PRACTICAL CH

(For B.Sc. I, II & III Year Students of All Indian Universities)

Dr. O.P. PANDEY

M.Sc. Ph.D.
Principal
B.B.N. Degree College
KANPUR

D.N. BAJPAI

M.Sc.
Reader, Chemistry Department
B.S.N.V. Post Graduate College
Lucknow University
LUCKNOW

Dr. S. GIRI

M.Sc. Ph.D.
Retd. Professor and Head
Department of Chemistry
University of Gorakhpur
GORAKHPUR

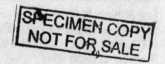
S. CHAND & COMPANY LTD.
(AN ISO 9001: 2000 COMPANY)
RAM NAGAR, NEW DELHI-110 055

S. CHAND & COMPANY LTD.

(An ISO 9001 : 2000 Company)

Head Office: 7361, RAM NAGAR, NEW DELHI - 110 055
Phone: 23672080-81-82, 9899107446, 9911310888
Fax: 91-11-23677446
Shop at: **schandgroup.com**; e-mail: **info@schandgroup.com**

Branches :

AHMEDABAD	: 1st Floor, Heritage, Near Gujarat Vidhyapeeth, Ashram Road, **Ahmedabad** - 380 014, Ph: 27541965, 27542369, ahmedabad@schandgroup.com
BENGALURU	: No. 6, Ahuja Chambers, 1st Cross, Kumara Krupa Road, **Bengaluru** - 560 001, Ph: 22268048, 22354008, bangalore@schandgroup.com
BHOPAL	: Bajaj Tower, Plot No. 243, Lala Lajpat Rai Colony, Raisen Road, **Bhopal** - 462 011, Ph: 4274723. bhopal@schandgroup.com
CHANDIGARH	: S.C.O. 2419-20, First Floor, Sector - 22-C (Near Aroma Hotel), **Chandigarh** -160 022, Ph: 2725443, 2725446, chandigarh@schandgroup.com
CHENNAI	: 152, Anna Salai, **Chennai** - 600 002, Ph: 28460026, 28460027, chennai@schandgroup.com
COIMBATORE	: No. 5, 30 Feet Road, Krishnasamy Nagar, Ramanathapuram, **Coimbatore** -641045, Ph: 0422-2323620 coimbatore@schandgroup.com **(Marketing Office)**
CUTTACK	: 1st Floor, Bhartia Tower, Badambadi, **Cuttack** - 753 009, Ph: 2332580; 2332581, cuttack@schandgroup.com
DEHRADUN	: 1st Floor, 20, New Road, Near Dwarka Store, **Dehradun** - 248 001, Ph: 2711101, 2710861, dehradun@schandgroup.com
GUWAHATI	: Pan Bazar, **Guwahati** - 781 001, Ph: 2738811, 2735640 guwahati@schandgroup.com
HYDERABAD	: Padma Plaza, H.No. 3-4-630, Opp. Ratna College, Narayanaguda, **Hyderabad** - 500 029, Ph: 24651135, 24744815, hyderabad@schandgroup.com
JAIPUR	: A-14, Janta Store Shopping Complex, University Marg, Bapu Nagar, **Jaipur** - 302 015, Ph: 2719126, jaipur@schandgroup.com
JALANDHAR	: Mai Hiran Gate, **Jalandhar** - 144 008, Ph: 2401630, 5000630, jalandhar@schandgroup.com
JAMMU	: 67/B, B-Block, Gandhi Nagar, **Jammu** - 180 004, (M) 09878651464 **(Marketing Office)**
KOCHI	: Kachapilly Square, Mullassery Canal Road, Ernakulam, **Kochi** - 682 011, Ph: 2378207, cochin@schandgroup.com
KOLKATA	: 285/J, Bipin Bihari Ganguli Street, **Kolkata** - 700 012, Ph: 22367459, 22373914, kolkata@schandgroup.com
LUCKNOW	: Mahabeer Market, 25 Gwynne Road, Aminabad, **Lucknow** - 226 018, Ph: 2626801, 2284815, lucknow@schandgroup.com
MUMBAI	: Blackie House, 103/5, Walchand Hirachand Marg, Opp. G.P.O., **Mumbai** - 400 001, Ph: 22690881, 22610885, mumbai@schandgroup.com
NAGPUR	: Karnal Bag, Model Mill Chowk, Umrer Road, **Nagpur** - 440 032, Ph: 2723901, 2777666 nagpur@schandgroup.com
PATNA	: 104, Citicentre Ashok, Govind Mitra Road, **Patna** - 800 004, Ph: 2300489, 2302100, patna@schandgroup.com
PUNE	: 291/1, Ganesh Gayatri Complex, 1st Floor, Somwarpeth, Near Jain Mandir, **Pune** - 411 011, Ph: 64017298, pune@schandgroup.com **(Marketing Office)**
RAIPUR	: Kailash Residency, Plot No. 4B, Bottle House Road, Shankar Nagar, **Raipur** - 492 007, Ph: 09981200834, raipur@schandgroup.com **(Marketing Office)**
RANCHI	: Flat No. 104, Sri Draupadi Smriti Apartments, East of Jaipal Singh Stadium, Neel Ratan Street, Upper Bazar, **Ranchi** - 834 001, Ph: 2208761, ranchi@schandgroup.com **(Marketing Office)**
SILIGURI	: 122, Raja Ram Mohan Roy Road, East Vivekanandapally, P.O., **Siliguri**- 734001, Dist., Jalpaiguri, (W.B.) Ph. 0353-2520750 **(Marketing Office)**
VISAKHAPATNAM	: Plot No. 7, 1st Floor, Allipuram Extension, Opp. Radhakrishna Towers, Seethammadhara North Extn., **Visakhapatnam** - 530 013, (M) 09347580841, visakhapatnam@schandgroup.com **(Marketing Office)**

© *1972, Dr. O.P. Pandey, D.N. Bajpai and Dr. S. Giri*

First Edition 1972
Subsequent Editions and reprints 1974, 76, 79, 81, 83, 85, 87, 88, 89, 90, 91, 92, 95, 96, 98, 99, 2000, 2001 (Twice), 2002, 2003, 2004, 2005 (Twice), 2006, 2007 (Twice), 2008, 2009 (Twice)
Revised Edition 2011
Reprint 2012

ISBN : 81-219-0812-4   **Code :**   04A 033

PRINTED IN INDIA

By Rajendra Ravindra Printers Pvt. Ltd., 7361, Ram Nagar, New Delhi -110 055 and published by S. Chand & Company Ltd., 7361, Ram Nagar, New Delhi -110 055.

PREFACE TO THE REVISED EDITION

This revised edition of the book **Practical Chemistry (For B.Sc. I, II and III year students)** has been updated to meet suitably the needs of undergraduate students of various Indian Universities. It also covers UGC model Curriculum. The salient features of the book are :

1. A significant haul up comprises important additions in the chapter 'Few Inorganic **Preparations.**

2. Totally fresh chapters *viz.* '**Conductometric Titrations**', '**Potentiometric Titrations**' and '**Determination of Molecular Weight**' have been added keeping in view the latest examination pattern.

3. Simultaneously, induction of 'Viva-Voce Questions' makes the book highly useful and informative, making the difficult concepts clear to students in a lucid manner.

Sincere thanks are expressed to Mrs. Nirmala Gupta, CMD ; Mr. Himanshu Gupta, JMD ; Mr. Amit Gupta, CEO ; Mr. Navin Joshi, EVP ; and Mr. S. Bhatnagar, Manager (Editorial & Pre-Press) for helping in bring out this edition.

Suggestions for further improvement, along with constructive critcism, from teachers and students will be gratefully acknowledged.

Dr. O.P. Pandey
D.N. Bajpai
Dr. S. Giri

PREFACE TO THE FIRST EDITION

The present book is intended for the use of under-graduate students preparing for Chemistry Practical examination of the various Indian Universities. Though much literature is available on the subject yet the authors during their experience in teaching profession have felt the need of concise and comprehensible book providing simple and easy approach to the subject matter.

The book aims at developing the habit of scientific reasoning in the students, making them familiar with technicalities involved in analytical procedures and thus providing the promising results with utmost clarity. Efforts have been made to remove every possible difficulty arising during practical work. Care has been taken not to make the treatment too exhaustive as it often leaves the students with initiative of their own.

The arrangement of topics have been made on rational ground. In the beginning Qualitative Analysis of Inorganic mixtures at macro, semi-micro and micro scales has been dealt with. The subject matter has been discussed in sufficient details to meet all requirements of students. Next the topic of quantitative Inorganic Analysis has been taken up : the Volumetric Analysis is followed by Gravimetric Analysis. The difficult points have been fully explained by means of questions and answers to make the subject sufficiently clear. Qualitative Organic Analysis comes next. It has been dealt with in a very systematic and logical manner. Last of all Inorganic and Organic Preparations and practicals of Physical Chemistry have been described and explained. At the end, a number of important appendices have been added to make the book as useful and informative as possible to the reader.

It is hoped that the book will provide a better and explicit understanding of the subject matter to both teachers and students.

Our thanks are due to publishers, M/s S. Chand & Company Ltd., New Delhi, for their co-operation in bringing out the books in such a short time.

Any suggestion towards the improvement of the book will be gratefully acknowledged.

April, 1972   **AUTHORS**

CONTENTS
INORGANIC QUALITATIVE ANALYSIS.

APPENDICES

1. Preparation of the Chemical Reagents used In Laboratory
2. Preparation of Some Special Reagents
3. Atomic Weights and other Constants
4. List of Equivalent Weights of Substances commonly used in Volumetric Analysis
5. Data on the Specific Gravity, Percentage Composition, Normality, *etc.* of some Reagents
6. Ionization Constants of weak Bases at Room Temperature (25°C)
7. Ionization Constants of Weak Acids at Room Temperature
8. Ionization (instability) Constant of Complex Ions at Room Temperature
9. Solubility Products at Room Temperature
10. Transition Temperatures
11. Viscosity of Liquids (in Centipoise)
12. Surface Tension of Liquids (in Dynes/cm)
13. Surface Tension of Water (in Dynes/cm)
14. Densities at 25°C (in gm/c.c.)
15. Density and Specific Volume of Water at Different Temperatures
16. Solubility of certain substances
17. Viscosity (in centipoise) of Different Aqueous solutions
18. Surface Tension of Different Aqueous solutions
19. Hints for Preparing Solution for Gravimetric Estimations
20. Laboratory First Aid, Safety and Treatment of Fires etc.
21. Consolidated List of some Organic Compounds
    Log Tables

1

Elementary Idea in Analysis

A general knowledge of some fundamental concepts is very essential in analytical proceedings. While analysing a sample, it must be clear that the ingredients present in the sample may be salts, acids, bases, metals and their oxides. When brought into solution, they often dissociate into two types of charged species, the positively charged ions are known as cations or *basic radicals* and negatively charged ion as anions or *acid radicals*. The tests in solution are actually the tests of these ions.

Solubility product and precipitation. At times, when a reagent is added to a clear solution, a solid mass separates. This is known as precipitate. Initially, it is voluminous and covers almost full portion of the solution but gradually it settles at the bottom.

For a given electrolyte, the product of the ionic concentration, in the saturated solution at a constant temperature is known as *solubility product*. For a binary electrolyte, if ionic product exceeds solubility product, precipitation occurs. As long as the product of ionic concentration remains less than or equal to solubility product of that electrolyte, it remains in solution. An alternation in the ionic concentration can be made by adding a suitable electrolyte. In this way, we may precipitate one compound retaining the other in soluble form.

Oxidising and reducing flames. Students generally think that two different burners are used to produce such flames. In fact, the burner used in laboratory has both the types of flames and it requires technique to use these selectively.

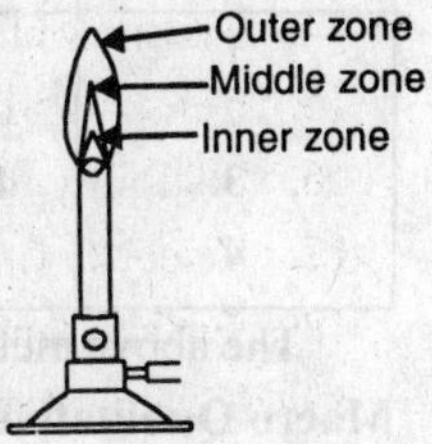

When the air hole of the burner is open, the flame contains three zones as shown in Fig. 1.1. The inner zone which is cold, does not involve any combustion. The middle zone has partial combustion and hence it contains carbon particles of the fuel and can reduce a substance coming in contact with it. This is regarded as *Reducing Flame*. The outer part of the flame is hottest and shows complete combusion. It is known as *oxidising flame*.

Outer zone
Middle zone
Inner zone

Fig. 1.1.

To heat the substance in reducing flame, we keep the mouth of the blowpipe at a short distance from the flame and blow so as to cause inner zone to come in contact with substance. The tip of the flame works as reducing flame (Fig. 1.2).

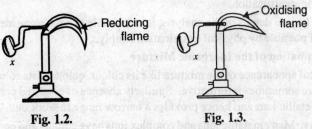

Reducing flame

Oxidising flame

Fig. 1.2. **Fig. 1.3.**

x — Rubber bladder with a blowpipe to supply the current of air.

The *oxidising flame* is the name of the uppermost part. But this can not be used everywhere straight away such as in cavity tests. For this purpose, we put the mouth of blowpipe inside the flame and blow it slowly. In this way, whole of the flame may be directed towards desired side and the upper tip of the flame is utilised (Fig. 1.3)

Solubilities of Electrolytes in aqueous solution. These are goverened by following rules :

1.   All nitrates, acetates, nitrites, chlorates and perchlorates are generally soluble. Silver acetate is an exception which is only moderately soluble.

2.   Sulphates are generally soluble. Exceptions are $PbSO_4$, $BaSO_4$ and $SrSO_4$ which are insoluble. $CaSO_4$ and Ag_2SO_4 are only slightly soluble. .

3.   All chlorides, bromides and iodides are generally soluble. Insoluble halides are of Ag^+, Pb^{2+}, Hg_2^{2+} and of Cu^+. $PbCl_2$ is soluble in hot water but almost insoluble in cold water.

4.   Compounds of alkali metals and ammonium salts are generally soluble.

5.   Carbonates, hydroxides, chromates, phosphates and sulphites are found generally insoluble leaving those of alkali metals and ammonium ions which are soluble.

6.   Sulphides are also in general insoluble leaving those of alkali metals and ammonium sulphide which are soluble. Sulphides of alkaline earth metals and of aluminium are decomposed in aqueous solution.

Qualitative Analysis

The inorganic qualitative analysis involves the identification of individual components present in a sample. The basis of identification is the observed chemical behaviour of the components.

On the basis of the amount of the sample to start with, the qualitative analysis has been categorized as follows :

S.No.	Amount of Substance	Method of Analysis
1.	0.5 gm—1gm, or solution (20 –50 ml).	Macro method
2.	25 mg — 40 mg, or 1 ml solution.	Semi-micro method
3.	5 mg—10 mg or one tenth of a ml. of solution	Micro method
4.	Less than 1 mg	Ultramicro method

The above methods differ somewhat in process also.

Macro Qualitative Analysis

This was introduced by Robert W. Bunsen nearly 150 years ago and this is regarded as *Classical Method of Analysis*. In it, the mixture is put into solvent such as water or acids where the dielectric constant of the medium overcomes the interionic forces of attraction and in this way putting the ions a bit separately. This is now made to react with the ions of various reagents to bring out the characteristic changes such as colour, appearance, precipitation and complex formation etc. which serve the purpose for their identification.

Before we go into details of the analysis, it is desirable to have some idea about the constituents of the sample if possible by physical examination simply.

Physical Examination of the Inorganic Mixture

The physical appearance of the mixture like its colour, colour of its solution prepared and smell of mixture prove sometimes informative. Similarly absence of coloured constituents in mixture excludes many metallic ions and hence provides a narrow range to work out.

1.   **Colour.** Many metallic ions and complex ions have colours and consequently their salts are also coloured. The following table shows the possible colours of the mixture in regular course of analysis along with constituents suspected.

S.No.	Colour	Compound Suspected
1.	Black	Powdered metals, sulphides of Ag(I), Cu(I), Cu(II), Fe(II), Co(II), Hg(II), Ni(II) and Pb(II) ; MnO_2, Fe_3O_4, FeO, CuO, Co_3O_4, Ni_2O_3.
2.	Blue	Hydrated copper salts, anhydrous cobalt salts.

Cont.

3.	Brown	Sn (II) Sulphides.
4.	Dark brown	PbO_2, Ag_2O, CdO, Fe_2O_3, Bi_2S_3, $CuCrO_4$
5.	pale brown	$MnCO_3$
6.	Green	Nickel salts, hydrated ferrous salts, some Cu (II), Chromium (III) salts K_2MnO_4
7.	Orange	Sb_2S_3, some dichromates and ferricyanides
8.	Pink	
	(a) Light pink	Mn Salts in hydrated form
	(b) Reddish pink	Hydrated cobalt (ous) salts
9.	Purple	Permanganates and a few Cr (III) Salts
10.	Red	As (II), Hg (II) and Sb (III) Sulphides, Pb_3O_4, HgI_2
11.	Yellow	Chromates ; ferric salts ; As (III) and As (V) Cd and Sn (IV) Sulphides ; some iodides and ferrocyanides.

Note. *Numerals in brackets such as (I) and (II) etc. represent their oxidation states.*

(2) Colour-change on heating. Certain substances change their colour when they are hot and revert to their original colour in cold as is clear below :

S.No.	Observation	Compound Suspected
1.	Yellow when hot and white in cold again	Zn oxide and some Zn salts
2.	Yellowish brown in hot and yellow in cold	Bi_2O_3 and SnO_2
3.	Black or red in hot and brown in cold	Fe (III) oxide
4.	Yellow in hot and yellow in cold	PbO and some Pb salts

(3) Colour of the solution. In certain cases, we get coloured solutions. The usual colours of the solution with their possible ions are given below :

S.No.	Colour of the solution	Ions Suspected
1.	Yellow	CrO_4^{2-}, Fe^{3+}, $[Fe(CN)_6]^{4-}$
2.	Green or blue	Ni^{2+}, Fe^{2+}, Cr^{3+} and Cu^{2+}
3.	Pink	Co^{2+} and Mn^{2+}
4.	Orange or purple	Dichromates (orange), permanganates (purple)

(4) Smell of original mixture. Sometimes, the original mixture gives characteristic smell. This also serves as an information as shown below :

S.No.	Smell	Inference
1.	Vinegar or acetic acid type	Acetates
2.	Ammoniacal smell	Amm. carbonate and some other ammonium salts
3.	Chlorine gas smell	Hypochlorites
4.	Bitter almond type smell	Cyanides.

Analysis of Acid Radicals
(Macro Method)

The complete qualitative analysis of any inorganic mixture covers the identification of both, the cations and anions. The analysis of cations is simple as it is systematic in approach while analysis of anions is slightly independent.

It is not only important but necessary too, to identify the acid radicals first followed by the analysis of basic radicals as the presence of some acid radicals (interfearing radicals) create unwanted problems and, therefore, their removal becomes essential at appropriate place. We shall discuss the analysis of both the classes individually ahead.

2·1. Preliminary test with dilute H_2SO_4

Take a small amount of mixture in a testtube and add about 2 ml dil. H_2SO_4. Note the change in cold. Now warm gently, observe the change and draw inference as follows :

Observation	Radical Suspected	Confirmatory Test
1. Effervescence with evolution of colourless gas in cold.	Carbonate (CO_3^{2-})	1. Pass the gas into lime water. It turns *milky*. (When the gas is passed in excess, the milky colour disappears).
2. Reddish brown fumes in cold.	Nitrite (NO_2^-)	1. Mix. + dil. H_2SO_4 + starch soln. (fresh) + KI soln. (fresh) $\longrightarrow$ Blue colour
3. Colourless gas (in hot) smelling ing like rotton eggs (H_2S smell).	Sulphide (S^{2-})	1. Paper dipped in lead acetate solution turns black on exposure to gas.
		2. Add a few drops of sodium nitroprusside solution to 2 ml. of sodium carbonate extract—violet or purple colour.
4. Colourless gas (in hot) with choking smell of burning sulphur	Sulphite (SO_3^{2-})	1. Filter paper moistened with acidified dichromate solution *turns green* on exposure to gas.
		2. Decolourises $KMnO_4$ solution.
5. White or yellowish white turbidity on warming.	Thiosulphate $(S_2O_3^{2-})$	See the confirmatory tests given ahead.
6. Solution smells like vinegar (acetic acid or gas) $\longrightarrow$. No fumes visible.	Acetate (CH_3COO^-)	1. Aqueous extract of the mixture + a few drops of *neutral $FeCl_3$ solution $\longrightarrow$ blood red colour. The colour is destroyed by adding an acid.
		2. In a watch glass, rub the mixture with a little solid oxalic acid and a few drops of water $\longrightarrow$ vinegar smell.

* **Preparation of neutral $FeCl_3$.** Take a little $FeCl_3$ solution (say 3 ml) from shelf in a test tube and add dil. NH_4OH dropwise with shaking until a slight permanent turbidity appears. Filter and use the filtrate as neutral $FeCl_3$. Ordinary $FeCl_3$ solution reacts acidic due to hydrolysis and turns reddish brown :

$$FeCl_3 + 3H_2O \longrightarrow Fe(OH)_3 + 3HCl$$

All the above tests can be performed with dilute HCl as well. Sulphate radical can also be tested if dilute HCl is used.

2·2. Preliminary Test with Conc. H_2SO_4

Few salts are not decomposed by dilute mineral acids and for these concentrated acid is used. It must be kept in mind that salts decomposed by dilute acid, will also be decomposed by concentrated acid. Take small amount of the mixture in a test tube, add about 2 ml. of conc. H_2SO_4 and warm gently. Observe the changes and draw the inference as follows :

Observation	Radical Suspected	Confirmatory Test
1. Colourless pungent smelling gas comes out. Bring a rod moistened with conc. NH_4OH near the mouth of the test tube ⟶ dense fumes	Chloride (Cl^-)	1. Add a pinch of MnO_2 in the solution ⟶ pale green gas. 2. Take a small amount of mixture, add 1–2 ml. dil. HNO_3 and heat. If no clear solution, filter. To the filtrate, add $AgNO_3$ solution ⟶ curdy white ppt. soluble in NH_4OH. 3. *Chromyl chloride test* (see ahead).
2. Reddish-brown vapours which intensify on addition of MnO_2. Vapours passed in water, make it yellow.	Bromide (Br^-)	1. Heat a little mixture with dil. HNO_3 and filter. To the filtrate, add $AgNO_3$ solution → pale yellow ppt. sparingly soluble in NH_4OH. 2. Heat a small amount of the mixture with dil H_2SO_4. Cool, add 1 ml. chloroform or carbon tetrachloride and then excess of chlorine water with shaking. The chloroform layer becomes orange-brown. [**N.B.** *The chloroform and chlorine water may be added to sodium carbonate extract to obtain the same results.*]
3. Violet pungent vapours evolved. The vapours passed over starch paper turn it blue.	Iodide (I^-)	1. Acidify sodium carbonate extract with dil HNO_3 and add $AgNO_3$ solution ⟶ yellow ppt. insoluble in NH_4OH. 2. Add chloroform and chlorine water to sodium carbonate extract ⟶ violet colour in chloroform layer.
4. Light brown gas evolved and this increases on addition of Cu-turnings in solution.	Nitrate (NO_3^-)	1. *Ring test.* See ahead.
5. Colourless gas which turns lime water milky and burns with a blue flame.	Oxalate ($C_2O_4^{2-}$)	1. Dissolve a small amount of the mixture in dilute HCl or H_2SO_4. Add a few drops of $KMnO_4$ solution ⟶ decolorisation occurs (this test may be given by other reducing agents also hence perform test No. 2 also). 2. Take 2 ml. sodium carbonate extact and acidify with acetic acid. Add $CaCl_2$ solution in excess and allow it to stand for sometime. Filter the white ppt. and

		extract it with hot dil. H_2SO_4. Now add to it a few drops of dilute $KMnO_4$ solution. The pink colour of $KMnO_4$ disappears.
6. Colourless pungent smelling fumes. The test tube becomes greasy.	Fluoride (F^-)	See test (b) of interfering acid radicals.
7. Yellow gas with cracking sound and the solution turns orange-yellow.	Chlorate (ClO_3^-)	Dissolve the mixture in dil. HNO_3, add a little sodium nitrite and warm. Now add $AgNO_3$ solution → a curdy white ppt. soluble in NH_4OH.
8. On heating the mixture with conc. H_2SO_4 charring occurs (solution becomes black) and smell of burnt sugar is obtained.	Tartarate ($C_4H_4O_6^{2-}$)	See the test ahead.

Nitrate. (a) **Ring Test.** Take water extract or soda extract (2 ml) just acidified with dil H_2SO_4, add to it 4 ml of conc. H_2SO_4 slowly and mix the two liquids thoroughly. Cool the test tube under a stream of water and now hold the test tube in a slanting position. Add freshly prepared saturated solution of ferrous sulphate slowly from the side of test tube without pouring it at a time and keep the test tube once again straight. Now allow to stand for ~3 minutes. A sharp brown ring will be formed at the junction of two layers.

Alternative. Add ~3 ml of freshly prepared saturated solution of ferrous sulphate to ~2 ml of water extract (or soda extract acidified with dil. H_2SO_4) of mixture and pour 3-5 ml conc. H_2SO_4 slowly from the side of test tube so that acid forms a layer beneath the solution. Brown ring will be formed on cooling where liquids meet.

Brown ring is due to formation of a complex, $[Fe(NO)]^{2+}$

$$NaNO_3 + H_2SO_4 \longrightarrow NaHSO_4 + HNO_3$$
$$2HNO_3 + 6FeSO_4 + 3H_2SO_4 \longrightarrow 3Fe_2(SO_4)_3 + 2NO + 4H_2O$$
$$FeSO_4 + NO \longrightarrow [Fe(NO)]SO_4$$

Limitations. This test should not be preformed for nitrate in presence of nitrite, bromide and iodide. This ring test is also not reliable in presence of chlorate, chromate, sulphite, thiosulphate, ferrocyanide and ferricyanide. The interference of bromide and iodide is due to halogen liberated with conc. H_2SO_4. Since nitrites are decomposed by conc. H_2SO_4, nitrite will also give the same ring test even in absence of nitrate.

2·3. Reactions involved with dilute H_2SO_4

Carbonate. A carbonate is decomposed by dilute acid to give carbon dioxide. Evolution of this gas is the cause of effervescence in solution e.g.,

$$Na_2CO_3 + H_2SO_4 \longrightarrow Na_2SO_4 + H_2O + CO_2 \uparrow$$

The carbon dioxide when passed into lime water, it turns milky due to the formation of suspension of $CaCO_3$. The excess of CO_2 converts $CaCO_3$ into soluble bicarbonate and hence the colour disappears later.

$$Ca(OH)_2 + CO_2 \longrightarrow \underset{\substack{\text{milky} \\ \text{appearance}}}{CaCO_3} + H_2O$$

$$CaCO_3 + H_2O + CO_2 \longrightarrow \underset{\text{colourless}}{Ca(HCO_3)_2}$$

Sulphite. Dilute acid decomposes a sulphite to give sulphur dioxide gas which has smells like burning sulphur.

$$Na_2SO_3 + H_2SO_4 \longrightarrow Na_2SO_4 + H_2O + SO_2 \uparrow$$

SO_2, reduces acidic $K_2Cr_2O_7$ to green $Cr_2(SO_4)_3$ and acidic $KMnO_4$ to colourless $MnSO_4$ as follows :

(i)
$$K_2Cr_2O_7 + 4H_2SO_4 \longrightarrow K_2SO_4 + Cr_2(SO_4)_3 + 4H_2O + 3O$$
$$SO_2 + H_2O + O \longrightarrow H_2SO_4 \qquad [\times 3]$$
$$\overline{K_2Cr_2O_7 + H_2SO_4 + 3SO_2 \longrightarrow K_2SO_4 + Cr_2(SO_4)_3 + H_2O}$$

(ii)
$$2KMnO_4 + 2H_2O + 5SO_2 \longrightarrow K_2SO_4 + 2MnSO_4 + 2H_2SO_4$$

Sulphide. Action of dilute acid decomposes a sulphide salt to form H_2S gas which smells like rotten eggs.
$$Na_2S + H_2SO_4 \longrightarrow Na_2SO_4 + H_2S \uparrow$$
The lead acetate paper turns black on exposure to this gas due to formation of lead sulphide.
$$Pb(CH_3COO)_2 + H_2S \longrightarrow PbS \downarrow 2CH_3COOH$$
$$\text{black}$$

H_2S like SO_2, is a good reducing agent ; it reduces (i) acidified solution of $KMnO_4$, (ii) acidified $K_2Cr_2O_7$ solution and (iii) iodine solution.

(i)
$$2KMnO_4 + 5H_2S + 6HCl \longrightarrow 2KCl + 2MnCl_2 + 8H_2O + 5S$$
(ii)
$$K_2Cr_2O_7 + 3H_2S + 8HCl \longrightarrow 2KCl + 2CrCl_3 + 7H_2O + 3S$$
(iii)
$$H_2S + I_2 \longrightarrow 2HI + S$$

Some sulphides are not ordinarily decomposed by dilute acid. In such cases, add a pinch of zinc dust in solution. A brisk evolution of H_2S now takes place. For example,
$$Zn + H_2SO_4 \longrightarrow ZnSO_4 + 2H$$
$$HgS + 2H \longrightarrow Hg + H_2S$$

Nitrite. By treatment with dilute acid (dil. HCl, dil. H_2SO_4 or dil. CH_3COOH), a nitrite salt decomposes to ive nitric oxide. This in contact with air forms reddish-brown NO_2.
$$2NaNO_2 + H_2SO_4 \longrightarrow Na_2SO_4 + 2HNO_2$$
$$3HNO_2 \longrightarrow HNO_3 + 2NO + H_2O$$
$$\text{colourless}$$
$$2NO + O_2 \longrightarrow 2NO_2$$
$$\text{(air)} \qquad \text{nitrogen dioxide}$$
$$\text{(brown)}$$

The nitrous acid liberates iodine from potassium iodide which gives blue colour with starch
$$2KI + 2HNO_2 + H_2SO_4 \longrightarrow K_2SO_4 + 2H_2O + 2NO + I_2$$
$$I_2 + \text{starch} \longrightarrow \text{Blue coloured iodo-starch complex.}$$
starch iodide has no colour.

Acetate. Action of dilute acid on an acetate salt liberates acetic acid which smells like vinegar.

Aqueous extract of acetate gives red coloured ferric acetate when treated with neutral ferric chloride. On ution and heating, basic ferric acetate is produced.
$$3CH_3COONa + FeCl_3 \longrightarrow (CH_3COO)_3Fe + 3NaCl$$
$$\text{blood red colour}$$
$$Fe(CH_3COO)_3 + 2H_2O \longrightarrow CH_3COO(OH)_2Fe \downarrow + 2CH_3COOH$$
Rection with oxalic acid also liberates acetic acid.
$$2CH_3COONa + H_2C_2O_4 \longrightarrow Na_2C_2O_4 + 2CH_3COOH$$

Note : In $FeCl_3$ test for acetate, the solution must be neutral and must not contain anions which precipitate ferric ions: CO_3^{2-}, SO_3^{2-}, PO_4^{3-}, $[Fe(CN)_6]^{4-}$, etc. (the CNS^- and I^- ions should also be absent. Iodide is oxidised by Fe^{3+} ions to I_2 which produces a red brown colour similar to that expected in the reaction). All these ions can be removed by addition of $AgNO_3$ or Ag_2SO_4 solution to the neutral solution. However, this involves the use of considerable amount of silver salts, and, therefore, the test should not be used unless very necessary.

Reactions involved with conc. H_2SO_4

Chloride. Upon warming with conc. H_2SO_4, a chloride salt is decomposed to give colourless HCl gas h forms dense fumes of NH_4Cl when comes in contact with NH_4OH.

$$NaCl + H_2SO_4 \longrightarrow NaHSO_4 + HCl \uparrow$$
$$NH_4OH + HCl \longrightarrow NH_4Cl + H_2O$$

The HCl is oxidised by manganese dioxide to give pale green, chlorine gas.

$$MnO_2 + 4HCl \longrightarrow MnCl_2 + 2H_2O + Cl_2 \uparrow$$

The chloride salt gives white ppt. of AgCl with silver nitrate solution. This precipitate is soluble in ammonia due to formation of a complex.

$$NaCl + AgNO_3 \longrightarrow AgCl + NaNO_3$$
$$\text{White ppt.}$$

$$AgCl + 2NH_4OH \longrightarrow Ag[NH_3]_2Cl + 2H_2O$$
$$\text{Diammine silver chloride}$$
$$\text{(complex)}$$

Bromide. A bromide salt is decomposed by conc. H_2SO_4 to form reddish-brown vapours of bromine.

$$NaBr + H_2SO_4 \longrightarrow NaHSO_4 + HBr$$
$$2HBr + H_2SO_4 \longrightarrow Br_2 + SO_2 + 2H_2O$$

Addition of MnO_2 to the reaction mixture causes the evolution of bromine only hence the fumes intensify.

$$2NaBr + MnO_2 + 3H_2SO_4 \longrightarrow Br_2 + 2NaHSO_4 + MnSO_4 + 2H_2O$$

From the solution, bromide is precipitated as pale yellow AgBr upon addition of $AgNO_3$. This is soluble in excess of ammonium hydroxide.

$$NaBr + AgNO_3 \longrightarrow AgBr \downarrow + NaNO_3$$
$$AgBr + 2NH_4OH \longrightarrow Ag[NH_3]_2Br + 2H_2O$$
$$\text{soluble}$$

Bromine is liberated from bromide salt by addition of chlorine water. Free bromine dissolves in chloro form and gives orange-brown colour.

$$2NaBr + Cl_2 \longrightarrow 2NaCl + Br_2$$
$$Br_2 \text{ in Chloroform} \longrightarrow \text{orange-brown colour.}$$

Iodide. The salt is decomposed by conc. H_2SO_4 to liberate free iodine which is seen as violet vapours.

$$NaI + H_2SO_4 \longrightarrow NaHSO_4 + HI$$
$$2HI + H_2SO_4 \longrightarrow 2H_2O + I_2 + SO_2$$

Sodium iodide present in sodium carbonate extract gives yellow precipipate of AgI with $AgNO_3$ solutio This does not dissolve in ammonia.

$$NaI + AgNO_3 \longrightarrow AgI + NaNO_3$$
$$\text{Yellow ppt.}$$

Chlorine liberates iodine from NaI. Iodine is soluble in chloroform to give violet colour.

$$2NaI + Cl_2 \longrightarrow 2NaCl + I_2$$
$$I_2 \text{ in Chloroform} \longrightarrow \text{Violet colour}$$

Nitrate. The nitrate salt is decomposed by conc. H_2SO_4 to give reddish brown fumes of nitrogen dioxi

$$NaNO_3 + H_2SO_4 \longrightarrow NaHSO_4 + HNO_3$$
$$4HNO_3 \longrightarrow 4NO_2 \uparrow + O_2 \uparrow + 2H_2O$$

Oxalate. The oxalates are decomposed by conc. H_2SO_4 to give carbon monoxide and carbon dioxide. latter when passed into lime water, it turns it milky. The carbon monoxide burns with a blue flame at the mo of the test tube.

$$Na_2C_2O_4 + H_2SO_4 \longrightarrow Na_2SO_4 + H_2C_2O_4$$
$$\text{Oxalic acid}$$

$$H_2C_2O_4 + H_2SO_4 \longrightarrow CO + CO_2 + H_2O + H_2SO_4$$

The sodium oxalate present in sodium carbonate extract reacts with $CaCl_2$ to give white ppt. of cal oxalate which is insoluble in acetic acid.

$$Na_2C_2O_4 + CaCl_2 \longrightarrow CaC_2O_4 \downarrow + 2NaCl$$
$$\text{White ppt.}$$

In presence of dil. H_2SO_4 the oxalate reduces permanganate solution hence decolorisation occurs.

$$2KMnO_4 + 3H_2SO_4 \longrightarrow K_2SO_4 + 2MnSO_4 + 3H_2O + 5[O]$$
$$CaC_2O_4 + H_2SO_4 + [O] \longrightarrow CaSO_4 + H_2O + 2CO_2$$

Fluoride. Action of conc. H_2SO_4 on a fluoride salt liberates hydrogen fluoride. This reacts with the silica of glass to give silicon tetrafluoride. The hydrogen fluoride gives fumes in moist air. Its corrosive action on the glass gives a greasy appearance.

$$NaF + H_2SO_4 \longrightarrow NaHSO_4 + HF$$
$$SiO_2 + 4HF \longrightarrow SiF_4 + 2H_2O$$

Chlorate. A chlorate salt is decomposed by conc. H_2SO_4 to give chlorine dioxide gas which dissolves in the acid to give an orange-yellow solution.

$$3\,KClO_3 + 3H_2SO_4 \longrightarrow 3KHSO_4 + 2ClO_2 + HClO_4 + H_2O$$

The chlorate is reduced to chloride by nitrite salt and this gives a white ppt. of AgCl with $AgNO_3$ solution.

$$KClO_3 + 3NaNO_2 \longrightarrow 3NaNO_3 + KCl$$
$$KCl + AgNO_3 \longrightarrow AgCl \downarrow + KNO_3$$

Tartarate. A tartrate salt when heated with conc. H_2SO_4 decomposes to give various products. One of them is carbon which imparts black colour.

$$Na_2C_4H_4O_6 + H_2SO_4 \longrightarrow H_6C_4O_6 + Na_2SO_4$$
$$H_2C_4H_4O_6 \longrightarrow CO + CO_2 + 2C + 3H_2O$$
$$C + 2H_2SO_4 \longrightarrow 2SO_2 + CO_2 + 2H_2O.$$

·5. Specific tests in Solution

1. **Sulphate (SO_4^{2-}).** (*i*) Dissolve a little amount of mixture in dil. HCl and add $BaCl_2$ solution. A white precipitate insoluble in conc. HNO_3 is obtained.

If an insoluble sulphate is given in the mixture, it may not dissolve in dil. HCl. In that case, test for sulphate should be done as follows with sodium carbonate extract.

(*ii*) To about 1 ml sodium carbonate extract, add sufficient dil. HCl until effervescence ceases and the solution becomes distinctly acidic. Now add $BaCl_2$ solution. A white ppt. insoluble in nitric acid is obtained.

$$Na_2SO_4 + BaCl_2 \longrightarrow BaSO_4 \downarrow + 2NaCl$$

2. **Thiosulphate ($S_2O_3^{2-}$).** If sulphide, sulphite or sulphate is present, shake a portion of sodium carbonate extract with $CdCO_3$ or $PbCO_3$ and then with $Sr(NO_3)_2$ solution. Filter and discard if any ppt. forms (which may be due to S^{2-}, SO_3^{2-} or SO_4^{2-}). From the filtrate, thiosulphate can be tested by any of the following tests. In absence of sulphide, sulphite or sulphate, use the water extract for testing thiosulphate.

(*i*) **$AgNO_3$ test.** With thiosulphate, $AgNO_3$ produces a white ppt. of silver thiosulphate which is unstable and decomposes on warming or on addition of HCl into black Ag_2S through a colour change from yellow to brown and finally black.

$$Na_2S_2O_3 + 2AgNO_3 \longrightarrow Ag_2S_2O_3 \downarrow + 2NaNO_3$$
White

$$Ag_2S_2O_3 + H_2O \xrightarrow{\text{Warm}} Ag_2S \downarrow + H_2SO_4$$
Black

Silver thiosulphate is soluble in excess of thiosulphate ions. At first, a sparingly soluble complex $AgS_2O_3]$ is formed which dissolves in excess of sodium thiosulphate forming a soluble complex $Ag_2(S_2O_3)_3]$

$$Ag_2S_2O_3 + Na_2S_2O_3 \longrightarrow 2Na[AgS_2O_3]$$
$$2Na[AgS_2O_3] + Na_2S_2O_3 \longrightarrow Na_4[Ag_2(S_2O_3)_3]$$
soluble

On warming the dilute solution, Ag_2S is precipitated.

$$Na_4[Ag_2(S_2O_3)_3] \longrightarrow Ag_2S\downarrow + S\downarrow + SO_2\uparrow + Na_2S_2O_3 + Na_2SO_4$$
$$2Na[AgS_2O_3] \longrightarrow Ag_2S\downarrow + S\downarrow + SO_2\uparrow + Na_2SO_4$$

(*ii*) **Blue ring test.** A solution of thiosulphate is mixed with an equal volume of 10% ammonium molybdate soln. The resulting mixture is then added slowly down the side of a test tube containing conc. H_2SO_4. A blue ring is formed at the junction of two layers.

(*iii*) **Dilute HCl test.** Acidify thiosulphate soln. with dil. HCl and warm the soln. It becomes white or yellowish turbid due to release of free sulphur. On warming, SO_2 is evolved which can be detected by its characteristic smell and by acidified $K_2Cr_2O_7$ paper.

$$Na_2S_2O_3 + 2HCl \xrightarrow{\text{Warm}} SO_2\uparrow + S\downarrow + H_2O + 2NaCl$$

(*iv*) **KCN Test.** Make the thiosulphate solution alkaline by NaOH and then boil with KCN solution. On acidifying this solution with dil. HCl and adding $FeCl_3$ solution, the blood red colour o ferric thiocyanate is obtained (difference from sulphide).

$$S_2O_3^{2-} + CN^- \longrightarrow SCN^- + SO_3^{2-}$$
$$Fe^{3+} + SCN^- \longrightarrow [Fe(SCN)]^{2+}$$
$$\text{Blood red}$$

(*v*) **FeCl$_3$ Test.** It produces short lasting deep violet colouration probably due to formation c sodium ferri-thiosulphate, $Na[Fe(S_2O_3)_2]$ which disappears on warming leaving an almost colourles solution forming $FeCl_2$ and sodium tetrathionate.

$$2Na_2S_2O_3 + FeCl_3 \longrightarrow Na[Fe(S_2O_3)_2] + 3NaCl$$
$$Na[Fe(S_2O_3)_2] + NaCl + FeCl_3 \longrightarrow 2FeCl_2 + Na_2S_4O_6$$
$$\text{Ionically}: 2S_2O_3^{2-} + 2Fe^{3+} \longrightarrow 2Fe^{2+} + S_4O_6^{2-}$$

3. Interfering acid radical. (*a*) **Phosphate.** Take a little mixture in a test tube, add ~3 ml conc. HNO_3 and boil. Now add excess of ammonium molybdate solution and heat again to abc 40°C. If crystalline canary yellow ppt. is produced, phosphate is present. This ppt. is soluble NH_4OH and NaOH.

$$Na_3PO_4 + 3HNO_3 \longrightarrow 3NaNO_3 + H_3PO_4$$
$$H_3PO_4 + 12(NH_4)_2\,MoO_4 + 21HNO_3 \longrightarrow (NH_4)_3[PMo_{12}O_{40}] + 12H_2O + 21NH_4NO_3$$
$$\text{Canary yellow ppt.}$$

[**Note : 1.** Since arsenic also responds to the above test, presence of phosphate should always be tes with the II group filtrate after boiling off H_2S.

2. Excess of HCl interferes with the above test and should preferably be removed by evaporation small volume with excess of conc. HNO_3.

3. The ppt. of ammonium phosphomolybdate dissolves in excess of phosphate to form complex ani The reagent amm. molybdate must, therefore, be taken in excess.

4. In absence of phosphate, a white ppt. often results, due to decomposition of amm. molybdate reag Discard such a ppt.

5. Reducing agents such as SO_3^{2-}, $S_2O_3^{2-}$, S^{2-} etc. interfere as they produce "molybdenum b $(MoO_8.xH_2O)$. The solution, therefore, turns blue. These should be oxidised before performing the above for phosphate. If mixture is boiled with 3-4 ml conc. HNO_3 before adding ammonium molybdate reagent, are decomposed.]

(*b*) **Fluoride.** Mix a little sand with mixture in a dry test tube and add to it 1-2 ml conc. H_2 Now warm the test tube gently. Bring the rod moistened with water inside the test tube wit touching the test tube and solution, a white deposition on the rod shows the presence of fluorid greasy appearance in the test tube is also seen.

$$NaF + H_2SO_4 \longrightarrow NaHSO_4 + HF$$

$$SiO_2 + 4HF \longrightarrow SiF_4 + 2H_2O$$
$$3SiF_4 + 4H_2O \longrightarrow 2H_2[SiF_6] + H_4SiO_4 \quad \text{(Silicic acid)}$$
$$\text{white}$$

The gas HF being corrosive, attacks silica present in the glass of test tube to produce oily drops. The corrosive action of HF is also responsible for the greasy appearance. The fluosilicic acid which is known only in aqueous solution, is decomposed into silicon tetrafluoride and hydrofluoric acid on evaporating the aqueous solution.

$$H_2[SiF_6] \longrightarrow SiF_4 \uparrow + 2HF \uparrow$$
$$\text{(Fluosilicic acid)}$$

(c) **Borate.** (i) In a porcelain dish, take a little mixture, add to it 1 ml. conc. H_2SO_4 and 1 ml. ethyl alcohol. Heat slightly and bring a burning paper on the dish. Green flame shows the presence of *Borate*.

[**Note :** Since Cu and Ba both respond to the above test, perform this test in a test tube. In this way, the vapours of only ethyl borate are able to reach the mouth of test tube and metallic copper and barium are left behind.]

$$2Na_3BO_3 + 3H_2SO_4 \longrightarrow 3Na_2SO_4 + 2H_3BO_3$$
$$H_3BO_3 + 3C_2H_5OH \longrightarrow B(OC_2H_5)_3 + 3H_2O$$
$$\text{Ethyl borate}$$

(ii) **Test with CaF_2.** This test is quite reliable because metallic ions like copper and barium do not interfere.

In a watch glass or porcelain dish, mix CaF_2 and mixture in equal amounts and make the thick paste of it with conc. H_2SO_4. Keep a little paste on the tip of glass rod and bring it close to (but don't let it touch) the flame. Volatile BF_3 is formed and flame appears green which confirms borate.

$$CaF_2 + H_2SO_4 \longrightarrow CaSO_4 + 2HF$$
$$B_2O_3 + 6HF \longrightarrow 2BF_3 \quad + 3H_2O$$
$$\text{Volatile, burns}$$
$$\text{with green flame}$$

(d) **Tartarate.** (i) Heat the mixture with conc. H_2SO_4. If charring occurs immediately accompanied by the smell of burnt sugar, tartarate is present. In the charring, carbon is produced and solution in the test tube becomes black.

(ii) *Silver mirror test.* Generally salts of tartaric acid are water soluble.

Take water extract or soda extract (neutralized with dil. HNO_3) in a test tube. In another clean test tube take 5—10 ml. of $AgNO_3$ solution (10%) and add 2 drops of dil. NaOH solution. Now add dil. NH_4OH drop by drop until the ppt. redissolves. Now introduce 0.5—1.0 ml of neutral tartarate solution. Place this test tube in a beaker containing water and then warm. After sometime, if a *bright silver mirror* is formed, tartarate is present.

(e) **Oxalate.** Take 5 ml. of soda extract in a test tube and acidify with excess of dil. CH_3COOH. Add $CaCl_2$ solution in excess. It gives white ppt. soluble in mineral acids. Allow the ppt. to settle. Filter and wash the ppt. with hot water. Take this ppt. in a test tube and extract with a little quantity of hot dil. H_2SO_4. Now add to it dil. $KMnO_4$ drop by drop and shake well after each addition. If the colour of $KMnO_4$ disappears, oxalate is present.

$$Na_2C_2O_4 + CaCl_2 \longrightarrow 2NaCl + CaC_2O_4 \downarrow$$
$$\text{white ppt.}$$
$$CaC_2O_4 + H_2SO_4 \longrightarrow CaSO_4 + H_2C_2O_4$$
$$2KMnO_4 + 3H_2SO_4 + 5H_2C_2O_4 \longrightarrow K_2SO_4 + 2MnSO_4 + 8H_2O + 10CO_2$$
$$\text{violet} \qquad\qquad\qquad\qquad\qquad\qquad \text{colourless}$$

4. Chromate CrO_4^{--}. This can be tested by following methods :

(i) To an aqueous solution, add barium chloride solution. A pale yellow precipitate insoluble in acetic acid indicates the presence of chromate.

(ii) To an aqueous solution or solution in acetic acid, add silver nitrate solution. A brownish red precipitate of silver chromate is obtained.

(iii) Treat its acidic solution with H_2O_2. This gives deep blue solution of chromium pentoxide which fades quickly on standing.

(iv) In acidic solution, pass H_2S gas. This turns the solution green.

(v) Acidify the solution with dilute H_2SO_4 or acetic acid and add diphenyl carbazide reagent. A deep blue colouration is produced.

5. Permanganate, MnO_4^-. All permanganates are, as a rule, soluble in water imparting purple solution. Following tests can be applied to ascertain the presence of permanganate radical.

(i) To an aqueous solution, add concentrated HCl and boil. Chlorine gas is evolved which turns starch–iodide paper blue.

(ii) To concentrated aqueous solution, add NaOH solution and warm the mixture, A green solution is obtained with the liberation of oxygen gas. The green solution when poured into a beaker full of water and contents acidified, purple colours of permanganate is restored.

6. Arsenate, AsO_4^{3-} (i) To a neutral solution of arsenate, add silver nitrate solution. A brownish red precipitate insoluble in acetic acid but soluble in mineral acid and in ammonia indicates the presence of arsenate.

(ii) To an aqueous solution, add nitric acid and ammonium molybdate. Boil the mixture. A yellow crystalline precipitate of ammonium arsenomolybdate, $(NH_4)_3AsMo_{12}O_{40}$ indicates the presence of arsenate. This precipitate is soluble in ammonia and in alkalies. The ionic equation for the above test is written as :

$$AsO_4^{3-} + 12MoO_4^{2-} + 3NH^{4+} + 24H^+ \longrightarrow (NH_4)_3AsMo_{12}O_{40} \downarrow + 12H_2O$$
$$\text{precipitate}$$

(iii) Acidify the solution with large amount of concentrated HCl and to it, add KI solution. This will precipitate iodine which can be tested as usual. The ionic equation is :

$$AsO_4^{3-} + 2H^+ + 2I^- \rightleftharpoons AsO_3^{3-} + I_2 \downarrow + H_2O$$

2·6. Tests with sodium carbonate extract

A large number of acid radicals can be confirmed by using sodium carbonate extract of the mixture. This is especially useful when the mixture is less soluble in dilute acids. For example, many halides and sulphates do not readily dissolve in acids and hence they may not be tested.

When mixture is heated strongly in the minimum quantity of water with excess of Na_2CO_3, the double decomposition takes place i.e., the partner radicals are changed. Sodium joins with anions of mixture and carbonate attaches with basic radicals of mixture. After heating, it is mixed with water where sodium salts which contain anions of mixture dissolve (as sodium salts are soluble) and carbonates of the metals remain as insolubles. In this way, we separate the anions from the cations of the mixture after filtration.

Preparation of Sodium Carbonate Extract

Take one part of mixture and four parts of pure Na_2CO_3 in a porcelain dish. Add 15-20 m distilled water and heat while stirring with the glass rod. Heat it to boiling gently for about 5-1 minutes, add water to make up the loss by evoparation and filter. The filtrate is the desired sodiu carbonate extract. Use 1-2 ml of this extract for each test.

Precautions

1. Certain substances may remain undecomposed by boiling with Na_2CO_3. They include certain phosphates, fluorides and silver halides. Hence if corresponding anions are not found in the prepared soda extract, they should be tested in the residue left by action of Na_2CO_3 or in certain instances, in separate portion of original mixture.

2. During neutralization of soda extract with acids, sometimes there appears ppt. though it becomes soluble in excess of acid. It is desirable to filter off such a ppt. This is because the cations forming amphoteric oxides may partially pass into soda extract (*e.g.* AlO_2^-). The copper ion forms a complex compound and also partially passes in soda extract colouring it bluish. On subsequent neutralization, all these compounds are decomposed and corresponding cations are precipitated, *e.g.*,

$$NaAlO_2 + CH_3COOH + H_2O \longrightarrow Al(OH)_3 \downarrow + CH_3COONa$$

Similarly, copper is precipitated as basic carbonate.

3. Before adding a particular reagent to test acid radicals, extract must be made acidic because carbonate should be decomposed otherwise confusion may arise due to precipitation of the carbonates of the metals. For example, in test for sulphate, we add $BaCl_2$ solution to acidified sodium carbonate extract. The appearance of white ppt. indicates the presence of SO_4^{2-} radical. If sufficient acid is not added prior to the addition of $BaCl_2$, some undecomposed sodium carbonate will remain in solution. This will give a white ppt. of $BaCO_3$ even in absence of SO_4^{2-}. This will obviously create a confusion.

Hence in testing various radicals by the sodium carbonate extract, acidify according to the following chart:

Radicals to be tested	Acid used for neutralization	Reagent for precipitation
Chloride	dil. HNO_3	$AgNO_3$
Bromide	"	"
Iodide	"	"
Sulphate	dil. HCl	$BaCl_2$
Oxalate	CH_3COOH	$CaCl_2$

(4) Do neutralization very carefully, shaking the liquid thoroughly after each addition and avoid an excess of acid otherwise certain anions (*e.g.*, S^{2-}, NO_2^- etc.) may also be lost.

2·7. Combination Tests of Acid Radicals

Generally the following combinations are given :

1. Sulphide, Sulphite, Sulphate and Thiosulphate
2. Nitrite and Nitrate
3. Nitrite and Iodide
4. Nitrite and Sulphide
5. Nitrate in presence of Bromide and Iodide
6. Chloride, bromide and Iodide
7. Oxalate and Fluoride
8. Carbonate and oxalate
9. Carbonate in presence of Sulphite
10. Sulphite and Fluoride
11. Phosphate, Arsenate and Arsenite

All the combinations are not essential to be tested ; the radicals which respond test in preliminary amination, their combinations are to be seen and carried out.

1. Sulphide, sulphite, sulphate and thiosulphate present

Systematic analysis is required for detection of S^{2-}, $S_2O_3^{2-}$, SO_3^{2-} and SO_4^{2-} ions when present gether, because some of these interfere in test of others. For example, if the solution contains S^{2-}

and SO_3^{2-}, addition of acid liberates H_2S and SO_2 simultaneously and the two immediately interact.

$$SO_2 + 2H_2S \longrightarrow 3S\downarrow + 2H_2O$$

As a result of this reaction, only the gas present in excess will be detected. If this gas is SO_2, addition of acid would be accompanied by the same effects (formation of SO_2 and S) as would be observed if the solution contained $S_2O_3^{2-}$. Thiosulphate would, therefore, be wrongly found. It is obvious that a mixture of SO_3^{2-} and $S_2O_3^{2-}$ would behave on acidification in the same way as $S_2O_3^{2-}$ by itself.

$$Na_2SO_3 + 2HCl \longrightarrow 2NaCl + SO_2\uparrow + H_2O$$
$$Na_2S_2O_3 + 2HCl \longrightarrow 2NaCl + SO_2\uparrow + S\downarrow + H_2O$$

Finally in presence of S^{2-}, the test for SO_3^{2-} with acidified $K_2Cr_2O_7$ can not be used (H_2S also reduces acidified $K_2Cr_2O_7$ to green chromium salt). Similarly, detection of SO_4^{2-} is difficult in presence of $S_2O_3^{2-}$ or mixture of S^{2-} and SO_3^{2-}. In both cases, a white ppt. of sulphur is formed which being insoluble in acids, it may be mistaken for $BaSO_4$.

Procedure of analysis

The systematic analysis of mixture of S^{2-}, $S_2O_3^{2-}$, SO_3^{2-} and SO_4^{2-}, described in detail below, is based on (a) precipitation of S^{2-} by the action of $CdCO_3$, when SO_3^{2-}, SO_4^{2-} and $S_2O_3^{2-}$ remain in solution ; (b) precipitation of SO_3^{2-} and SO_4^{2-} by $BaCl_2$ (or $SrCl_2$) or $Sr(NO_3)_2$, which does not precipitate $S_2O_3^{2-}$; (c) the fact that $BaSO_4$ (or $SrSO_4$) is almost insoluble while $BaSO_3$ (or $SrSO_3$) is soluble in dilute acids.

Take the sodium carbonate extract and add excess of freshly prepared $CdCO_3$, shake well and filter.

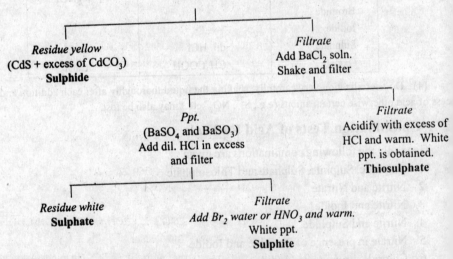

Residue yellow
(CdS + excess of $CdCO_3$)
Sulphide

Filtrate
Add $BaCl_2$ soln.
Shake and filter

Ppt.
($BaSO_4$ and $BaSO_3$)
Add dil. HCl in excess
and filter

Filtrate
Acidify with excess of
HCl and warm. White
ppt. is obtained.
Thiosulphate

Residue white
Sulphate

Filtrate
Add Br_2 water or HNO_3 and warm.
White ppt.
Sulphite

Reactions :

$$Na_2S + CdCO_3 \longrightarrow CdS\downarrow + Na_2CO_3$$

On addition of $BaCl_2$, SO_3^{2-} and SO_4^{2-} both are precipitated but HCl dissolves $BaSO_3$.

1. $$Na_2SO_3 + BaCl_2 \longrightarrow BaSO_3\downarrow + 2NaCl$$
2. $$Na_2SO_4 + BaCl_2 \longrightarrow BaSO_4\downarrow + 2NaCl$$

Br_2 water converts SO_3^{2-} to SO_4^{2-} on boiling

$$Br_2 + H_2O \longrightarrow 2HBr + [O]$$
$$BaSO_3 + [O] \longrightarrow BaSO_4\downarrow$$

(i) **Thiosulphate.** Boil the mixture with water and filter. Cool the filtrate and add to it a drops of ferric chloride solution. Dark violet colour appears which fades slowly.

$$2Na_2S_2O_3 + FeCl_3 \longrightarrow Na[Fe(S_2O_3)_2] + 3NaCl$$
$$\text{Dark violet}$$
$$Fe^{3+} + [Fe(S_2O_3)_2]^- \longrightarrow S_4O_6^{2-} + 2Fe^{2+}$$

(*ii*) **Sulphide.** This is tested by addition of $PbCO_3$ to soda extract solution which gives black ppt. of PbS.

Sulphide can also be tested by a sodium nitroprusside, $Na_2[Fe(CN)_5NO]$. Add 1 ml. of this reagent to 1 ml soda extract. A red-violet colour, due to formation of complex $Na_4[Fe(CN)_5(NOS)]$ is obtained.

2. Nitrite and Nitrate. Nitrite and nitrate, both respond to ring tests, it is, therefore, difficult to detect when these are present together.

The combination of both these radicals recommends *starch iodide test*. This is based on the fact that iodine (not iodide) with starch gives blue colour.

Acidify the soda extract with dil. H_2SO_4 (or acetic acid) and heat it gently. Add KI and starch solutions (both freshly prepared). A blue colour appears—*Nitrite*.

When nitrite is present, to test nitrate, the nitrite is decomposed first and nitrate is then converted into nitrite. Finally, the above test is applied. Therefore, proceed as below.

To test *nitrate*, acidify soda extract with acetic acid. Boil this solution with excess of NH_4Cl so long as it gives blue colour with KI and starch (test it by taking out few drops of solution separately). Now add Zn pieces and dil. CH_3COOH and boil the solution (to convert NO_3^- to NO_2^-). Add KI and freshly prepared starch solution. Blue colour appear shows *Nitrate*.

Reactions :

Nitrite :

$$2NaNO_2 + H_2SO_4 \longrightarrow Na_2SO_4 + 2HNO_2$$
$$H_2SO_4 + 2HNO_2 + 2KI \longrightarrow 2NO + I_2 + K_2SO_4 + 2H_2O$$
$$I_2 + \text{starch} \longrightarrow \text{Adsorption complex of blue colour}$$

Nitrate :

(*i*) *Removal of* NO_2^-

$$NaNO_2 + NH_4Cl \longrightarrow NH_4NO_2 + NaCl$$
$$NH_4NO_2 \xrightarrow{\text{boiling}} N_2 \uparrow + 2H_2O$$

(*ii*) *Reduction of* NO_3^-

$$Zn + 2CH_3COOH \longrightarrow (CH_3COO)_2Zn + 2H$$
$$NaNO_3 + 2H \longrightarrow NaNO_2 + H_2O$$

(*iii*) *Confirmation*

$$2NaNO_2 + 2KI + 4CH_3COOH \longrightarrow 2NO + I_2 + 2CH_3COOK + 2CH_3COONa + 2H_2O$$
$$I_2 + \text{starch} \longrightarrow \text{Blue colour}$$

3. Nitrite and Iodide. Acidify the soda extract with dil. H_2SO_4 (in excess) and add only starch (not KI as I^- may be present). If it gives blue colour—*Iodide and Nitrite both*. If it requires freshly prepared KI to develop the colour—*Nitrite alone*. However, the Iodide may also be identified as follows:

Add dil. HNO_3 or dil. H_2SO_4 to soda extract till it becomes distinctly acidic. Add $CHCl_3$ or CCl_4 solution and then Cl_2 water slowly while shaking. This gives violet colour in chloroform layer. The iodide is oxidised to I_2 which in organic layer, appears violet.

4. Nitrite and Sulphide. The nitrite is tested by starch iodide test.

To test the sulphide, we acidify the soda extract with CH_3COOH and add $AgNO_3$. A brown precipitate indicates *sulphide*.

Note. When sulphide is present, the nitrite does not give brown fumes with dil. H_2SO_4 rather precipitates sulphur.

$$2NaNO_2 + H_2SO_4 \longrightarrow Na_2SO_4 + 2HNO_2$$
$$H_2S + 2HNO_2 \longrightarrow 2H_2O + 2NO + S\downarrow$$

5.   Nitrate in presence of Bromide and Iodide. As bromide and iodide are also decomposed by conc. H_2SO_4, the brown vapours for nitrate are difficult to be recognised. Ring test for nitrate can not also be applied in presence of bromide and iodide.

However, we can proceed to test NO_3^- as follows :

Acidify a portion of soda extract with dil. H_2SO_4 and then add zinc dust and heat for 2-3 minutes. Add KI and starch solutions. (both freshly prepared). Blue colour confirms *Nitrate*.

The *bromide* and *iodide* can be tested by chloroform test [see combination 6 (*b*)].

6.   Chloride, Iodide and Bromide : (*a*) **Chloride.** The chloride radical is tested by *chromyl chloride test* : mix the powdered mixture thoroughly with almost three times of its weight of solid $K_2Cr_2O_7$ in a hard glass test tube and to it, add conc. H_2SO_4. Heat it carefully. The dark-red vapours of chromyl chloride are evolved. These are passed into dil. NaOH. The NaOH solution becomes yellow. Acidify it, with dil. CH_3COOH and then add lead acetate solution. Yellow ppt. of $PbCrO_4$ is obtained. $PbCrO_4$ ppt. is soluble in alkali hydroxides and dil. HNO_3 but insoluble in acetic acid or dil. NH_4OH. *If the yellow ppt. so formed dissolves in NaOH and reappears when excess of acetic acid is added, chloride is confirmed. Final step is the complete confirmation of the chloride ion.*

Reactions :

$$4NaCl + K_2Cr_2O_7 + 6H_2SO_4 \longrightarrow 4NaHSO_4 + 2KHSO_4 + 3H_2O + 2CrO_2Cl_2$$
$$\text{(Chromyl chloride)}$$

$$CrO_2Cl_2 + 4NaOH \longrightarrow \underset{\text{yellow solution}}{Na_2CrO_4} + 2NaCl + 2H_2O$$

$$Na_2CrO_4 + (CH_3COO)_2Pb \longrightarrow \underset{\text{yellow ppt.}}{PbCrO_4 \downarrow} + 2CH_3COONa$$

Note. Acetic acid is used to neutralize excess of NaOH as the ppt. of $PbCrO_4$ is soluble in NaOH :

$$\underset{\text{yellow ppt.}}{PbCrO_4} + 4NaOH \longrightarrow \underset{\text{soluble}}{Na_2CrO_4} + \underset{\text{souble}}{Na_2PbO_2} + 2H_2O$$

Limitations of Chromyl chloride test. 1. This test fails with chlorides of mercury and very poorly responds with the chlorides of Pb, Ag, Sb and Sn. Thus, when any of these metals are found present in the mixture, take 10 ml of soda extract in a procelain dish and evaporate it to solid residue and perform the chromyl chloride test with this residue.

2.   Fluoride too must be removed prior to application of chromyl chloride test because of similarity in the behaviour. Fluoride changes to volatile CrO_2F_2 which interferes in the chromyl chloride test.

3.   The chlorates are decomposed with conc. H_2SO_4 even in cold ; the greenish yellow and explosive chlorine dioxide, ClO_2, gas is formed and gentle warming may cause a violent explosion. Hence never perfom chromyl chloride test in presence of chlorates.

$$3KClO_3 + 3H_2SO_4 \longrightarrow 2ClO_2 + HClO_4 + 3KHSO_4 + H_2O$$

4.   If the mixture contains iodide, the ratio of iodide to chloride should never be greater than 1 : 16 otherwise the chromyl chloride test fails. Therefore, in presence of iodide proceed as follows :

Acidify the soda extract with dil. HNO_3 and boil off CO_2. Add an excess of $AgNO_3$ solution and filter. Reject the filtrate. Wash the ppt. with distilled water. Now shake the ppt. with dil. NH_4OH and filter. To the filtrate add an excess of dil. HNO_3. White ppt. or milkiness confirmes *Chloride*.

(*b*) **Iodide and Bromide.** Acidify 1 ml soda extract with dil. HCl or dil. H_2SO_4 and add few drops of $CHCl_3(CCl_4$ or $CS_2)$. Add freshly prepared chlorine water drop by drop while shaking. If $CHCl_3$ layers turns violet $\longrightarrow$ Iodide confirmed. Add excess of chlorine water with vigorous shaking. If orange-brown colour appears in $CHCl_3$ layer $\longrightarrow$ Bromide confirmed.

Reactions

Since out of three halogens, iodine is least electronegative, iodide is liberated first by adding chlorine water. Liberated iodine being soluble in $CHCl_3$, CCl_4 and CS_2, produces its violet colour in organic layer.

$$2NaI + Cl_2 \longrightarrow 2NaCl + I_2$$

Chlorine also displaces bromine from bromide salts. The iodide which imparts violet colour to organic layer, obscures the orange-brown colour of bromine (liberated from the bromide salt). The excess of chlorine water oxidises the free iodine to iodic acid which is colourless and then the bromine colour can be noticed.

$$I_2 + 5Cl_2 + 6H_2O \longrightarrow 2HIO_3 + 10HCl$$
$$\text{(Iodic acid)}$$

On further addition of chlorine water, more bromine is liberated from bromide and dissolves into organic layer and a clear orange-brown colour is developed.

$$2NaBr + Cl_2 \longrightarrow 2NaCl + Br_2$$
$$Br_2 + CHCl_3 \longrightarrow \text{Orange-brown colour}$$

[Note : In presence of reducing agents S^{2-}, SO_3^{2-} or $S_2O_3^{2-}$ (which are stonger reducing agents than I^- and Br^-), chlorine water is first consumed for oxidation of these ions. However, as the concentration of free Cl_2 in chlorine water is low, a large volume may be required for this oxidation. Therefore, S^{2-}, SO_3^{2-} and $S_2O_3^{2-}$ must first be oxidised by $KMnO_4$ in acid solution ($KMnO_4$ is a convenient oxidising agent because it can be obtained in much more concentrated solution than Cl_2. Moreover, change of colour indicates the course of oxidation). Acidify 1 ml. soda extract strongly with dil. H_2SO_4 and add $KMnO_4$ solution drop by drop with continuous shaking until a purple (excess $KMnO_4$) or yellowish-brown (liberation of I_2 or Br_2) colour appears and persists even after shaking. To remove the colour add very-very dilute $Na_2S_2O_3$ solution drop by drop, shaking continuously ; avoid the slightest excess of $Na_2S_2O_3$ as it interferes with test for I^- and Br^-. The solution is now free from reducing agents ; test it for I^- and Br^- as described above.

This method for removing reducing anions is rapid and gives very good results but only with careful and precise working. If I^- and Br^- are not detected, the following procedure may be recommended to confirm that the operations have been carried out correctly. Put one or two drops of KI solution into the test tube in which the test for I^- and Br^- was performed and which, therefore, already contains CCl_4 and Cl_2 water. If the violet colour characteristic of iodine appears on shaking, the solution does not contain anything which can interfere in detection of I^- and Br^-, and their absence is confirmed. On the other hand, if violet colour does not appear, it follows that either reducing agent has not been completely removed or an excess of $Na_2S_2O_3$ had been used. In that case the oxidation procedure must be repeated with fresh portion of the sodium carbonate extract].

7. **Oxalate and Fluoride.** This may be confirmed by either of these methods :

First method. Add CH_3COOH to soda extract till it becomes distinctly acidic. Add $CaCl_2$ in excess where oxalate as CaC_2O_4 and fluoride as CaF_2, both as white ppt. are obtained.

Add dil. HCl

(a) If the precipitate is soluble, the *oxalate* is only present.

(b) If the ppt, is not at all soluble, it should be *fluoride*.

However, if both are present, it fails and we proceed to another method.

Second Method. Neutralise the soda extract with dil. HNO_3 and add $AgNO_3$. Filter the precipitate and do the following treatment ;

(a) *Residue treatment.* Add to it, dil H_2SO_4 and few drops of $KMnO_4$ and warm. The colour of $KMnO_4$ is discharged $\longrightarrow$ *Oxalate.*

(b) *Filtrate treatment.* Add Barium nitrate to this filtrate. This gives white precipitate of Barium fluoride $\longrightarrow$ *Fluoride.*

8.   Carbonate and oxalate. Both are decomposed by conc. H_2SO_4 but carbonate is decomposed also by dil. acids. Hence the oxalate can be tested only after destroying the carbonate.

Take little mixture in a test tube and add dil. H_2SO_4 and warm. If effervescence take place, pass the gas into lime water. If lime water turns milky, *carbonate* is present. Heat the solution in test tube till no effervescence occurs. Now add a pinch of MnO_2. Fresh effervescence confirms the *presence of oxalate*. Oxalate can also be confirmed by its usual *sodium carbonate extract test.*

9.   Carbonate in presence of Sulphite. The carboante can not be tested by a soda extract as it has already some Na_2CO_3.

Both, sulphite and carbonate are decomposed by dil. acid (HCl, H_2SO_4) and gases SO_2 and CO_2 both turn lime water milky. The dichromate test for sulphite is, however, not affected by presence of carbonate but sulphite interferes with carbonate. Hence to test carbonate in presence of sulphite, proceed as follows :

Take little mixture in a test tube and add dil. H_2SO_4 and some solid $K_2Cr_2O_7$, warm and pass the gas into lime water. If lime water still turns milky, *carbonate* is present.

Reactions :

1. $$CO_2 + Ca(OH)_2 \longrightarrow CaCO_3 + H_2O$$
$$\text{(Milky)}$$

2. $$Na_2SO_3 + 2HCl \longrightarrow 2NaCl + H_2O + SO_2 \uparrow$$
$$SO_2 + Ca(OH)_2 \longrightarrow CaSO_3 + H_2O$$
$$\text{(Milky)}$$

On passing excess of gases

CO_3^{2-} case : $$CaCO_3 + H_2O + CO_2 \longrightarrow Ca(HCO_3)_2$$
$$\text{(Colourless)}$$

SO_3^{2-} case : $$CaSO_3 + H_2O + SO_2 \longrightarrow Ca(HSO_3)_2$$
$$\text{(Colourless)}$$

In presence of dichromate, sulphite which produces SO_2 is used up as follows :
$$K_2Cr_2O_7 + H_2SO_4 + 3SO_2 \longrightarrow K_2SO_4 + Cr_2(SO_4)_3 + H_2O$$
$$\text{(Green)}$$

10.   Sulphate in presence of fluoride. Fluoride is tested in usual way (*i.e.*, mix. + sand + conc. $H_2SO_4 \longrightarrow$ white deposit on the rod moistened with water).

However, presence of fluoride interferes with $BaCl_2$ test for sulphate. By addition of $BaCl_2$ solution to soda extract, acidified with dil. HCl, a white ppt. of barium fluoride is obtained which is insoluble in HCl even if sulphate is absent. Hence to test SO_4^{2-} in presence of F^-, we proceed as follows :

Acidify soda extract with excess of acetic acid, add lead acetate solution and warm. A white ppt. is produced which confirms presence of sulphate.

Reactions :
$$Na_2SO_4 + (CH_3COO)_2Pb \longrightarrow PbSO_4 + 2CH_3COONa$$
$$\text{(insoluble)}$$

$$2NaF + (CH_3COO)_2Pb \longrightarrow PbF_2 + 2CH_3COONa$$
$$\text{(soluble)}$$

The ppt. of $PbSO_4$ is insoluble whereas PbF_2 is soluble in acetic acid. In this way, fluoride fails to interfere in the above test.

11.   Phosphate, Arsenate and Arsenite in presence of each other.

Acidify a portion of sodium carbonate extract with dil. HCl and then pass H_2S ; immediate precipitation of yellow As_2S_3 confirms *arsenite*. Continue to pass H_2S until no more precipitation occurs. Filter off the ppt. Boil off H_2S from the filtrate then add little dil. HCl and 1-2 pinch of Na_2SO_3 and heat (SO_2 produced will reduce arsenate to arsenite), boil off excess of SO_2 (test with $K_2Cr_2O_7$ paper) and then pass $H_2S \longrightarrow$ appearance of yellow ppt. of As_2S_3 confirms *arsenate* ; continue to pass H_2S until no more ppt. forms. Filter off the ppt.

Evaporate the filtrate to a small volume, add 2-3 ml conc. HNO_3 and heat. Now add excess of ammonium molybdate solution and warm $\longrightarrow$ appearance of canary yellow ppt. confirms *phosphate*.

[**Note.** Arsenate may also be reduced by adding 1 ml of 10% NH_4I solution and 5 drops of conc. HCl followed by boiling.]

Analysis of Basic Radicals as Cations

(Macro Method)

General

Unlike the analysis of acid radicals, this analysis requires the preparation of solution of the original mixture. This is mainly because once the solution is prepared, it can be taken to undergo various chemical changes which precipitate the similar type of ions at one time and other types at different times. This way, we divide the analysis of cations into six groups. The characteristics of the true solutions are :

1. It should be transparent having no visible particles of the mixture. This may have, ofcourse, any colour depending upon the ions present in it.

2. This should not undergo sedimentaion, i.e., its constituents should not settle on the bottom of the vessel.

3·1. Merits and Demerits of Acid Solvents

1. Hydrochloric Acid

Merits. Since the chlorides are more volatile than nitrates, cations dissolved in HCl impart the flame test better than in HNO_3. HCl can be easily evaporated off. The water-insoluble oxidising agents, e.g., SnO_2, MnO_2, etc. are insoluble in HNO_3 but dissolve in HCl. HCl is not an oxidising agent hence it is frequently used for preparing the solution of mixture where H_2S is empolyed as precipitant

Demerits. The sulphides of Cu, Pb, Bi, As, Sb, Co and Ni are insoluble in conc. HCl but dissolv in conc. HNO_3. In certain cases, Cl^- ions in high concentration work as complex forming agent e.g. $HgCl_2$ forms $HgCl_4^{2-}$ ion with excess of Cl^- ion and this complex ion is too stable to detect Hg^{2+} ion unless the complex is destroyed. Certain chlorides e.g., arsenic and mercuric chlorides are vapourise from hot HCl solution.

2. Nitric Acid

Merits. The nitrates are less volatile than chlorides. When the solution is evaporated to remo excess of HNO_3, there are less chances for losing nitrates. All nitrates are soluble in water ar generally more soluble than chlorides. HNO_3 behaves as solvent and as an oxidising agent. T nitrate ions in presence of hydronium ions (H_3O^+) oxidise certain substances to a soluble form whi are unaffected by hydronium ions alone.

Demerits. When the solution has been prepared, the oxidising action of HNO_3 is undesirab throughout the analysis of cations. HNO_3 oxidises sulphide ions into colloidal sulphur and to son extent into sulphate ions. The precipitated sulphur creates unnecessary disturbance and confusion precipitation tests for certain cations ; the sulphate ion produced by oxidation of H_2S, will cause precipitation of V group cations and lead as sulphates in II group where they are unwanted and the is no provision to identify them in II group. HNO_3 will not dissolve water-insoluble oxidising age such as MnO_2, SnO_2 and PbO_2.

3. Sulphuric Acid.
For the purpose of solubility tests H_2SO_4 is least desirable of the th acids HCl, HNO_3 and H_2SO_4. Several sulphates are insoluble, or may be only partly soluble in wa or in acids.

Note. As a general rule, it is preferred to use a single acid if possible.

3·2. Preparation of Solution

Selection of the Solvent. To *prepare the solution, the selection of the suitable solvent is* neces-sary. This may be done by using one of the following reagents. Mixture should be first tried with reagents in cold, if not found soluble then contents should be heated. The following solvents are tried one by one in the order given below :

1. Water.
2. Dil. HCl.
3. Conc. HCl alone (add a few drops of Br_2 water if needed).
4. Dil. HNO_3 ⎫ (Avoid as far as possible to prepare
5. Conc. HNO_3 ⎭ solution with HNO_3).

1. Water. As it is clear from above, first of all take 5 ml of water and a very small quantity of the mixture, *i.e.*, 0.1 gm is put into it. Shake it well. If the substance disappears in the solution, we assume it to be soluble. When it is not soluble in cold, we heat the contents and observe its solubility. A general idea about water soluble compounds is given in chapter 1.

2. Dil. HCl. Finding insolubility of mixture in water, take again the same quantity of mixture, add 2 ml of dil. HCl to it and do similarly, *i.e.*, first in cold then in hot.

Sometimes, it has been observed that while preparing the solution in dil. HCl, we get *white precipitate.* This indicates the presence of first group in the mixture. Similarly, when HCl dissolves the mixture completely giving a transparent solution, we assume the absence of first group members. The white precipitate by addition of HCl does not always ensure the presence of first group members.Certain salts such as NaCl, $BaCl_2$ and some other chlorides when present substantially, precipitate down in HCl but this precipitate disappears on dilution with water.

3. Conc. HCl. When the dilute acid fails to dissolve, we take, 1-2 ml of conc. HCl to dissolve 0.1 gm of mixture in similar way. If the mixture is soluble, we prepare the solution and dilute the solution with water finally.

If it is not soluble in conc. HCl, add few drops of Br_2 water, boil and observe its solubility.

It is necessary to dilute this solution with water otherwise some of the second group radicals will not be precipitated.

4. Dil. HNO_3. The third solvent with which we try is dil. HNO_3. As earlier, here also take 0.1 m of the mixture in test tube and add 2 ml of HNO_3. Shake it well to dissolve. If it does not, heat it and observe.

5. Conc. HNO_3. Finally try with conc. HNO_3 *i.e.*, first in cold then in hot. Mostly the mixtures re soluble in it. Boil off excess of HNO_3. Like the solution obtained by conc. HCl, it must also be diluted. Conc. HNO_3 solution will be highly oxidising in its nature and when H_2S is passed to pre-pitate second group radicals, it will take hydrogen from H_2S leaving sulphur precipitated or will ve yellow white suspension. This creates complication. Hence avoid to prepare solution in HNO_3.

After finding out the suitable solvent, take sufficient amount (0.5 to 0.1 gm) of the mixture and ake its solution.

Proper Procedure for Preparation of Starting Solution for Cation Analysis. In a boiling be, take about 1/3rd amount (0.5—1.0 gm) of the original mixture, add 5 ml of water, boil and filter. llect the filtrate (say A). Transfer the undissolved portion to a boiling tube, add 5 ml of dil. HCl, il and observe.

(a) If transparent solution then mix with filtrate A. Proceed with this for group precipitation.

(b) If not transparent, add again 1-2 ml of dil. HCl to the same solution, heat and observe. If the ution does not become clear, add 2-3 ml of water to the same solution. If it becomes clear barium or sodium or both are indicated ; otherwise filter. Collect the filtrate (say B).

Transfer the undissolved portion in step (b) in a boiling tube, add 2-3 ml conc. HCl and heat till almost half volume is left. Dilute by adding 2-3 ml water, boil and filter. Collect the filtrate (say C). If still any undissolved portion is left, reject it. Mix filtrates A, B and C. Proceed with this for group precipitation. *On mixing, if any white ppt. appears which does not dissolve by adding more quantity of dil. HCl, I group is taken present. Filter and use the filtrate for II group precipitation. When I group is absent, use the original solution for further group precipitation.*

3·3. Group separation

The flow-sheet given below reveals in brief the total procedure of cationic separation. NH_4^+ is tested from original solution.

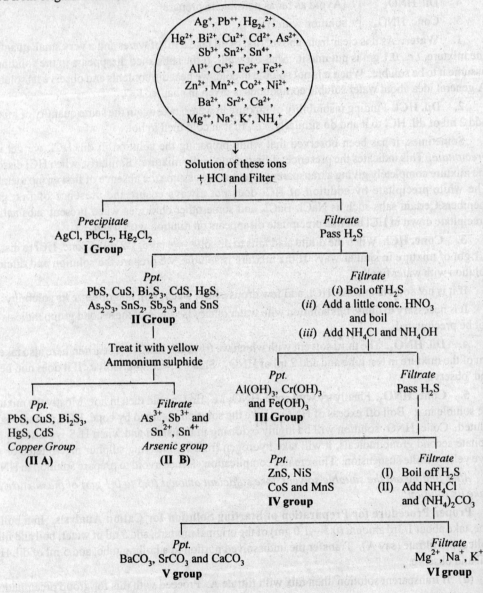

3·4. Basis of Group Separation

Unlike acid radicals, the analysis of basic radicals is quite systematic. Once the original solution is prepared, various radicals from it are precipitated groupwise in the form of suitable compounds. The ionic concentration of precipitating reagent is kept controlled using common ion effect.

The first group radicals are precipitated in the form of *their chlorides* and the reagent used is dilute HCl acid. The K_{sp} value of first group chloride is very low and, infact, the lowest of chlorides of all analytical groups.

The second group radicals are precipitated from the filtrate of the first group in form of *their sulphides* using H_2S gas as precipitating reagent. Likewise, the K_{sp} value of II group sulphides is lowest of II to VI group sulphides. However, these values of IV group sulphides are not much higher and are next to II group sulphides.

Since HCl is already in filtrate, it suppresses the concentration of S^{--} ions due to H^+ ion common with that of H_2S.

$$H_2S \rightleftharpoons \boxed{2H^+} + S^{--} \qquad \text{(overall ionisation)}$$
$$HCl \longrightarrow \boxed{H^+} + Cl^-$$
$$\text{common ion}$$

From the filtrate of II group, the third group radicals are precipitated in the form of their hydroxides. The concentration of hydroxide ions should also be quite low lest the hydroxides of higher group should precipitate. NH_4OH is used as a precipitating reagent and NH_4Cl is used to suppress the concentration of OH^- ions.

The fourth group radicals are precipitated as their sulphides from filtrate of III group and these need large concentration of S^{--} ions. The alkaline medium helps in doing so :

$$NH_4OH \rightleftharpoons NH_4^+ + OH^-$$
$$H_2S \rightleftharpoons 2H^+ + S^{--} \qquad \text{(overall ionisation)}$$

H^+ and OH^- ions join together to form non-electrolyte water. To make up the loss of H^+ ions. H_2S ionises more and more and thus provides a high concentration of S^{--} ions.

The fifth group radicals are precipitated in form of their carbonates from the filtrate of the fourth group. Ammoniacal medium is already here to control the concentration of CO_3^{--} ions by NH_4^+ ion common.

Finally in sixth group, Mg^{2+}, Na^+ and K^+ are tested independently.

3·5. Analysis of Group I

1. Radicals—Ag^+, Pb^{2+}, Hg_2^{2+}.
2. Group Reagent—Dil. HCl.

Add dilute HCl to the original solution. If first group is present, a white precipitate of the chlorides of first group ($PbCl_2$, AgCl and or Hg_2Cl_2) is obtained. Filter it and keep the filtrate for subsequent groups. Boil the residue with water and filter it hot.

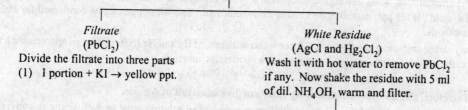

Filtrate	White Residue
($PbCl_2$)	(AgCl and Hg_2Cl_2)
Divide the filtrate into three parts	Wash it with hot water to remove $PbCl_2$
(1) 1 portion + KI → yellow ppt.	if any. Now shake the residue with 5 ml
	of dil. NH_4OH, warm and filter.

(2) II portion + K_2CrO_4
     $\rightarrow$ yellow ppt.
     soluble in NaOH

Filtrate
$[Ag(NH_3)_2]Cl$
Divide into 3 parts
(a)   Acidify 1st portion with
     dil. $HNO_3 \rightarrow$ white ppt.
     (AgCl)

(3) III portion + dil. H_2SO_4
     $\rightarrow$ white ppt. ($PbSO_4$)

(b)   2nd portion + KI
     solution $\rightarrow$ pale yellow
     ppt. (AgI)
(c)   3rd portion + K_2CrO_4
     $\rightarrow$ Brick red ppt. (Ag_2CrO_4)

Lead

Silver

Residue (Black)
$[Hg \,\&\, Hg(NH_2)Cl]$

It gradually turns black. Dissolve this
residue in Aqua regia (1 part conc. HNO_3
+ 3 parts conc. HCl) and add $SnCl_2$ solution $\rightarrow$ A white ppt. turning to grey
colour.

Mercurous, Hg_2^{2+}

[**Note.** If it is mercuric salt, it will not be precipitated in the first group because $HgCl_2$ is soluble in HCl. It is then precipitated in the second group.]

Notes

(i)   If the solution is not filtered while being hot, $PbCl_2$ which should have come in filtrate, is retained in residue (as it is insoluble in cold) and the filtrate contains nothing but water.

(ii)   The white ppt. sometimes disappears from the solution on adding water to it. This shows the absence of I group chlorides. This may be due to higher group radicals which precipitate in large concentration of Cl^- ions as their chlorides.

(iii)   Lead is not completely precipitated in the first group because its precipitate ($PbCl_2$) is partially soluble. It passes slightly in II group also.

(iv)   The solution obtained in HCl when diluted with H_2O, sometimes gives white ppt. This may also be due to the formation of BiOCl and SbOCl. Such precipitation disappears by adding 2–3 drops of conc. HCl.

(v)   Silver and lead should be carefully distinguished as both with KI add K_2CrO_4 give coloured precipitate. Silver chromate is soluble in dil. HNO_3.

(vi)   To confirm Pb^{2+}, KI should not be added in excess as it makes lead soluble K_2PbI_4.

(vii)   The large concentration of HCl is not desirable as it assists the formation of the complex ion, $AgCl_3^{2-}$ and $PbCl_4^{2-}$ which are soluble.

(viii)   When the solution is prepared in conc. HCl and a white ppt. appears which dissolves on adding water, the presence of Ba^{2+} and Na^+ is indicated.

(ix)   Sometimes a white ppt. is obtained by adding dil. HCl to the mixture. The ppt. may also be due to boric acid from borate. This ppt. dissolves in cold on adding water. Hence test the borate before the separation of basic groups.

(x)   Sometimes, when the original solution of mixture is prepared in dil. HCl white cyrstalline ppt. is ob-tained in cold. If the ppt. dissolves on heating Pb is indicated. The white ppt. can be confirmed for Pb^{2+} by usual methods.

(xi)   If the ammoniacal solution is left in contact with ppt. of Hg and Hg (NH_2) Cl for an appreciable length of time, the soluble silver complex ion may react with metallic mercury.

$$2Hg + 2Ag(NH_3)_2^+ \longrightarrow Hg_2^{2+} + 2Ag + 4NH_3$$

Therefore, a small concentration of Ag^+ may not give usual test of Ag^+ ion.

(xii)   Test the acidity of the solution with litmus paper. The solution must be well acidic to convert the $Ag(NH_3)_2^+$ to AgCl, otherwise, even though silver ion be present, no precipitate will be formed.

(xiii) If mercurous ions are present, the residue left after treatment with ammonia must be black because of precipitation of colloidal mercury.

$$Hg_2Cl_2 + 2NH_3 \longrightarrow HgNH_2Cl + Hg + NH_4^+ + Cl^-$$

A white ppt. at this point may be either lead oxychloride formed by ammonia-action upon some undissolved lead chloride, or excess silver chloride undissolved by the ammonia.

Specific Tests

Lead. (1) This gives *red* or *reddish violet colouration* with *p*-dimethyl amino ber.zilidine rhodamine. (2) To the original solution, add 2 drops of dithiozone and shake. *Red colour.*

Silver. Ammoniacal solution of it, when treated with aqueous solution of Resorcinol, gives *blue colour* slowly.

Mercurous. Solution when treated with alcoholic solution of gallic acid gives an *orange* or *yellow* colour.

3·6.   Analysis of Group II

Group Reagent—H_2S in the presence of HCl.

Take the filtrate from first group or the original solution + HCl if first group is absent. Divide it in two parts and heat them. In first part, pass H_2S directly. Dilute the second part to double of its volume with water and then pass H_2S. It is done because

1.   Arsenic is most readily precipitated in moderate amount of HCl as in very low concentration of HCl arsenic sulphide forms a soluble complex AsS_2^- (due to excess of S^{--} ions).

2.   Cadmium is completely precipitated in very low concentration of HCl as CdS is soluble in excess of HCl.

Note : The following points are to be kept in mind while passing H_2S :

(i) The solution should be hot when H_2S is passed.

(ii) If mixture contains NO_3^-, NO_2^- or SO_3^-, solution should be boiled for few minutes to ensure their decomposition which otherwise will precipitate sulphur (milky solution appears on passing H_2S) because HNO_3, HNO_2 and SO_2 are oxidising agents and they oxide H_2S to free sulphur.

$$2HNO_3 + H_2S \longrightarrow S\downarrow + 2H_2O + 2NO_2 \uparrow$$
$$2HNO_2 + H_2S \longrightarrow S\downarrow + 2H_2O + 2NO \uparrow$$
$$SO_2 + 2H_2S \longrightarrow 3S\downarrow + 2H_2O$$

The precipitate so obtained in dilute and conc. solution is mixed. The following table contains all the second group radicals, their precipitated forms and colours.

Radicals	Precipitated as	Colour of the ppt.
1. Hg^{2+}	HgS	White, changing to yellow and finally to black
2. Pb^{2+}	PbS	Black, may be Red
3. Bi^{3+}	Bi_2S_3	Dark Brown or Black
4. Cu^{2+}	CuS	Black
5. Cd^{2+}	CdS	Yellow
6. As^{3+}	As_2S_3	Bright yellow
7. Sn^{2+}	SnS	Dark brown
8. Sb^{3+}	Sb_2S_3	Orange
9. Sn^{4+}	SnS_2	Dull or dirty yellow

Filter the precipitate and keep the filtrate for subsequent groups. Take its residue in a porcelain ish, add 5 ml of yellow ammonium sulphide $(NH_4)_2S_x$ and heat it to nearly 60°C with constant irring for 5 minutes and filter.

(a) Residue contains copper group, IIA (HgS, PbS, Bi_2S_3, CuS and CdS).

(b) Filtrate contains arsenic group, IIB (nitrates of As, Sb and Sn).

Copper Group Analysis (IIA Group)

Wash the residue with hot water and then boil it with 5-10 ml HNO_3 (one part conc. HNO_3 + one part water) for 3-4 minutes and filter

residue (Black)

dissolve it in aqua-regia. Boil the solution, dilute it with water and divide :

I portion + $SnCl_2$ in excess → white ppt. turning grey.

II portion + Cu turning → white deposite

Mercuric, Hg^{2+}

Filtrate (1)

may contain

$[Pb(NO_3)_2, Bi(NO_3)_3, Cu(NO_3)_2$ and $Cd(NO_3)_2]$ Take small portion of it and add dil. H_2SO_4. If white ppt. comes, Pb^{2+} is present. In such a case, add acid to rest of the filtrate (1). Filter and test the residue for Pb^{2+} as in first group. Keep this filtrate (2) to proceed. If Pb^{2+} is absent, need not add dilute H_2SO_4 to the filtrate (1) and proceed. Add NH_4OH solution while shaking until it smells ammonia. A white precipitate shows the presence of Bi^{3+}.

Residue

$Bi(OH)_3$

Dissolve it in least amount of dilute HCl and divide.

I portion + excess of water → white turbidity

II portion + sodium stannite solution in excess and cool ⟶ Black ppt.

BISMUTH

Filtrate

$[Cu(NH_3)_4^{2+}$ and $Cd(NH_3)_4^{2+}]$

If we get blue colour after adding NH_4OH, this shows presence of copper but cadmium may or may not be present. In such a case, divide blue soln. into two parts and test these as follows :

(a) **Test for Copper.** To one part add CH_3COOH to acidify and then $K_4Fe(CN)_6$ soln. ⟶ Chocolate brown ppt. **Copper.**

(b) **Test for Cadmium.** Divide second part into two :

(i) To one part, add KCN (poison) soln. drop by drop until blue colour disappears. Pass H_2S ⟶ yellow ppt. shows *cadmium.*

Note. *Pass H_2S for a short time.*

(ii) To second part, add excess of dil. H_2SO_4 to make soln. strongly acidic. Now add a little zinc dust, boil and filter. To the filtrate add NH_4OH to neutralize excess of acid, then add 1 ml, dil. HCl and pass H_2S ⟶ yellow ppt. **Cadmium**

Note : *Iron powder can also be used in place of zinc dust.*

Analysis of Arsenic Group (II B Group)

Warm the filtrate (obtained after the addition of yellow ammonium sulphide) with dilute HCl. I white precipitate (of sulphur) comes, filter it off. Now add 10 ml of conc. HCl and boil it for 2- minutes. Dilute it with about 5 ml of hot water and filter :

Residue   *Filtrate*

[Arsenic as As_2S_3 or Ar_2S_5 or   [$SnCl_4$, $SbCl_3$]
both and sulphur]

Divide and dissolve the residue separately   Boil the filtrate to remove H_2S completely
in   and divide the solution in two parts.

(*a*) Conc. HNO_3 and add excess of amm.   (*a*) I Part + NH_4OH till ammoniacal. Add
      molybdate and heat $\longrightarrow$ yellow ppt.   solid oxalic acid, boil and pass $H_2S \longrightarrow$
                                                                            orange ppt. (Sb_2S_3).

                                                                                    Antimony

(*b*) Hot amm. carboante. Acidify with dil.   (*b*) II part + 1-2 iron nails (or 2 cm iron wire
      HCl. Yellow ppt. is obtained with or   or little iron powder or 2 zinc pieces). Boil
      wihtout passing H_2S.   and filter.

                  Arsenic   (*i*)   If black residue $\longrightarrow$ **Antimony.**

                                                           (*ii*) Filtrate + $HgCl_2$ soln. $\longrightarrow$ white ppt.
                                                                  ($HgCl_2$) changing to grey (Hg) on standing.

                                                                                       Tin

Notes

1. (*a*) If the mixture contains NO_2^- or NO_3^- or SO_3^{2-}, the original solution prepared in HCl is boiled for a long time in order to decompose these radicals.

(*b*) If the original solution is prepared in HNO_3, evaporate the solution to dryness, add. dil. HCl, boil and filter if necessary and then pass H_2S.

The above operations are necessary because if S^{2-} and SO_3^{2-} are present in mixture, they are oxidised to sulphate. If 5th group basic radicals are also present, they will precipitate as sulphates (white ppt.) along with II group sulphides.

2. If nitrate is present in mixture, it should also be destroyed by boiling a little mixture with 1 gm solid NH_4Cl and water. Then add dil. HCl to it and prepare the clear solution, dilute it and pass H_2S.

3. For complete precipitation of II group basic radicals, the solution before passing H_2S must be properly acidic, *i.e.*, it should contain about 0.3 N-HCl. For this purpose we should proceed as follows :

(*a*) If the original solution of mixture is prepared in dil. HCl (which is about 4–5 N), dilute the solution by adding 10–12 times water.

(*b*) If concentrated acid is required to prepare the original solution, take a little mixture in a boiling tube and 2-3 ml conc. HCl and boil off almost completely. Now add to it 5 ml dil. HCl, boil and filter if necessary and dilute by adding nearly 50 ml of water and then pass H_2S.

The reasons for maintaining proper acidity are :

(*i*) In highly acidic medium (*i.e.*, solution contains too much HCl) $\longrightarrow$ Cd, Pb, Sn are difficult to be precipitated as sulphides. Sometimes in too much acidic medium if Pb is present, a red ppt. of $PbCl_2$, PbS is obtained when H_2S is passed in the solution.

(*ii*) If solution is less acidic (*i.e.*, it contains very little HCl) $\longrightarrow$ ZnS etc. are precipitated along with II group sulphides and arsenic does not completely precipitate.

To ensure complete precipitate of II group basic radicals, the filtrate should be repeatedly diluted by adding water and H_2S is passed so long as ppt. appears. When the filtrate does not give ppt., the precipitation is complete and use this filtrate for III group analysis.

4. While passing H_2S the white precipitate may be seen, which may be due to mercury but this is for a moment and such ppt. converts to yellow and finally black. This gives an idea about the presence of mercury. The white precipitate has probably a composition of $HgCl_2.HgS$. A long standing white precipitate is of only sulphur and this should be discarded.

5. A white suspension formed after continued H_2S treatment is the result of oxidation of H_2S to produce free sulphur. This suspension passes out through filter paper and thus it does not indicate the presence of any II group sulphides.

6. In case, the mixture contains arsenate, a yellow ppt. is presistently formed by repeatedly passing H_2S in hot solution and complete precipitation takes more time. In order to hasten the precipitation, add NH_4I to the original solution and then pass H_2S. In the solution NH_4I produces HI due to hydrolysis. HI and H_2S will reduce arsenate to arsenite and hence will hasten the precipitation of As_2S_3.

7. Pentavalent As is precipitated only when the solution is hot and strongly acidic.

8. Test the complete precipitation of II group sulphide by filtering off 1-2 ml of liquid and adding an equal volume of freshly prepared H_2S water (instead of passing H_2S). If ppt. is formed, pass H_2S again through the whole solution after diluting it again with water. When precipitation is complete, filter off the ppt. and wash it with H_2S water containing small amount of NH_4Cl. The H_2S in wash water retards oxidation of precipitated sulphides to sulphates. Another advantage of using H_2S water is that it avoids error if incomplete precipitation is due to excessive acidity of the solution. However, it must be freshly prepared and must smell strongly of H_2S. NH_4Cl is added to help in breaking up any colloidal dispersion which may be present.

9. Treatment with 1 : 1 HNO_3 in the beginning of II group analysis may convert sometimes black H_2S to white $Hg(NO_3)_2.2HgS$ and this gives white colour. Therefore, the lack of black ppt. of HgS at this stage does not confirm the absence of mercuric iron.

10. Aqua regia dissolves HgS by two different types of reactions : the oxidation of sulphide ion to free sulphur by nitrate ion and the complexing action of chloride ion to produce the $HgCl_4^{2-}$ ion. A small insoluble residue after the aqua regia treatment, is free sulphur.

$$3HgS + 2NO_3^- + 8H_3O^+ + 12Cl^- \longrightarrow 3HgCl_4^{2-} + 2NO + 3S + 12H_2O.$$

11. Aqua regia liberate chloride which, if not removed, will oxidise stannous ion to stannic ions. Only stannous ion will give the characteristic test for mercuric ions.

12. Before confirming lead by K_2CrO_4, the solution should be made acidic with CH_3COOH otherwise, bismuth which might be present will also precipitate as chromate.

13. To prepare sodium stannite, add NaOH solution, drop by drop to $SnCl_2$ solution until white ppt. of $Sn(OH)_2$ first formed redissolves. The equations for these reactions are : $Sn^{2+} + 2(OH)^- \longrightarrow Sn(OH)_2$; and $Sn(OH)_2 + OH^- \longrightarrow Sn(OH)_3^-$.

14. The KCN solution is added to form the stable complex ion $Cu(CN)_4^{3-}$, by decomposing the cupric ammonia complex, $Cu(NH_3)_4^{2+}$, to prevent interference with cadmium test. In Cu and Cd separation while passing H_2S gives sometimes brown ppt. This indicates incomplete separation.

Specific Tests

1. **Arsenic.** The residue obtained to test the arsenic is dissolved in NH_4OH (throw the undissolved portion). Acidify the solution with dil. HCl. It will give a *yellow precipitate.*

2. **Bismuth.** (*i*) Bismuth may be confirmed by cinchonine potassium iodide complex which with the solution mixture gives *red precipitate.*

[To prepare the reagent, we dissolve 1 gm of cinchonine in 100 ml of water. This is acidified with conc. HNO_3. At last 2 or 3 gms KI is also added to it.]

(*ii*) Add dil. HNO_3 to original solution and some thio-urea. A yellow colouration confirms the presence of bismuth.

3. **Copper.** When aldoxine is added to acidified solution (made by CH_3COOH) of mixture, a *green precipitate* shows the presence of copper.

4. **Lead.** The ammoniacal solution of mixture when treated with 5–10% aqueous solution of resorsinol, the appearance of *blue colour* after few minutes indicates the presence of lead.

5. **Cadmium.** It may be confirmed by thio-cinnamine. The solution is made alkaline by NaOH and 2 or 3 drops of the reagent is added to it. On heating, it gives yellow precipitate.

6. **Antimony.** Pentavalent antimony when added to (.01 to 5%) solution of Rhodamine B, bright red colour is developed to confirm antimony.

7. **Mercury.** When alcoholic solution of diphenyl carbazide is added to mercuric salt solution and little dil. HNO_3 is added, a *violet blue colour* is developed.

In case mixture contains mercurous salts, an alcoholic solution of gallic acid gives an *orange* or *yellow colour*.

8.   **Tin.** Tin in bivalent (ous state) gives light yellow colour with cacothelin. The importance of the test can be extended to stannic salts also, but first of all these are to be reduced by zinc and HCl.

3·7. Interfering Acid Radicals, and their Removal

Cause of interference. Oxalate, borate, fluoride, tartarate and phosphate are known as interfering acid radicals because of their interference in the regular course of analysis of basic radicals. *In fact their salts are soluble in acid medium and become insoluble in alkaline medium.* As long as the analysis remains under the grip of acid medium, they remain soluble and get no chance to interfere. After second group, the solution is made alkaline and interfering radicals are also precipitated along with the third group hydroxides. That is to say that radicals of IV, V and VI groups have fair chances to precipitate as phosphates, oxalates, borates, fluorides or tartarates. This really creates a problem in the systematic analysis. The reason for the interference may be explained as follows :

In acid medium these remain in the following equilibria.

$$HCl \rightleftharpoons H^+ + Cl^-$$

1.	Borate	H_3BO_3	$\rightleftharpoons 3H^+ + BO_3^{3-}$	(overall ionisation)
2.	Oxalate	$H_2C_2O_4$	$\rightleftharpoons 2H^+ + C_2O_4^{2-}$	
3.	Phosphate	H_3PO_4	$\rightleftharpoons 3H^+ + PO_4^{3-}$	
4.	Fluoride	HF	$\rightleftharpoons H^+ + F^-$	
5.	Tartarate	$H_2C_4H_4O_6$	$\rightleftharpoons 2H^+ + C_4H_4O_6^{2-}$	

It is clear from ionic forms that H^+ ions become the subject of *Common ion effect* thus suppressing the ionisation of interfering acid. Less ionisation brings only less ions and the product of concentration of interfering ions and its components never exceeds the solubility produced (solubility phenomenon) of their salts. In this way, they do not precipitate and remain in solution.

The alkaline medium produces OH^- ions and thus give following equilibria :

$$NH_4OH \rightleftharpoons NH_4^+ + OH^-$$

(F⁻ for example) $$HF \rightleftharpoons H^+ + F^-$$

At this stage, the H^+ ion of interfering acid and the OH^- ion of NH_4OH join together to produce water. H_2O being feebly ionising substance does not furnish H^+ ions further and the interfering acid has to ionise more to maintain the equilibrium. This enhances the ionisation of interfering acid and the product of concentration of ions exceeds their solubility product. In this way, these are precipitated.

The discussion held above already points out that the analysis of the basic radicals beyond second group is not safe and the removal of such radicals become necessary before proceeding to third group and onwards.

1.   **Removal of the oxalate or tartarate.** (*a*) *From the filtrate of the second group.* If second group is present, take the filtrate from the second group in a porcelain dish and evaporate it until it becomes paste type. Add 2-3 ml of conc. HNO_3 and then evaporate it to dryness (stir with a glass rod to avoid spurting). Repeat the process for 3-4 times. Extract the dry mass with hot dilute HCl and filter off the residue if any and proceed with the filtrate for third group analysis.

(*b*) *From the mixture.* If first and second group are absent, take 2-3 gms of the original mixture in a porcelain dish, add 2-3 ml of conc. HNO_3, evaporate it to dryness (stir with glass rod to avoid spurting). Repeat the process for 3-4 times by adding 2-3 ml of HNO_3. Extract the dry mass with hot dilute HCl. Filter off the residue, if any, and proceed for third group analysis with the filtrate.

2.   **Removal of the fluoride and borate.**

If second group is present, take the filtrate from second group in a porcelain dish and evaporate the solution to almost paste. If first and second group are absent, take 2-3 gms of original mixture.

Add 2-3 ml of conc. HCl and evaporate it to dryness (Stir with glass rod to avoid spurting). Repeat the process 3-4 times. Finally, extract the dry mass with hot dilute HCl. Filter off the residue, if any, and proceed with the filtrate for third group analysis.

3. **Removal of Phosphate.** Phosphate radical is not decomposed by heating with mineral acids. It is, therefore, removed by the following way :

If second group is present, take the filtrate of the second group and boil for 5 minutes to remove H_2S completely (test with lead acetate paper). If first and second group radicals are absent, take the original solution. In either way, proceed further as follows :

Add 4-5 drops of conc. HNO_3 and boil for 2-3 minutes and cool. Add 1-2 gms of solid NH_4Cl and excess of NH_4OH solution while shaking until turbidity comes and the solution smells ammonia. To it, add 5 gms of CH_3COONa or CH_3COONH_4. Add acetic acid while shaking until the solution smells vinegar. Now filter the solution and use the residue, if any, for third group analysis and note the colour of the filtrate.

(a) If the filtrate obtained so is not reddish, add few drops of neutral ferric chloride solution. It will give buff coloured precipitate. Continue adding $FeCl_3$ until the solution just acquires tea colour. Add water to equal volume and boil for 2-3 minutes. Filter and discard the residue and with filtrate proceed for IV group analysis.

(b) If the filtrate is reddish, sufficient iron is already present. In such a case, dilute it with equal volume of water and boil for 2-3 minutes. Filter and discard the residue and with filtrate, proceed for IV group analysis.

3·8. | Analysis of group III

$(Al^{3+}, Fe^{3+}$ and $Cr^{3+})$

Group Reagent—NH_4Cl and NH_4OH (in excess).

Radicals	Precipitated as	Colour
Fe^{3+}	$Fe(OH)_3$	Reddish brown
Cr^{3+}	$Cr(OH)_3$	Dirty green
Al^{3+}	$Al(OH)^3$	White gelatinous

(a) **If no interfearing radical is present.** Boil the filtrate of the second group to remove H_2S completely (test it with lead acetate paper). Add few drops of conc. HNO_3 and boil to oxidise ferrous into ferric. Now add 0.5-1 gm of solid NH_4Cl and then excess of NH_4OH solution. This will precipitate the third group radicals. Filter and use residue for analysis of individual radicals.

(b) **If oxalate is present.** Use the mentioned filtrate after oxalate removal. Add to it solid NH_4Cl and NH_4OH solution. This will precipitate the third group radicals. Filter and use residue.

(c) **If borate or fluoride is present.** Use the mentioned filtrate after its removal. Add HNO_3, NH_4Cl and NH_4OH as mentioned above in 'a' part. This will precipitate the third group radicals. Filter and use residue.

(d) **If phosphate is present.** Use the mentioned residue.

Procedure. Add excess of NaOH to the residue obtained as above and add few drops of Br. water. Boil it for 2-3 minutes. and filter.

---

*Sometimes, Mn is also precipitated as brown or black hydrous manganese oxide, $MnO_2. xH_2O$ along with III group hydroxides.

|
↓   ↓

Residue	*Filtrate*
$[Fe(OH)_3$ and $MnO_2.xH_2O]$	$[NaAlO_2$ and $Na_2CrO_4]$
[If it is black → Mn is present ;	(If the solution is yellow, Cr is present
if reddish brown → ferric only)	otherwise colourless soln.) Test Al and
	Cr as follows :

| |
                         ↓   ↓

Wash the residue with hot water Reject the washing. Dissolve in conc. HNO_3. Divide solution in three parts.	(*i*) I part + solid NH_4Cl and boil ⟶ white ppt. $[Al(OH)_3]$	Acidify with CH_3COOH. Divide in two parts.
(*i*) I part + $K_4[Fe(CN)_6]$ → Deep blue ppt. or colouration.	(*ii*) Add dil. HCl to II Part drop by drop until just acidic (test with blue lit-	(*i*) I part + lead acetate → yellow ppt. soluble in NaOH and reappears
Iron	mus paper). Then add	on adding CH_3COOH.
(*ii*) II part + pinch of PbO_2. Boil, cool and dilute with water. Allow it to stand for a few minutes → Violet or or purple colour (like $KMnO_4$ soln.) in upper layer of the solution.	dil. NH_4OH until just alkaline (test with red litmus paper). Warm → white gelatinous ppt. **Aluminium**	(*ii*) II Part + $AgNO_3$ soln. → Brownish red ppt. soluble in dil. HNO_3 and NH_4OH but insolu- ble in acetic acid.
Manganese		**Chromium**

Notes

1. H_2S must be completely driven off before adding conc. HNO_3 otherwise this may interfere in the following ways :

(*a*)   Conc. HNO_3 may precipitate sulphur of H_2S.

(*b*) H_2S may be converted to H_2SO_4 and that may precipitate barium and strontium as their sulphates.

$$H_2S + 8HNO_3 \longrightarrow H_2SO_4 + 8NO_2 + 4H_2O$$

2.   Avoid large excess of NH_4Cl. The reason is that NH_4^+ ions when present in excessive amount, exert solvent action and thus will reverse the precipitation of carbonates in V group precipitation. For example,

$$Ba^{2+} + (NH_4)_2CO_3 \rightleftharpoons BaCO_3 + 2NH_4^+.$$

That is why an excess of NH_4Cl in III group is avoided.

3. NH_4OH is added to hot solution and then boiled well because in cold soluble complex is likely to be formed and thus chromium may not be comletely precipitated and some of it may pass to filtrate. Hence solution should be boiled and excess of NH_4OH be avoided.

$$Cr(OH)_3 + 6NH_4OH \longrightarrow [Cr(NH_3)_6](OH)_3 + 6H_2O$$
Soluble

4.   Mn also gets precipitated in excess of HNO_3 and it converts $Mn(OH)_2$ to MnO_2 which precipitates down. This can be checked by adding the least amount of conc. HNO_3 which is just sufficient to convert Fe^{2+} to Fe^{3+}. However, complete conversion is necessary otherwise it will be precipitated partially.

Specific Tests

1.   **Iron.** Iron may be confirmed by KCNS, $K_3Fe(CN)_6$ and $K_4Fe(CN)_6$ easily as given in ferrous and ferric ions tests ahead.

2.   **Chromium.** The mixture is first boiled with Br_2 water to convert its salt to chromate. It is then treated with diphenylamine prepared in strong H_2SO_4. An intense *blue colour* of phenylene blue dye is obtained.

3.   **Aluminium.** It may be tested by aluminon (auric tricarboxylic acid) which gives *bright red colour* in slight acidic medium.

Distinction between Ferrous (Fe^{2+}) and Ferric (Fe^{3+}) Ions

This can be achieved by the following reagents :

1.   **Potassium Thiocyanate.** Dissolve the mixture in dil. HCl and and KCNS. Blood red colour shows the presence of Fe^{3+} ions. The blood red colouration is due to ferri-thiocyanate ion, $[Fe(SCN)]^{2+}$.

$$Fe^{3+} + SCN^- \rightleftharpoons [Fe(SCN)]^{2+}$$

2.   **Potassium Ferrocyanide.** Dissolve the mixture in HCl and add $K_4[Fe(CN)_6]$. A deep blue ppt. or colour (Prussian's blue) is produced with Fe^{3+} ions.

$$3K_4[Fe(CN)_6] + 4FeCl_3 \longrightarrow Fe_4[Fe(CN)_6]_3 + 12 KCl$$

Ferric ferrocyanide

(Prussian's blue)

Fe^{2+} ions produce a white ppt. of ferrous ferrocyanide in the complete absense of air.

$$K_4[Fe(CN)_6] + 2FeCl_3 \longrightarrow Fe_2[Fe(CN)_6] + 4KCl$$

Ferrous ferrocyanide

(white ppt.)

But in presence of air, a white blue ppt. is obtained due to partial oxidation of Fe^{2+} to Fe^{3+} ions.

$$3Fe_2[Fe(CN)_6] \xrightarrow{\text{oxidation}} Fe_4[Fe(CN)_6]_3 + 2Fe.$$

3.   **Potassium Ferricyanide.** Dissolve the mixture in HCl and add $K_3[Fe(CN)_6]$. A deep blue ppt. (Turnbull's blue) shows presence of Fe^{2+} ions.

$$3FeCl_2 + 2K_3[Fe(CN)_6] \longrightarrow Fe_3[Fe(CN)_6]_2 + 6KCl$$

Ferrous ferricyanide

(Turnbull's blue)

Fe^{3+} ions produce brown colour.

$$FeCl_3 + K_3[Fe(CN)_6] \longrightarrow Fe[Fe(CN)_6] + 3KCl$$

Ferric ferricyanide

(Brown)

Difference between Prussian's blue and Turnbull's blue. The ferrocyanide ions form dark blue ppt. or colour with Fe^{3+} ions. This ppt. has been assigned formula $Fe_4^{3+}[Fe^{2+}(CN)_6]_3^{4-}$ and is called prussian's blue.

The ferricyanide ions form a dark blue ppt. or colour with Fe^{2+} ions. This ppt. has been assigned the formula $Fe_3^{2+}[Fe^{3+}(CN)_6]_2^{3-}$ called Turnbull's blue.

There is some evidence that Turnbull's blue is identical with Prussian's blue. It is possible that ferricyanide ion oxidises ferrous ions to ferric.

$$[Fe(CN)_6]^{3-} + Fe^{2+} \longrightarrow [Fe(CN)_6]^{4-} + Fe^{3+}$$

The products of this redox reaction will give ferric ferrocyanide. That is

$$4Fe^{3+} + 3[Fe(CN)_6]^{4-} \longrightarrow Fe_4^{3+}[Fe(CN)_6]_3^{4-}.$$

Lately, the composition of both, Prussian blue and Turnbull's blue has been found to be $Fe_4[Fe(CN_6)]_3$.

3·9. | **Analysis of Group IV**

(Zn^{2+}, Mn^{2+}, Co^{2+} and Ni^{2+})

Group Reagent. H_2S in alkaline (made by NH_4OH) solution.

To the filtrate obtained after III group radicals precipitation, add NH_4Cl and more NH_4OH, warm and pass H_2S slowly. Filter off the ppt. Make the filtrate alkaline if needed with NH_4OH and pass H_2S repeatedly till no further precipitation takes place. Proceed with the ppt. for IV group analysis and keep the filtrate for subsequent group analysis.

Radicals	Precipitated as	Colour
Zn^{2+}	ZnS	White
Mn^{2+}	MnS	Buff coloured
CO^{2+}	CoS	Black
Ni^{2+}	NiS	Black

Procedure. Wash the ppt. with hot water and then shake it with cold dil. HCl (*dilute one vol. of conc. HCl with 2 vol. of water.*) for 3 minutes. *Do not heat.* Filter.

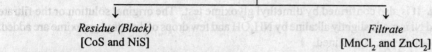

Residue (Black)	Filtrate
[CoS and NiS]	[$MnCl_2$ and $ZnCl_2$]

Residue (Black) [CoS and NiS]

Transfer the residue to a porcelain dish, add a little conc. HCl + 1-2 crystals of $KClO_3$ or a little conc. HNO_3 and then heat till the residue is dissolved. Evaporate the soln. just to dryness. Observe the colour of residue (if blue or bluish green → Co ; yellow residue → only Ni is indicated). Extract the residue with water. Cool. Divide the extract in three parts).

(A) I part + excess of saturated soln. (or solid) NaH-CO_3 + strong Br_2 water. Shake well. Note the colour change.

(*i*) Apple green in cold but no change on heating →
   Cobalt only

(*ii*) No green in cold but changes to black on heating
   →
   Nickel only

(*iii*) Apple green in cold and changed to black on heat-ing →
   Co and Ni both

(B) Divide II part into two.

(*i*) One part + NH_4Cl + excess of NH_4OH + K_3[Fe-$(CN)_6$] solution. Warm gently → Reddish brown ppt. or colour.
   Cobalt

Filtrate [$MnCl_2$ and $ZnCl_2$]

Boil off H_2S (test with lead acetate paper). Cool. Add NaOH soln. in excess. Boil for 3 mts. ar.d filter.

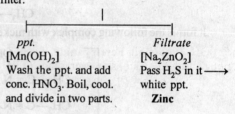

ppt.	Filtrate
[$Mn(OH)_2$]	[Na_2ZnO_2]

ppt. [$Mn(OH)_2$]
Wash the ppt. and add conc. HNO_3. Boil, cool. and divide in two parts.

(1) I part, dilute and cool. Add sod. bismu-thate. Stir well and wait → Supernatant liquid purple coloured.
   Manganese

(2) II part + pinch of PbO_2, boil for some-time and then allow to stand → purple colour in upper liquid.
   Manganese

Filtrate [Na_2ZnO_2]
Pass H_2S in it →
white ppt.
   Zinc

Notes

1. Do not use HCl in place of HNO_3 to test Mn otherwise the colour will be discharged by chloride ions.

2. Excess of dil. HCl should be avoided for separation of cobalt and nickel from zinc and manganese because CoS is slightly soluble in excess of dil. HCi.

3. In separation of Co and Ni, after treatment with conc. HNO_3 or $KClO_3$ if (*a*) Residue is blue or bluish green → Co is indicated ; (*b*) yellow residue → only Ni is indicated.

4. Use always strong solution of bromine wa⁌er.

5. *Dimethylglyoxime test for Ni.* In this test cobalt if present in excess produces some interference by reacting with dimethylglyoxime to form a dark coloured complex. But in excess of this test reagent, cobalt does not produce such complex. Hence proceed as follows :

Make the test solution just ammoniacal by NH_4OH. Filter if any ppt. To the filtrate, add excess of dimethylglyoxime → Voluminous rosy red ppt. confirms nickel.

6. In IV group, the solution is alkaline in which H_2S is readily absorbed. If too *much* H_2S is employed, NiS may partially form a colloidal solution which may pass through the filter paper. This is largely avoided by passing H_2S for a short time (30–60) seconds and testing for complete precipitation. It is also advised that H_2S should be passed into the hot alkaline solution of IV group (*i.e.*, filtrate of III group made alkaline by adding NH_4OH) acidified with CH_3COOH and heat the solution after passing H_2S. In presence of CH_3COOH, the colloidal NiS is coagulated.

Specific Tests

1. **Zinc.** To one drop of original solution, add 2 drops of $Na_2S_2O_3$. Now add KCN and 3-5 drops of diphenyl thiocarbazone (prepared in CCl_4). Shake it well → CCl_4 layer turns red.

2. **Cobalt.** Make original solution and add $Na_2S_2O_3$ solution and solid amm. thiocyanate. Add acetone. A *blue green colour* or deep blue colour of Ammonium tetrathiocyanato cobaltate, $(NH_4)_2[Co(CNS)_4]$ is obtained. This test is very sensitive but it is obscured by iron which forms soluble $Fe(CNS)^{2+}$. Ni does not interfere in it.

3. **Manganese.** Manganese is confirmed by sodium bismuthate solution (as given in alanysis scheme).

4. **Nickel.** It is also confirmed by dimethyl glyoxime test. The original solution or the filtrate meant for Co and Ni is made slightly alkaline by NH_4OH and few drops of dimethyl gyloxime are added. A rosy red crystalline ppt. is obtained.

[Dimethyl glyoxime has the structural formula :

$$CH_2 — C = N — O — H$$
$$|$$
$$CH_3 — C = N — O — H$$

It forms the following complex with nickel.

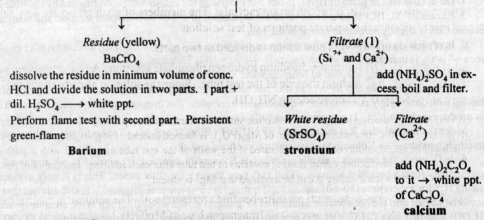

3·10.   Analysis of V Group

Members – Ba^{2+}, Sr^{2+} and Ca^{2+}

Group Reagent – Ammonium carbonate.

Colour of the precipitates – all white; $BaCO_3$, $SrCO_3$ and $CaCO_3$.

Procedure - Evaporate the filtrate from IV group and reduce it to nearly one third. Add NH_4OH slowly while shaking until the solution smells ammonia. Heat the solution to about 60°C and while being add saturated solution of ammonium carbonate. Allow it to stand for 5 minutes for precipitation. Filter and keep the filtrate for sixth group.

Dissolve the residue in minimum amount of hot dil. acetic acid and heat the solution to boiling. Add in hot K_2CrO_4 solution. If yellow precipitate comes, add more to precipitate it completely. Cool and filter.

|

Residue (yellow)   **Filtrate** (1)
$BaCrO_4$ (Sr^{2+} and Ca^{2+})

dissolve the residue in minimum volume of conc.   add $(NH_4)_2SO_4$ in ex-
HCl and divide the solution in two parts. I part +   cess, boil and filter.
dil. H_2SO_4 ⟶ white ppt.

Perform flame test with second part. Persistent   **White residue**   **Filtrate**
green-flame   ($SrSO_4$)   (Ca^{2+})

Barium   **strontium**

add $(NH_4)_2C_2O_4$
to it → white ppt.
of CaC_2O_4
calcium

*In absence of Ba^{2+}, treat the solution as you do with filtrate (1).

Notes

(1) The solution should not be highly acidic (*i.e.*, CH_3COOH) because this will not precipitate braium completely and this will surely interfere in strontium and calcium test.

(2) After the addition of amm. carboante, the solution should not be heated strongly otherwise amm. carbonate will undergo decomposition.

(3) After the addition of amm. sulphate, let the contents be kept for 10 minutes to ppt. strontium.

(4) The filtrate of IV group is boiled with a little CH_3COOH in order to coagulate any colloidal NiS present in IV group filtrate. The presence of colloidal NiS in IV group filtrate is indicated if filtrate has brownish colour otherwise addition of acetic acid is not essential.

(5) NH_4Cl should be sufficient otherwise Mg will also precipitate. Avoid too much of NH_4Cl otherwise V group will not be completely precipitated.

(6) The filtrate of V group should be tested for calcium before proceeding to VI group as sometimes calcium does not precipitate completely. Hence to the V group filtrate, add $(NH_4)_2CO_3$ and if any white ppt. comes, calcium is present. Filter and use the filtrate for VI group analysis.

(7) Before percipitating V group radicals, the filtrate of IV group should be boiled off to remove H_2S completely and bulk of filtrate should be reduced to about one-third and then proceed for V group precipitation.

(8) Sometimes, when Sr^{2+} is present in high concentration or CrO_4^{2-} added excessively, it is likely to precipitate along with $BaCrO_4$. However, the use of $K_2Cr_2O_7$ in buffer soln. ($CH_3COOH + CH_3COONa$) will prevent Sr^{2+} from precipitating as $SrCrO_4$.

Specific Tests. The fifth group radicals are confirmed by flame test. In all the three tests, dip the platinum wire in conc. HCl, touch with precipitate and hold in the flame. Characteristic flame colours are observed for their identity. (See page 51)

3·11 | Analysis of Group VI |

[Mg^{2+}, Na^+, K^+ and NH_4^+]

Group reagent : No common reagent.

Like other analytical groups, we are not able to precipitate all the sixth group radicals by any single reagent. Hence this group may also be regarded as a *soluble group*. First, we ensure complete removal of 5th group radicals and then we test all these radicals individually.

Procedure. Reduce the bulk of V group filtrate to one-third by evaporation. To this, add a little quantity of amm. sulphate and amm. oxalate solutions. Boil for 3 minutes. Any ppt. if comes, is filtered and is discarded as this may be of 5th group radicals. The members of sixth group are now subsequently identified with the separate portions of test solution.

1. Magnesium (Mg^{2+}). The solution is divided in two parts.

(*a*) I part + NH_4OH solution + disodium hydrogen phosphate solution. A fine *crystalline white precipitate* appears on scratching the side of the test tube. Since the solution already contains NH_4^+ ions, no need to add NH_4Cl before adding NH_4OH.

[Note. NH_4OH solution is added until solution smells of ammonia (*i.e.*, pH ≈ 9). Do not add excess of NH_4OH, as at pH > 10 a noncharacteristic ppt. of $Mg_3(PO_4)_2$ is formed instead of $MgNH_4PO_4$. The precipitation is delayed in dilute solutions. It is accelerated if the walls of the test tube are rubbed with a glass rod. Na_2HPO_4 should be added drop by drop shaking contents of test tube after each addition. In absence of ppt. the solution must be left to stand before it can be assumed that Mg^+ is absent.]

(*b*) II part + magneson reagent (paranitrobenzine azoresorcinol). The solution is made alkaline by sodium hydroxide. A *sky blue precipitate* [magneson blue of $Mg(OH)_2$] is obtained. Ammonium salts should be absent as they prevent the precipitation of $Mg(OH)_2$.

by sodium hydroxide. A *sky blue precipitate* [magneson blue of $Mg(OH)_2$] is obtained. Ammonium salts should be absent as they prevent the precipitation of $Mg(OH)_2$.

This test, however, can be better preformed by precipitate itself obtained by the addition of Na_2HPO_4. To proceed, we wash the ppt. with hot water and discard washings. Magneson is added to it followed by NaOH slowly. This gives *sky blue ppt.* as soon as the medium becomes alkaline.

[**Note.** It is essential to perform blank test with the reagent as it frequently gives blue colouration. For this reason a blue ppt. should be looked for.]

(c) This may also be confirmed by cobalt test also (see Index).

2. Ammonium Ion (NH_4^+). The ammonium ion should not be tested from the group filtrate because during the anaylsis from first group to fifth group, we have supplied NH_4^+ ions many a time. It must be, therefore, tested from original solution. The confirmation may be achieved by the following ways :

(a) Heat the mixture with NaOH. The characteristic smell of ammonia is obtained.

(b) To the original solution, add NaOH and few drops of Nesseler's reagent—A *brown precipitate* is obtained.

If we replace Nesseler's reagent by mercurous nitrate, the precipitate is black

[**Note.** Salts of Hg interfere in test of NH_4^+ with NaOH. Thus when Hg is present in mixture, we should proceed as follows :

Take a part of filtrate of II group, boil off H_2S completely. Evaporate the filtrate to dryness in a procelain dish. Now add NaOH to this solid residue and boil $\longrightarrow$ smell of ammonia indicates the presence of NH_4^+].

3. Sodium. This may be confirmed by flame test.

Golden yellow flame—Sodium.

Alternatively, it may be tested as follows :

Take the filtrate of the V group, precipitate off magnesium by Na_2HPO_4. Now, evaporate this filtrate to dryness to remove NH_4^+ ions. Lixivate it with water and add *sodium reagent* (Magnesium acetate and uranyl acetate). Shake and scratch the side of the test tube by glass rod. Leave the solution to rest for 10 minutes. A *crystalline yellow precipitate* of composition $NaMg(UO_2)_3 (C_2H_3O_2)_9$ $9H_2O$ is obtained.

4. Potassium. (i) The potassium can be confirmed alone by its flame test which gives fleeting *pale violet flame* but this demands the absence of those radicals which give colour to flame. This would otherwise make it confusing.

(ii) Remove NH_4^+ as it interferes with the test. For this, take the filtrate after the precipitation of V group radicals and evaporate it to dryness. Lixivate it with water and add some acetic acid and finally sodium cobaltinitrite in excess—*yellow precipitate*.

[**Note.** In conc. solution, the ppt. is immediately formed and slowly in dilute solution ; precipitation may be accelerated by warming. Iodide and other reducing agents must be removed, as they interfere before applying the test.]

3·12.   Reactions in Basic Radicals Analysis

Chemical reactions invovled in 1 Group Analysis : Ag^+, Pb^{2+} and Hg_2^{2+} ions present in the solution react with Cl^- ions from HCl, to give white ppt.

1. $Pb(NO_3)_2 + 2HCl \longrightarrow PbCl_2 \downarrow + 2HNO_3$

2. $AgNO_3 + HCl \longrightarrow AgCl \downarrow + HNO_3$

3. $Hg_2(NO_3)_2 + 2HCl \longrightarrow Hg_2Cl_2 \downarrow + 2HNO_3$

Separation and Identification of Lead (filtrate) : On boiling the ppt. with water, $PbCl_2$ becomes soluble thus separated by filtration. This when treated with KI, K_2CrO_4 and H_2SO_4 separately, following reactions take place.

I Portion—　　　　　$PbCl_2 + 2KI \longrightarrow PbI_2 \downarrow + 2KCl.$

II Portion—　　　　$PbCl_2 + K_2CrO_4 \longrightarrow PbCrO_4 \downarrow + 2KCl.$

III Portion—　　　$PbCl_2 + H_2SO_4 \longrightarrow PbSO_4 \downarrow + 2HCl.$

Treatment of residue with NH_4OH : (*i*) Silver chloride dissolves in NH_4OH forming a soluble complex.

$$AgCl + 2NH_4OH \longrightarrow [Ag(NH_3)_2]Cl + 2H_2O$$
$$\text{diammine silver}$$
$$\text{chloride}$$

(*ii*) NH_4OH converts Hg_2Cl_2 to a *black residue* which consists of *white* amino-mercuric chloride and *black* finely divided Hg. In this case internal redox reaction takes place whereby Hg_2^{2+} ions of Hg_2Cl_2 are reduced to metallic Hg and others are oxidised to mercuric state.

$$Hg_2Cl_2 + 2NH_4OH \longrightarrow Hg(NH_2)Cl \downarrow + Hg \downarrow + NH_4Cl + 2H_2O$$
$$\underbrace{\quad\quad\text{white}\quad\quad\quad\quad\text{black}\quad\quad}_{\text{Black}}$$

Reactions of Ag. The filtrate containing soluble silver complex is treated in three ways separately.

(*a*)　I portion— $[Ag(NH_3)_2]Cl + 2HNO_3 \longrightarrow AgCl \downarrow + 2NH_4NO_3$

(*b*)　II portion—　　$[Ag(NH_3)_2]Cl + KI \longrightarrow AgI \downarrow + KCl + 2NH_3$

(*c*)　III portion—$2[Ag(NH_3)_2]Cl + K_2CrO_4 \longrightarrow Ag_2CrO_4 + 2KCl + 4NH_3$

Reductions of Hg_2^{2+} (ions). Aquaregia dissolves the black residue forming soluble $HgCl_2$.

$$HNO_3 + 3HCl \longrightarrow NOCl + 2Cl + 3H_2O$$
$$Hg + 2Cl \longrightarrow HgCl_2$$
$$Hg(NH_2)Cl + 3Cl_2 \longrightarrow 2HgCl_2 + 4HCl + N_2$$

Addition of $SnCl_2$ reduces $HgCl_2$ to white Hg_2Cl_2.

Further addition of $SnCl_2$ reduces white Hg_2Cl_2 to black finely divided Hg. Usually *grey* mixture of Hg_2Cl_2 and Hg is obtained.

$$2HgCl_2 + SnCl_2 \longrightarrow Hg_2Cl_2 \downarrow + SnCl_4$$
$$\text{white}$$

$$Hg_2Cl_2 + SnCl_2 \longrightarrow 2Hg \downarrow + SnCl_4$$
$$\text{black}$$

Chemical Reactions Involved in II Group

Precipitation of Group II with H_2S. The cations of Group II are precipitated as sulphides in a solution containing 0.3 N HCl with H_2S.

Mercuric ions give a white ppt. turning to yellow, brown and finally black with H_2S.

$$3Hg^{2+} + 2Cl^- + 2H_2S \longrightarrow Hg_3S_2Cl_2 + 4H^+$$
$$\text{White}$$

$$HgCl_2.2HgS + H_2S \longrightarrow 3HgS \downarrow + 2HCl$$
$$\text{Black}$$

Copper ions precipitate as black CuS with H_2S

$$CuCl_2 + H_2S \longrightarrow CuS \downarrow + 2HCl$$

Bismuth ions precipitate as dark-brown Bi_2S_3 with H_2S

$$2BiCl_3 + 3H_2S \longrightarrow Bi_2S_3 \downarrow + 6HCl.$$

Lead ions give black ppt. of PbS with H_2S

$$PbCl_2 + H_2S \longrightarrow PbS \downarrow + 2HCl$$

Sometimes, in highly acidic medium, a red ppt. is obtained.

$$PbS + PbCl_2 \longrightarrow PbCl_2.PbS \downarrow$$
$$\text{Red}$$

Cadmium ions give ppt. with H_2S. The composition and colour of ppt. depend upon the acidity and temperature of precipitating solution. In a cold and slightly acidic solution the ppt. is yellow, but in a hot acidic solution the colour of ppt. may be from orange to red.

$$CdCl_2 + H_2S \longrightarrow CdS \downarrow + 2HCl.$$

Trivalent arsenic is readily precipitated in acidic solution as yellow As_2S_3.

$$2AsCl_3 + 3H_2S \longrightarrow As_2S_3 \downarrow + 6HCl$$

$$\underset{\text{(Arsenious acid)}}{2H_3AsO_3 + 3H_2S} \longrightarrow As_2S_3 \downarrow + 6H_2O$$

$$\underset{\text{(Sod. arsenite)}}{2Na_3AsO_3} + 6HCl + 3H_2S \longrightarrow As_2S_3 \downarrow + 6NaCl + 6H_2O$$

Pentavalent. As does not precipitate immediately in low conc. of HCl (0.3 N) but continuous passing of H_2S precipitates very slowly As_2S_3 and sulphur. The precipitation is enhanced in hot solution. The reason for slow precipitation is that H_2S is first used to reduce pentavalent As to trivalent As and then it will precipitate trivalent As to As_2S_3. In hot solution, the reduction is enhanced hence more rapid precipitation takes place.

$$Na_2H.AsO_4 + 2HCl \longrightarrow \underset{\text{(Arsenic acid)}}{H_3AsO_4} + 2NaCl$$

$$H_3As^{5+}O_4 + H_2S \longrightarrow \underset{\text{(Arsenious acid)}}{H_3As^{3+}O_3} + H_2O + S \downarrow$$

$$2H_3As^{3+}O_3 + 3H_2S \longrightarrow As_2S_3 \downarrow + 6H_2O$$

Even in low conc. of HCl, arsenic from arsenates can be precipitated completely if before passing H_2S ammonium iodide is added or SO_2 is passed in the solution. This will reduce arsenate to arsenite and when H_2S is passed in this warm reduced solution, arsenic as As_2S_3 is immediately precipitated.

$$NH_4I + H_2O \longrightarrow HI + NH_4OH$$

$$\underset{\substack{\text{(arsenate} \\ \text{ion)}}}{AsO_4^{3-}} + 2HI \longrightarrow \underset{\substack{\text{(arsenite} \\ \text{ion)}}}{AsO_3^{3-}} + I_2 + H_2O$$

The liberated iodine is again reduced to HI by H_2S, hence HI acts as catalyst

$$I_2 + H_2S \longrightarrow 2HI + S$$

In presence of high concentration of HCl (*above 2N*) arsenate ion is rapidly precipitated as As_2S_5 when H_2S is passed in cold, whereas in hot solution H_2S precipitates a mixture of As_2S_3 and As_2S_5.

$$AsO_4^{3-} + H_2S \longrightarrow AsO_3^{3-} + S \downarrow + H_2O.$$

Thus arsenic present as arsenite or arsenate or both is completely and rapidly precipitated in highly acidic medium (*above 2N*) *without using an iodide or* SO_2.

Antimonous and Antimonic ions with H_2S give orange red ppt. of Sb_2S_3 and Sb_2S_5 respectively. The two antimony ions probably exist in HCl solution as complex ions, $SbCl_4^-$ and $SbCl_6^-$ which hydrolyse upon dilution to give white ppt. of antimony oxychloride.

$$SbCl_4- + 3H_2O \longrightarrow SbOCl \downarrow + 2H_3O^+ + 3Cl^-$$

$$SbCl_6^- + 6H_2O \longrightarrow SbO_2Cl \downarrow + 4H_3O^+ + 5Cl^-$$

The reaction of $SbCl_4^-$ and $SbCl_6^-$ with H_2S are as follows :

$$2SbCl_4^- + 3H_2S + 6H_2O \longrightarrow Sb_2S_3 \downarrow + 6H_3O^+ + 8Cl^-$$

$$2SbCl_6^- + 5H_2S + 10H_2O \longrightarrow Sb_2S_5 \downarrow + 10H_3O^+ + 12Cl^-$$

Stannous ions with H_2S give brown ppt. of SnS

$$SnCl_2 + H_2S \longrightarrow SnS \downarrow + 2HCl$$

Stannic ion in dil. HCl solution can exist only as complex chlorostannate ion, $SnCl_6^{2-}$ which gives a dull or light yellow ppt. of SnS_2 with H_2S.

$$SnCl_6^{2-} + 2H_2S + 4H_2O \longrightarrow SnS_2 \downarrow + 4H_3O^+ + 6Cl^-$$

Separation of II Group into II-A and II-B. The separation is affected by using warm yellow ammonium sulphide, $(NH_4)_2S_x$ which makes As, Sb and tin sulphides soluble as follows :

1. $$3(NH_4)_2S_2 + As_2S_3 \longrightarrow 2(NH_4)_3.AsS_4 + S$$

2. $$3(NH_4)_2S_2 + As_2S_5 \longrightarrow \underset{\substack{\text{Amm. thioarsenate} \\ \text{(soluble)}}}{2(NH_4)_3.AsS_4} + 3S$$

3. $3(NH_4)_2S_2 + Sb_2S_3 \longrightarrow 2(NH_4)_3SbS_4 + S$

Amm. thioantimonate

4. $3(NH_4)_2S_2 + SbS_5 \longrightarrow 2(NH_4)_3SbS_4 + 3S$

Amm. thioantimonate

5. $(NH_4)_2S_2 + SnS \longrightarrow (NH_4)_2SnS_3$

6. $(NH_4)_2S_2 + SnS_2 \longrightarrow (NH_4)_2.SnS_3 + S$

Amm. thiostannate

Yellow amm. sulphide is unable to dissolve copper group and hence these remain undissolved. The subsequent reactions in the copper group are as follows :

Copper Group Reactions

The sulphides are boiled with 1 : 1 HNO_3, whereby these are converted into their corresponding soluble nitrates.

1. $3CuS + 8HNO_3 \longrightarrow 3Cu(NO_3)_2 + 2NO + 3S + 4H_2O$

2. $3CdS + 8HNO_3 \longrightarrow 3Cd(NO_3)_2 + 2NO + 3S + 4H_2O$

3. $3PbS + 8HNO_3 \longrightarrow 3Pb(NO_3)_2 + 2NO + 3S + 4H_2O$

4. $Bi_2S_3 + 8HNO_3 \longrightarrow 2Bi(NO_3)_3 + 2NO + 3S + 4H_2O$

Reactions of Mercuric. The mercuric sulphide being insoluble is treated with aqua regia (3 parts conc. HCl + 1 part conc. HNO_3) and it goes into solution :

$$3HgS + 2HNO_3 + 6HCl \longrightarrow 3HgCl_2 + 2NO + 3S + 4H_2O$$

(a)   This mercuric chloride is treated with $SnCl_2$:

$$2HgCl_2 + SnCl_2 \longrightarrow Hg_2Cl_2 \downarrow + SnCl_4$$

white ppt.

$$Hg_2Cl_2 + SnCl_2 \longrightarrow 2Hg + SnCl_4$$

Black or

grey

(b)   On adding copper turning, Hg is deposited on copper :

$$HgCl_2 + Cu \longrightarrow CuCl_2 + Hg \text{ (deposited on Cu)}$$

Treatment of filtrate with NH_4OH. The filtrate containing $Pb(NO_3)_2$, $Bi(NO_3)_2$, $Cu(NO_3)_2$ and $Cd(NO_3)_2$ is treated with conc. NH_4OH, whereby Pb and Bi are precipitated as their hydroxides and Cu and Cd pass into soln. in the form of soluble complex cations, $[Cu(NH_3)_4]^{2+}$ and $[Cd(NH_3)_4]^{2+}$.

$$Pb(NO_3)_2 + 2NH_4OH \longrightarrow Pb(OH_2) \downarrow + 2NH_4NO_3$$

$$Bi(NO_3)_3 + 3NH_4OH \longrightarrow Bi(OH)_3 \downarrow + 3NH_4NO_3$$

$$Cu(NO_3)_2 + 4NH_4OH \longrightarrow [Cu(NH_3)_4](NO_3)_2 + 4H_2O$$

Deep blue

Tetrammine copper (II)

nitrate

$$Cd(NO_3)_2 + 4NH_4OH \longrightarrow [Cd(NH_3)_4](NO_3)_2 + 4H_2O$$

Tetrammine cadmium (II)

nitrate

Reactions of Pb and Bi. The residue containing $Pb(OH)_2$, $Bi(OH)_2$ is warmed with NaOH whereby Pb is changed into soluble sodium plumbite and $Bi(OH)_3$ remains undissolved.

Test for Pb. The filtrate containing $Na[PbO_2]$ when acidified with acetic acid and K_2CrO_4 is added, a yellow ppt. of $PbCrO_4$ is obtained.

$$Na[PbO_2] + 4CH_3COOH \longrightarrow (CH_3COO)_2Pb + 3CH_3COONa + 2H_2O$$

$$(CH_3COO)_2Pb + K_2CrO_4 \longrightarrow PbCrO_4 \downarrow + 2CH_3COOK$$

Yellow

The yellow ppt. is insoluble in acetic acid and NH_4OH but soluble in alkali hydroxide and HNO_3.

$$PbCrO_4 + 4NaOH \longrightarrow Na_2[PbO_2] + Na_2CrO_4 + 2H_2O$$

Test for Bi :

$$Bi(OH)_3 + 3HCl \longrightarrow BiCl_3 + 3H_2O$$

ppt.   Soluble

In this solution Bi can be confirmed by any of the following reactions :

(*i*) $BiCl_3 + H_2O \longrightarrow BiOCl + 2HCl$

                 (excess)   white

                          turbidity

(*ii*) $2BiCl_3 + 6NaOH + 3Na_2[SnO_2] \longrightarrow 2Bi\downarrow + 3NaSnO_3 + 3H_2O + 6NaCl$

                                             Black

Reactions with Cu. The ionisation of tetra-ammine cupric cation

$$[Cu(NH_3)_4]^{2+} \rightleftharpoons Cu^{2+} + 4NH_3$$

is small in presence of excess of ammonia, even then the Cu^{2+} ions thus produced are sufficient for their identification. Hence with $K_4[Fe(CN)_6]$ a reddish brown ppt. insoluble in CH_3COOH but soluble in NH_4OH is produced.

$$2Cu^{2+} + [Fe(CN)_6]^{4-} \longrightarrow Cu_2[Fe(CN)_6] \downarrow$$

Reddish-brown

Cupric ferrocyanide

Reactions with Cd :

$$[Cd(NH_3)_4]^{2+} + S^2 \longrightarrow CdS\downarrow + 4NH_3\uparrow$$

Yellow

Reactions in Cu and Cd separation. *When filtrate is blue* : Addition of excess of KCN forms soluble colourless complexes with Cu and Cd respectively.

$$2[Cu(NH_3)_4](NO_3)_2 + 10KCN + 8H_2O \longrightarrow 2K_3Cu(CN)_4 + 8NH_4OH + (CN)_2 + 4KNO_3$$

$$[Cd(NH_3)_4](NO_3)_2 + 4KCN + 4H_2O \longrightarrow K_2Cd(CN)_4 + 2KNO_3 + 4NH_4OH$$

Cyanide complex of copper is more stable than that of cadmium. The primary ionisation of both complexes are as under :

$$K_3Cu(CN)_4 \rightleftharpoons 3K^+ + [Cu(CN)_4]^{3-}$$

$$K_2Cd(CN)_4 \rightleftharpoons 2K^+ + [Cd(CN)_4]^{2-}$$

The secondary ionisation and their ionisation constans are :

$$[Cu(CN)_4]^{3-} \rightleftharpoons Cu^+ + 4CN^- \quad ; \quad K = 5 \times 10^{-28}$$

$$[Cd(CN)_4]^{2-} \rightleftharpoons Cd^{2+} + 4CN^- \quad ; \quad K = 1.4 \times 10^{-17}$$

It is thus clear that secondary ionisation of cupro-cyanide complex is so low that no appreciable quantity of Cu^+ ions is available in the solution and when H_2S is passed, no black ppt. of Cu_2S is produced ; whereas secondary ionisation for cadmicyanide complex is comparatively high, therefore, enough Cd^{2+} ions are avail- able in the solution and when H_2S is passed, a yellow ppt. of CdS is produced. In this way, cadmium is identified in the presence of copper.

Arsenic Group Reactions

Treatment of soluble thio-salts with dil. HCl. Dil. HCl decomposes the soluble complex thio salts of As, Sb and Sn to corresponding sulphides (ppts.)

1. $2(NH_4)_3AsS_4 + 6HCl \longrightarrow As_2S_5\downarrow + 6NH_4Cl + 3H_2S$

2. $2(NH_4)_3SbS_4 + 6HCl \longrightarrow Sb_2S_5\downarrow + 6NH_4Cl + 3H_2S$

Sb_2S_5 formed initially, is partially decomposed to Sb_2S_3

$$Sb_2S_5 \longrightarrow Sb_2S_3 + 2S$$

3. $(NH_4)_2SnS_3 + 2HCl \longrightarrow 2NH_4Cl + 2nS_2\downarrow + H_2S$

Treatment of above ppts. with conc. HCl. Conc. HCl converts sulphides of Sb and Sn to the respective chlorides but it does not dissolve arsenic sulphide.

$$Sb_2S_5 + 6HCl \longrightarrow 2SbCl_3 + 3H_2S + 2S$$

$$Sb_2S_3 + 6CHl \longrightarrow 2SbCl_3 + 3H_2S$$

$$SnS_2 + 4HCl \longrightarrow SnCl_4 + 2H_2S$$

Reactions of As :

(a)   Conc. HNO_3 and amm. molybdate (excess) produce :

$$3As_2S_5 + 10HNO_3 + 4H_2O \longrightarrow 6H_3AsO_4 + 10NO \downarrow + 15S$$
$$\text{(soluble)}$$

$$H_3AsO_4 + 12(NH_4)_2.MoO_4 + 21HNO_3 \longrightarrow (NH_4)_3[As(Mo_3O_{10})_4] + 21NH_4NO_3 + 12H_2O$$
$$\text{Amm. arsenomolybdate}$$
$$\text{(Yellow ppt.)}$$

Few authors suggest the composition of above complex as $(NH_4)_3[AsMo_{12}O_{40}]$

(b) $$As_2S_5 + 3(NH_4)_2CO_3 \longrightarrow (NH_4)_3AsS_4 + (NH_4)_3AsSO_3 + 3CO_2$$
$$\text{Amm. thoi-} \qquad \text{Amm. oxy.}$$
$$\text{arsenate} \qquad \text{thioarsenate}$$

$$As_2S_3 + 3(NH_4)_2CO_3 \longrightarrow (NH_4)_3AsS_3 + (NH_4)_3AsO_3 + 3CO_2$$
$$\text{Amm. arsenite}$$

Thus sulphides are dissolved forming soluble compounds as above. When solution is acidified with dil. HCl, a yellow ppt. is obtained.

$$2(NH_4)_3AsS_4 + 6HCl \longrightarrow As_2S_5 + 6NH_4Cl + 3H_2S$$
$$\text{Yellow ppt.}$$

$$2(NH_4)_3AsSO_3 + 12HCl \longrightarrow 2AsCl_3 + 6NH_4Cl + 6H_2O + 2S$$
$$\text{soluble}$$

When H_2S is passed :

$$2AsCl_3 + 3H_2S \longrightarrow As_2S_3 + 6HCl$$
$$\text{Yellow ppt.}$$

Reactions of Sb and Sn :

(a) $2NH_4OH + \begin{matrix} COOH \\ | \\ COOH \end{matrix} \longrightarrow (NH_4)_2C_2O_4 + 2H_2O$

$$SbCl_3 + 3(NH_4)_2C_2O_4 \longrightarrow (NH_4)_3[Sb(C_2O_4)_3] + 3NH_4Cl$$
$$\text{unstable}$$

$$SnCl_4 + 4(NH_4)_2C_2O_4 \longrightarrow (NH_4)_4[Sn(C_2O_4)_4] + 4NH_4Cl$$
$$\text{stable}$$

Since complex of antimony is unstable and gives sufficient Sb^{3+} ions which are precipitated when H_2S is passed whereas complex of tin does not give sufficient Sn^{4+} ions to be precipitated as sulphide when H_2S is passed.

(b)   (i) $$Fe + SbCl_3 \longrightarrow FeCl_3 + Sb \downarrow$$
$$\text{Black}$$

   (ii) $$Fe + 2HCl \longrightarrow FeCl_2 + 2H$$
$$2H + SnCl_4 \longrightarrow SnCl_2 + 2HCl$$
$$SnCl_2 + 2HgCl_2 \longrightarrow Hg_2Cl_2 + SnCl_4$$
$$\text{(white ppt.)}$$
$$Hg_2Cl_2 + SnCl_2 \longrightarrow SnCl_4 + 2Hg \downarrow$$
$$\text{grey}$$

Reactions in Removal of Interfering Radicals

1.   **Oxalate and Tartarate.** With conc. HNO_3, the oxalate and tartarate are converted into oxalic and artaric acids respectively. These acids are easily decomposed in presence of conc. HNO_3 as follows :

$$Na_2C_2O_4 + 2HNO_3 \longrightarrow 2NaNO_3 + H_2C_2O_4$$
$$H_2C_2O_4 + 2HNO_3 \longrightarrow 2NO_2 \uparrow + 2CO_2 \uparrow + 2H_2O$$
$$H_2C_4H_4O_6 + 8HNO_3 \longrightarrow 8NO_2 \uparrow + 3CO_2 \uparrow + CO \uparrow + 7H_2O$$
(Tartaric acid)

2. Borate and Fluoride. The decomposition takes as follows :

$$NaF + HCl \longrightarrow NaCl + HF \uparrow$$
$$\text{Volatile}$$
$$Ca_3(BO_3)_2 + 6HCl \longrightarrow 3CaCl_2 + 2H_3BO_3$$

The concentrated HNO_3 in place of HCl also serves the same purpose.

Precaution :

After the removal, take little portion of residue and test it for borate or fluoride as the case may be. If negative test, removal is complete otherwise repeat the process of removal with the rest of residue.

3. Phosphate radical

Theory. The principal involved in phosphate removal is very simple First of all by usual way of analysis, we precipitate third and all higher group phosphates. Amm. acetate and acetic acid is added to make medium buffer of constant pH which dissolves all the higher group phosphates leaving behind the third group phosphates only. In this way third group is separated. Since III group cations may not be in sufficient quantity to remove PO_4^{3-} ions completely, it is necessary to add neutral $FeCl_3$ to precipitate all PO_4^{3-} as $FePO_4$.

$$Fe^{3+} + PO_4^{3-} \longrightarrow FePO_4 \downarrow$$

The remaining Fe^{3+} ions react with acetate ions to give ferric acetate which on dilution and boiling precipitates as basic ferric acetate.

$$FeCl_3 + 3CH_3COONH_4 \longrightarrow (CH_3COO)_3Fe + 3NH_4Cl$$

On dilution and boiling

$$(CH_3COO)_3Fe + 2H_2O \rightleftharpoons Fe(OH)_2(CH_3COO) \downarrow + 2CH_3COOH$$
$$\text{Brown ppt.}$$

This reaction tends to reverse on cooling, hence filtration of solution while hot is desirable.

Comments on Acetate Buffer—$FeCl_3$ Method. The phosphate ions will be precipitated completely if trivalent ions are present in excess. For this purpose Fe^{3+} ions (as neutral $FeCl_3$) are preferred.

In the presence of considerable amounts of Fe and Al salts, chromium is coprecipitated with basic acetates of Fe and Al as an adsorption complex or as basic chromium acetate, $Cr(OH)_2.C_2H_3O_2$. If chromium is present in excess, only a fraction of all the III group metals will be precipitated as basic acetates, while some Al, Fe, Cr will remain in solution.

Reactions in III group Analysis

Ferrous is converted to ferric by conc. HNO_3

$$3Fe^{2+} + 4H^+ + NO_3^- \longrightarrow NO \uparrow + 2H_2O + 3Fe^{3+}$$

Finally, all members are precipitated as their hydroxides.

$$FeCl_3 + 3NH_4OH \longrightarrow Fe(OH)_3 \downarrow + 3NH_4Cl$$
$$AlCl_3 + 3NH_4OH \longrightarrow Al(OH)_3 \downarrow + 3NH_4Cl$$
$$CrCl_3 + 3NH_4OH \longrightarrow Cr(OH)_3 \downarrow + 3NH_4Cl$$

Bivalent manganese may also be converted to tetravalent manganese by HNO_3 and may be precipitated as follows :

$$Mn(OH)_2 + [O] \longrightarrow MnO_2 + H_2O$$

This finally becomes hydrated, *i.e.*, $MnO_2.xH_2O$.

Treatment of ppt. with excess of NaOH forming a soluble sodium aluminate, $NaAlO_2$

$$Al(OH)_3 + NaOH \longrightarrow NaAlO_2 + 2H_2O$$
$$\text{Soluble sod. aluminate}$$

In presence of $NaOH + Br_2$ water, $Cr(OH)_3$ is oxidised to soluble yellow sodium chromate.

$$Br_2 + NaOH \longrightarrow NaBr + HBr + [O]$$
$$2Cr(OH)_3 + 4NaOH + 3[O] \longrightarrow 2Na_2CrO_4 + 5H_2O$$
$$\text{Soluble and}$$
$$\text{yellow}$$

$Fe(OH)_3$ and $MnO_2.xH_2O$ remain unaffected on treatment with $NaOH + Br_2$ water.

Thus Al and Cr are tested from filtrate and Fe and Mn from residue.

Reactions of Al.

(i) *In presence of NH_4Cl*

$$NaAlO_2 + NH_4Cl + H_2O \longrightarrow Al(OH)_3 \downarrow + NaCl + NH_3 \uparrow$$
white
gelatinous

(ii) *In presence of HCl and NH_4OH*

$$NaAlO_2 + 4HCl \xrightarrow{boil} AlCl_3 + NaCl + 2H_2O$$

$$AlCl_3 + 3NH_4OH \longrightarrow Al(OH)_3 \downarrow + 3NH_4Cl$$

Reactions of Cr.

(i) $Na_2CrO_4 + Pb(CH_3COO)_2 \xrightarrow[gently]{heat} Pb\,CrO_4 \downarrow + 3CH_3COONa$

Yellow Soln.   Lead chromate

(Yellow)

$PbCrO_4$ is soluble in NaOH but insoluble in CH_3COOH.

$$PbCrO_4 + 4NaOH \longrightarrow \underbrace{Na_2PbO_2 + Na_2CrO_4}_{\text{Soluble}} + 2H_2O$$
(ppt.)

(ii) $Na_2CrO_4 + 2AgNO_3 \longrightarrow Ag_2CrO_4 \downarrow + 2NaNO_3$

(brownish red)

Reactions with Fe^{3+}.

$$Fe(OH)_3 + 3HNO_3 \longrightarrow Fe(NO_3)_3 + 3H_2O$$
$$Fe^{3+} + 3[Fe(CN)_6]^{4-} \longrightarrow Fe_4[Fe(CN)_6]_3$$
(Prussian's blue)

Reactions with Mn. In conc. HNO_3, $MnO_2.xH_2O$ dissolves.

(i) *Lead peroxide.*

$$2Mn(NO_3)_2 + 5PbO_2 + 6HNO_3 \longrightarrow 2HMnO_4 + 5Pb(NO_3)_2 + 2H_2O$$
(purple soln.)

(ii) *Sodium bismuthate.*

Manganese nitrate when reduced by sodium bismuthate gives following reactions.

$$2Mn(NO_3)_2 + 5NaBiO_3 + 16HNO_3 \longrightarrow 2HMnO_4 + 5Bi(NO_3)_2 + 5NaNO_3 + 7H_2O$$
(Purple soln.)

Chemical reactions involved in IV Group

The addition of ammonia (*i.e.*, conc. NH_4OH) converts these cations except Mn into soluble amino complexes.

$$Zn^{2+} + 4NH_3 \longrightarrow [Zn(NH_3)_4]^{2+}$$
$$Co^{2+} + 6NH_3 \longrightarrow [Co(NH_3)_6]^{2+}$$
$$Ni^{2+} + 6NH_3 \longrightarrow [Ni(NH_3)_6]^{2+}$$

The ammoniacal solution containing Mn^{2+} and these amino complexes gets converted into sulphides when H_2S is passed.

$$Mn^{2+} + S^{2-} \longrightarrow MnS \downarrow$$
(flesh coloured)

$$[Zn(NH_3)_4]^{2+} + S^{2-} \longrightarrow ZnS \downarrow + 4NH_3$$
white or dirty
white

$$[Co(NH_3)_6]^{2+} + S^{2-} \longrightarrow CoS \downarrow + 6NH_3$$
$$[Ni(NH_3)_6]^{2+} + S^{2-} \longrightarrow NiS \downarrow + 6NH_3$$
Black

Addition of very dil. HCl, dissolves only Zn and Mn sulphides.

$$ZnS + 2HCl \longrightarrow ZnCl_2 + H_2S$$
$$MnS + 2HCl \longrightarrow MnCl_2 + H_2S$$

Treatment of Zn and Mn Chlorides with excess of NaOH

At first Zn forms a white ppt. of $Zn(OH)_2$ but it dissolves in excess of NaOH forming a soluble complex sodium zincate.

$$ZnCl_2 + 2NaOH \longrightarrow Zn(OH)_2 \downarrow + 2NaCl$$
white

$$Zn(OH)_2 + 2NaOH \longrightarrow Na_2ZnO_2 + 2H_2O$$
soluble complex
(sod. zincate)

Mn is converted to a white ppt. of $Mn(OH)_2$ which is oxidised to black MnO_2 and finally becomes hydrated, i.e., $MnO_2.xH_2O$.

$$MnCl_2 + 2NaOH \longrightarrow Mn(OH)_2 \downarrow + 2NaCl$$
$$Mn(OH)_2 + O \longrightarrow MnO_2 \downarrow + H_2O$$
Brown or
Black

Thus filtrate contains Na_2ZnO_2 and residue contains $MnO_2.xH_2O$.

Test for Zn. On passing H_2S into the filtrate, a white ppt. of ZnS is produced.

$$Na_2ZnO_2 + H_2S \longrightarrow 2NaOH + ZnS \downarrow$$
white

Test for Mn. $Mn(OH)_2$ or hydrated manganese oxide, $MnO_2.xH_2O$ with conc. HNO_3 changes to $Mn(NO_3)_2$.

$$2MnO_2 + 4HNO_3 \longrightarrow 2Mn(NO_3)_2 + 2H_2O + O_2$$

(i) *Sodium bismuthate.*

$$5NaBiO_3 + 2Mn(NO_3)_2 + 16HNO_3 \longrightarrow 2HMnO_4 + 5Bi(NO_3)_3 + 5NaNO_3 + 7H_2O$$
Permanganic
acid (purple)

(ii) *Lead dioxide.*

$$2Mn(NO_3)_2 + 5PbO_2 + 6HNO_3 \longrightarrow 2HMnO_4 + 2H_2O + 5Pb(NO_3)_2$$

Treatment of Ni and Co sulphides. Conc. HCl along with conc. HNO_3 or $KClO_3$ produces nascent chlorine.

$$4KClO_3 + 12HCl \longrightarrow 9Cl + 4KCl + 3ClO_2 + 6H_2O$$

This nascent chlorine converts these sulphides back into their halides via nascent oxygen.

$$H_2O + 2Cl \longrightarrow 2HCl + [O]$$
$$CoS + 2HCl + [O] \longrightarrow CoCl_2 + H_2O + S$$
$$NiS + 2HCl + [O] \longrightarrow NiCl_2 + H_2O + S$$

Ni and Co can be separated by any of the following methods :

1. *Treatment with excess of $NaHCO_3$.* It converts $CoCl_2$ to sodium cobalto carbonate which is oxidise by Br_2 water to green sodium cobalticarbonate which is so stable that colour remains unchanged on heating.

$$2NaHCO_3 \longrightarrow Na_2CO_3 + H_2O + CO_2 \uparrow$$
$$Na_2CO_3 + CoCl_2 \longrightarrow CoCO_3 + 2NaCl$$
$$CoCO_3 + 2Na_2CO_3 \longrightarrow Na_4[Co(CO_3)_3]$$
Sodium cobalto
carbonate

$$2Na_4[Co(CO_3)_3] + 2NaHCO_3 + [O] \longrightarrow 2Na_3[Co(CO_3)_3] + 2Na_2CO_3 + H_2O$$
$$\text{(green) sodium cobalti-}$$
$$\text{carbonate}$$

$NiCl_2$ present at this stage, does not undergo complex carbonate formation and Br_2 water (in presence of $NaOH$ formed by hydrolysis of $NaHCO_3$) convertes it to black nickel oxide on heating.

$$NiCl_2 + 2NaHCO_3 \longrightarrow NiCO_3 + 2NaCl + H_2O + CO_2$$
$$2NiCO_3 + 4NaOH + [O] \longrightarrow Ni_2O_3 \downarrow + 2Na_2CO_3 + H_2O$$
$$\text{black}$$

2.   *Treatment with KCN.* With cobalt salt solution, *e.g.*, $CoCl_2$ and KCN give at first a reddish brown ppt. cobaltous cyanide, $Co(CN)_2$ but it dissolves in excess of KCN forming a yellowish brown solution of potassium cobaltocyanide, $K_4[Co(CN)_6]$.

$$CoCl_2 + 2KCN \longrightarrow Co(CN)_2 \downarrow + 2KCl$$
$$Co(CN)_2 + 4KCN \longrightarrow K_4[Co(CN)_6]$$
$$\text{(Soln.)}$$

On boiling, the yellowish brown solution of pot. cobaltocyanide, readily oxidises to bright yellow solution of pot. cobalticyanide $K_3[Co(CN)_6]$.

$$2K_4[Co(CN)_6] + H_2O + O \text{ (air)} \longrightarrow 2K_3[Co(CN)_6] + 2KOH$$

With nickel salt solution, *e.g.*, $NiCl_2$, KCN give a light green pot. of nickelous cyanide, $Ni(CN)_2$, soluble in excess of KCN forming soluble ppt. nickelocyanide $K_2[Ni(CN)_4]$.

$$NiCl_2 + 2KCN \longrightarrow Ni(CN)_2 \downarrow + 2KCl$$
$$2KCN + Ni(CN)_2 \longrightarrow K_2[Ni(CN)_4]$$

The secondary ionisation of pot. cobalticyanide is so small (*i.e.*, complex is very stable) that no cobalt is precipitated even by $NaOH$ and Br_2 water. The secondary ionisation of nickelocyanide is so high that Ni is precipitated as black $Ni(OH)_3$ by $NaOH + Br_2$ water, on gentle heating.

$$NaOH + Br_2 \longrightarrow NaBrO + HBr$$
$$2K_2[Ni(CN)_4] + 4NaOH + 9NaBrO + H_2O \longrightarrow 2Ni(OH)_3 \downarrow + 4KCNO + 9NaBr + 4NaCNO$$
$$\text{Black}$$

Thus nickel can be separated from cobalt.

Reactions Involved in V Group Analysis

The fifth group radicals are precipitated as their carbonates. The function of NH_4Cl is to suppress the onisation of NH_4OH and $(NH_4)_2CO_3$ (Common ion effect) which will otherwise precipitate magnesium.

The precipitation reactions take place as follows :

$$MCl_2 + (NH_4)_2CO_3 \longrightarrow MCO_3 \downarrow + 2NH_4Cl \qquad [M = Sr^{2+}, Ba^{2+} \text{ or } Ca^{2+}]$$

This white precipitate becomes soluble by the addition of hot acetic acid forming respective acetates.

$$MCO_2 + 2CH_3COOH \longrightarrow (CH_3COO)_2M + CO_2 \uparrow + H_2O$$

Reaction of Ba : With K_2CrO_4, Ba is converted to $BaCrO_4$ which being insoluble separates out. With conc. Cl, it is again converted to soluble $BaCl_2$ which by amm. sulphate is precipitated as barium sulphate.

$$2BaCrO_4 + 16HCl \longrightarrow 2BaCl_2 + 2CrCl_3 + Cl_2 + 8H_2O$$
$$BaCl_2 + (NH_4)_2SO_4 \longrightarrow BaSO_4 \downarrow + 2NH_4Cl$$

Reactions of Sr : The filtrate is treated with excess of $(NH_4)_2SO_4$ whereby Sr is precipitated as $SrSO_4$, SO_4 is also precipitated but it dissolves in excess of $(NH_4)_2SO_4$ forming a soluble complex salt.

$$(CH_3COO)_2Sr + (NH_4)_2SO_4 \longrightarrow SrSO_4 \downarrow + 2CH_3COONH_4$$
$$\text{white ppt.}$$

$$(CH_3COO)_2Ca + (NH_4)_2SO_4 \longrightarrow CaSO_4 \downarrow + 2CH_3COONH_4$$
$$\text{white ppt.}$$

But in excess of $(NH_4)_2SO_4$:

$$CaSO_4 + (NH_4)_2SO_4 \longrightarrow (NH_4)_2[Ca(SO_4)_2]$$

white ppt.   Soluble

Reaction of Ca. The filtrate containing Ca^{2+} ions when treated with amm. oxalate in presence of NH_4OH, white ppt. of calcium oxalate is obtained.

$$Ca^{2+} + (NH_4)_2C_2O_4 \longrightarrow CaC_2O_4 \downarrow + 2NH_4^+$$

White

CaC_2O_4 is dissolved in dil. HCl. The solution when treated with excess of $K_4[Fe(CN)_6]$, a white ppt. of calcium potassium ferrocyanide forms. The test is more sensitive in presence of excess of NH_4Cl.

$$CaC_2O_4 + 2HCl \longrightarrow H_2C_2O_4 + CaCl_2$$
$$CaCl_2 + K_4[Fe(CN)_6] \longrightarrow CaK_2[Fe(CN)_6] + 2KCl$$

(White ppt.)
Cal. Pot.
ferrocyanide

Reactions Involved in Sixth Group Analysis

1. *Magnesium* is precipitated as follows :

$$MgCl_2 + Na_2HPO_4 + NH_4OH \longrightarrow Mg(NH_4)PO_4 \downarrow + 2NaCl + H_2O$$

(white ppt.)

2. (a) To test the *ammonium ion*, we heat it with NaOH

$$NH_4Cl + NaOH \longrightarrow NH_3 \uparrow + NaCl + H_2O$$

(b) With Nesseler's reagent, it has to undergo according to :

$$NH_4Cl + NaOH \longrightarrow NH_3 + H_2O + NaCl$$

$$NH_3 + 3NaOH + 2K_2HgI_4 \longrightarrow 4KI + 3NaI + 2H_2O_2 + O\!\!<\!\!^{Hg}_{Hg}\!\!>\!\!NH_2I \downarrow$$

colouless   Brown

Oxydimercuri amm. iodide
(Iodide of Million's base)

(c) With mercurous nitrate –

$$2NH_3 + Hg_2(NO_3)_2 \longrightarrow \underbrace{Hg(NH_2)NO_2.Hg}_{Black} + NH_4NO_3$$

3. *Sodium* with Uranyl acetate

$$NaCl + 3(CH_3COO)_2UO_2 + (CH_3COO)_2Mg + CH_3COOH$$
$$\longrightarrow NaMg(UO_2)_3(CH_3COO)_9 + HCl$$

Sod. Mag. uranyl acetate

4. *Potassium with Cobaltinitrite* –

$$Na_3[Co(NO_2)_6] + 2KCl \longrightarrow K_2Na[Co(NO_2)_6] \downarrow + 2NaCl$$

Yellow ppt.
Pot. Sod. Cobaltinitrite

3·13. Special Tests for Some Basic Radicals

The students may also perform the following special tests and they may have some idea about so of the basic radicals which can be confirmed in the regular analysis described in preceding sections.

Copper

Prepare clear solution of mixture in HCl and add slowly NH_4OH solution $\longrightarrow$ *Bluish white* The ppt. dissolves by adding excess of $NH_4OH \longrightarrow$ *Deep blue solution* shows the presence copper.

Arsenic

1. Fleitmann's Test. The basis of this test is that AsH_3 is easily produced by treating an arsenic compound with nascent hydrogen. AsH_3 easily reduces a salt of Ag and Hg. Do not smell AsH_3 as it is *highly poisonous*. This test is performed as under :

All reagents used in this test must be free from arsenic. Take about 6–10 ml of NaOH solution in a test tube and add about 2 gms of granulated zinc. Introduce loose cotton plug in the mouth of this test tube and put over mouth of test tube a piece of filter paper moistened with $AgNO_3$ or $Hg_2(NO_3)_2$ solution. Heat the test tube and if there is *no stain* on filter paper, it is clear that reagent is free from arsenic. Now the substance under test is introduced and cotton plug is again fitted in. If any *black stain* on $AgNO_3$ or $Hg_2(NO_3)_2$ paper indicates presence of *arsenic*.

[**Note.** As there would be fast and energetic evolution or hydrogen throughout the experiment, it is better to take the substance in solution form.

Antimony salts do no respond to this test. Stain on $Hg_2(NO_3)_2$ paper is more dependable.]

2. Gutzeit's Test. The experimental details are same as in Fleitmann's test except that dil. H_2SO_4 is taken in place of NaOH solution. *Similar black stain* will be obtained on $AgNO_3$ paper.

[**Note.** Antimony salts also respond Gutzeit's Test. Hence this test is useful only when either Sb or As in present.]

Antimony

Perform Gutzeit's test (see above) in *absence of arsenic* and if black stain on $AgNO_3$ paper is obtained $\longrightarrow$ Sb in indicated.

Tin

Herbert-Meissner Test. In a porcelain dish, take zinc powder and the substance under test and add to it conc. HCl and let the reaction go on for sometime and simultaneously stir this mixture with a test tube filled with water. When the bottom of this test tube is brought very near to the non-luminous Bunsen flame, a bluish flame or fluorescence is seen at the bottom of the test tube to show that tin is present in the mixture. Probably the bluish-flame is produced due to burning of unstable tin hydride (probably SnH_4).

Iron (ferric)

1. Take the mixture, add to it dil. HCl and a few drops of conc. HNO_3 and boil for 2 minutes. To it, add a solution of $K_4Fe(CN)_6 \longrightarrow$ Deep blue colouration also called Prussian blue indicates presence of Fe^{3+}.

2. Take the mixture, add to it dil. HCl and a few drops of conc. HNO_3 and boil well. Dilute the solution by adding water. Now, add to it KCNS solution. Deep blood red colour indicates the presence of Fe^{3+}.

[**Note.** In KCNS test, a pink colour will be given by it in any solution. Only deep blood red colouration confirms the presence of Fe^{3+}].

Chromium and Manganese

Excess of Na_2CO_3 + KNO_3 are mixed with the substance under examination. This powdered mixture is fused on a piece of broken porcelain and then cooled.

If fused solid mass on cooling is

(1) Green $\longrightarrow$ Mn is indicated.

(2) Yellow $\longrightarrow$ Cr is indicated.

If both Mn and Cr are present, the green colour of manganate will mask yellow colour of chromate. Reactions during fusion are :

$$Cr_2(SO_4)_3 + 5Na_2CO_3 + 3KNO_3 \longrightarrow 2Na_2CrO_4 + 3KNO_2 + 3Na_2SO_4 + 5CO_2$$
$$\text{yellow}$$

$$MnSO_4 + 2KNO_3 + 2Na_2CO_3 \longrightarrow Na_2MnO_4 + 2KNO_2 + Na_2SO_4 + 2CO_2$$
green

The fused mass is extracted with boiling water. Acidify this extract with dil. $H_2SO_4 \rightarrow$ pink colour shows the presence of Mn. Boil this pink solution with alcohol $\rightarrow$ pink colour is destroyed and yellow or orange-yellow colour becomes distinct $\rightarrow$ Cr is present. When yellow solution is treated with acetic acid and lead acetate, a yellow ppt. confirms the presence of chromium.

Cobalt

Take the mixture in a test tube and add 1-2 ml of conc. HCl and boil $\rightarrow$ Dark green solution. To it add excess of water and boil $\rightarrow$ Light pink solution. Put a drop of this pink solution on a filter paper and dry it on low flame. Green spot on the paper shows presence of *cobalt*.

[Note. Most of the cobalt compounds respond to this test.]

Nickel

Since in dimethyl glyoxime test of nickel, Fe^{+2} and cobalt interfere, perform the dimethyl glyoxime test for Ni as follows :

Prepared the solution of mixture in HCl and boil it with 1-2 drops of conc. HNO_3. Cool and make solution alkaline by NH_4OH and filter out if any green ppt. Add dimethylglyoxime to the filtrate $\rightarrow$ rosy-red ppt. shows the presence of nickel.

Manganous Salts

The salt free from chlorides is boiled with a little of conc. HNO_3 in presence of a little lead dioxide diluted with water and then allowed to stand for 5-10 minutes. If supernatant liquid is purple (violet) in colour, Mn^{2+} is indicated. The purple colour is due to permanganic acid.

Since Pb_3O_4 on continued boiling with conc. HNO_3 changes to PbO_2, hence in above test Pb_3O_4 (red lead) may also be used in place of lead dioxide (PbO_2).

4

Analysis of Insoluble Compounds

The systematic analysis of cations, which has been discussed in preceding chapter is very simple in itself and promising but when we are unable to prepare the solution, we are deprived of subsequent treatment of analysis. The identification of such insoluble substances requires a different technique.

All those substances, which remain insoluble in by conc. HCl, conc. HNO_3 or aqua regia are practically regarded as insoluble residues. The use of aqua regia is not recommended as it creates many complications in analysis.

The insoluble substances which are under the scope of simple analysis are limited in number. Some of these are identified by their characteristic colours. The list of such insolubles is given below with their colours.

1. White — $BaSO_4$, $SrSO_4$, $PbSO_4$, CaF_2, AgCl, SnO_2, Sb_2O_4 and Al_2O_3
2. Yellow — AgBr and AgI
3. Green — Cr_2O_3, $Cr_2(SO_4)_3$ (anhydrous)
4. Black — HgS
5. Violet — $CrCl_3$ (mineral)
6. Dark red — Fe_2O_3
7. Brown — SnS_2 of mixture (soluble in auqa regia)

Analysis. That part of mixture which is not dissolved in conc. HCl, and conc. HNO_3 is taken out and washed with water. Now observe the colour carefully and proceed accordingly.

Case I — When residuce is white.

This residue is mixed with six times of Na_2CO_3 and distilled water is added to it. Finally, it is heated for sufficient time and filtered.

The filtrate is taken for acid radicals (SO_4^{2-}, Cl^- and F^-) analysis in the usual way. Further analysis rests on the radical confirmed.

(a) **Sulphate.** If the sulphate ion is present, there may be $BaSO_4$, $SrSO_4$ and $PbSO_4$. We take the residue to flame test.

(i) **$PbSO_4$.** This gives characteristic lemon blue flame. In such a case, the residue is dissolved in ammonium acetate by boiling. Add CH_3COOH and K_2CrO_4 solution. A *yellow precipitate* of lead chromate is obtained.

(ii) **$BaSO_4$.** This gives the characteristic green flame. The residue obtained after treatment with Na_2CO_3 is put in acetic acid to dissolve. Now add K_2CrO_4. A *yellow precipitate* of (Barium chromate) is obtained.

(iii) **$SrSO_4$.** If in the above treatment, no precipitate is obtained, we add amm. sulphate. A white precipitate of $SrSO_4$ comes.

(b) **Calcium Fluoride.** We take the residue obtained by the Na_2CO_3 treatment and dissolve it in acetic acid and add amm. oxalate—a white ppt. is obtained—**CaF_2**.

(c) **Silver Chloride.** In case of chloride ions, we take the insoluble residue and dissolve it in NH_4OH. Add HNO_3 till it becomes acidic. A white ppt. is obtained — **AgCl**.

(d) **No Acidic Radical.** If your insoluble white residue fails to give any test as above we can suspect the oxides of Sn, Sb or Al.

(*i*) **SnO$_2$ and Sb$_2$O$_4$.** Mix a portion of insoluble residue with Na$_2$CO$_3$ in almost equal amount and add to it powdered sulphur in excess. Fuse it in a porcelain dish by heating strongly. Lixivate the fused mass with water, add dil. HCl and observe.

(*a*) An orange precipitate is obtained — **Sb$_2$O$_4$.**

(*b*) In case, the precipitate is yellow or greyish yellow — **SnO$_2$.**

(*ii*) **Al$_2$O$_3$.** Test it by charcoal cavity test (see index). A *blue mass* is obtained.

Case II — When residue is yellow. Such residue is boiled with Zn and HCl which converts AgI or AgBr to metallic silver which is separated by filtration. This metallic residue is tested by dissolving it in HNO$_3$ and precipitating it as AgCl by addition of HCl. The filtrate so obtained (after Zn and HCl treatment) is tested for bromide and Iodide ions in the usual way.

Case III — When residue is green (Cr$_2$O$_3$) and Cr$_2$(SO$_4$)$_3$. The insoluble residue is mixed with Na$_2$CO$_3$ and KNO$_3$ in equal amount and fused by heating it in porcelain dish. The fused mass is lixivated with water and CH$_3$COOH is added to it. Divide it in two parts.

(*i*)   I part + Lead acetate ———→ yellow ppt. $\left.\begin{array}{c}\\ \\\end{array}\right\}$ → Cr$_2$(SO$_4$)$_3$
(*ii*)   II part + Barium chloride ———→ white ppt.

If it gives yellow precipitate with lead acetate but not white ppt. with BaCl$_2$ → Cr$_2$O$_3$

Case IV — When residue is black (HgS). This may be tested as follows :

(*i*) Boil a portion of residue with zinc and sulphuric acid. Pass the evolved gas in lead acetate solution. A black ppt. (PbS) is obtained → S^{2-} is confirmed.

(*ii*) To another portion of residue add aqua regia and boil it. Cool, add water to dilute and add SnCl$_2$. A white precipitate turning grey on heating with excess of SnCl$_2$ → Hg^{2+} is confirmed.

Case V — When residue is violet (CrCl$_3$). Proceed as in case III up to acetic acid addition and then divide it in two parts.

(*i*)   I part + Lead acetate ———→ yellow ppt. $\left.\begin{array}{c}\\ \\\end{array}\right\}$ **CrCl$_3$**
(*ii*) II part + dil. HNO$_3$ (excess) + AgNO$_3$ ———→ white ppt.   (Chromium chloride)

Case VI — When residue is dark red (Fe$_2$O$_3$). This may be confirmed by borax bead test (see index). Brown bead is obtained in oxidising flame.

Case VII — When residue is brown (SnS$_2$). The residue is dissolved in yellow amm. sulphide and then conc. HCl and iron fillings are added. It is then boiled for some time and filtered. To this, we add HgCl$_2$ solution. A white ppt. of mercurous chloride is obtained → **SnS$_2$.**

Some Useful Tests Beyond Systematic Analysis

5·1. Flame Test (volatilization reaction)

The vapours of certain elements and their salts impart characteristic colours when these are vapourised in the flame of burner. Since this property shows promising results, it has been found valuable in the analysis of cations. This analysis is more suitable to insoluble compounds where your systematic analysis fails. As this brings out results clearly even when the element is present in traces its value is further increased. The whole mixture, as such, can not be tested as it may have two or more cations capable of imparting colours and a multicoloured flame will be observed which may lead to confusion.

Theory. The metals whose outer orbital electrons are loosely held, readily achieve their excited conditions simply by heating them in flame. The heat first breaks the compounds to its ionic components and the metallic ions, may take the electrons which are frequently available in the flame. These electrons when fall in the valence orbital of the metal, there is emission of the light of characteristic wavelength which serves the purpose of their identification. The substance is supplied to a flame with the condition that

(1) the flame should be non-luminous (for details see chapter 1)

(2) the supporter on which the sample is put, should not give colour of its own to the flame.

For second condition, we are using platinum or nichrome wire as it does not give its own colour.

The flame colouration of any element is always same no matter whether the element is in free state or in combined state. The intense yellow colour of the sodium in flame test is always there whether we take NaCl, Na_2CO_3, Na_2SO_4 or sodium as such. But this is to note that it depends upon the volatile nature of the compound whether it would respond to the test quickly or will take a margin of time. For instance, barium takes sometime to impart its colour whereas Ca and Sr respond quickly.

The flame test can more successfully and conveniently be performed *when we convert elements to their chlorides.* As the chlorides of the elements are in general more volatile, these provide informations quickly.

Procedure. Heat the platinum wire or nichrome wire in the non-lumious flame of the burner. Observe that it must not give any colour to the flame even when red hot otherwise it *may not be clean.* If so dip it in conc. HCl and repeat.

Now touch the tip of the wire with the compound which has been made paste in conc. HCl and hold it in the edge of non-luminous flame. Observe the colour of the flame with naked eyes and then through cobalt glass. Observations give following results :

No.	Flame colour (Naked eye)	Intensity	Flame colour (cobalt glass)	Inference
1.	Persistent bright golden yellow	large	Not visible	Sodium
2.	Pale violet	slight	Pink or Crimson red	Potassium
3.	Deep bluish green	medium	Bluish green	Copper
4.	Brick red, flickering and non-persistent flame.	medium	Greenish grey	Calcium

5.	Crimson red (Steady red)	medium	Purple	Strontium
6.	Pale green very persistent	slight	Bluish green	Barium
7.	Pale blue grey	slight	Not visible	Lead

Precautions :

1.   Carry out the flame test in oxidising portion (upper portion) of colourless (non-luminous) flame.

2.   There are certain cations *e.g.* Ba^{2+} whose chlorides do not produce their characteristic colours to flame quickly. In such a case, to get the characteristic colour, dip the platinum wire (with paste of sample) in pure conc. HCl and heat again. Repeat this operation several times and make sure that test is not missed.

3.   In case of mixture containing Ca, Sr and Br, the calcium produces the colour first then Sr and lastly Ba. The colour of Ba develops after heating for sometime.

4.   If the test sample contains lead, it will spoil the platinum wire both, in oxidising and reducing flames because lead forms a brittle alloy with platinum. In such case, carry out flame test with iron wire.

5.   The sulphates of alkaline earths *e.g.* $CaSO_4$, $BaSO_4$ etc. do not easily react with conc. HCl. In such cases, first reduce these sulphates into sulphides (sulphides of Ca and Ba etc. easily react with conc. HCl) by heating on Pt. wire in reducing flame (lower portion of non-luminous flame) then dip in conc. HCl and heat again in oxidising flame and observe the colour.

6.   The life of Pt wire can be prolonged if the conc. HCl is free from HNO_3 and the substance on Pt wire is heated only in outer part of non-luminous flames, *i.e.* oxidising flame.

7.   Lead, As, Sb and Bi etc. corrode the Pt wire. In such a case, it is better to use asbestos fibre in place of Pt wire.

8.   The yellow flame is not always due to sodium. The sodium flame is bright yellow and in its light, skin of your hand will assume a peculiar corpse like appearance.

9.   Observe the colour of flame with naked eye and then through blue glass and draw the inference.

5·2. Borax Bead Test

Principle. When borax is heated, it loses the water molecules and then changes to colourless glassy transparent mass. When certain compounds are heated with this mass, the bead attains characteristic colour which serves the purpose of their identification. The borax undergoes following changes, producing glassy bead of composition $2NaBO_2.B_2O_3$.

$$Na_2B_4O_7.10H_2O \longrightarrow Na_2B_4O_7 + 10H_2O$$
$$\text{(anhydrous)}$$

$$Na_2B_4O_7 \longrightarrow \underbrace{2NaBO_2 + B_2O_3}_{\text{Transparent glassy bead}} \text{ (boric anhydride)}$$

In fact, this glassy bead, when in molten condition, can dissolve some metallic oxide forming a coloured *solid solution* which can be easily examined by taking it out of the flame. The sodium metaborate ($NaBO_2$) and boric anhydride (B_2O_3) present in borax bead react with metallic salt to form coloured metaborates.

The same metal can give different coloured beads in oxidising and reducing flames. It is because of the formation of higher metaborates (*i.e.* metal is present in higher valency state) in oxidising flame and lower metaborates (*i.e.* metal in lower valency state) or even metals in reducing flame.

Procedure. A loop is made at the end of platinum wire and heated to red hot. Immediately, it is touched with borax. It is again heated strongly where it swells and gives transparent bead. Now touch this hot bead to the solid to be tested and hold it now in oxidising and then in reducing flame. The bead gives characteristic colours which depends on the nature of compound. The following table discloses the colour of bead in both the flames and its inference.

Table — Borax Bead Test

S. No.	Colour of Bead in oxidising flame	Colour of Bead in reducing flame	Inference	Notes
1.	Greenish when hot and blue in cold.	Red and opaque	Copper	When Cu is in excess, the bead is red and opaque.
2.	Green in both hot and cold.	Green in both hot and cold	Chromium	The colour of the bead in both the flames is same because chromic metaborate formed by oxidising flame is not affected by reducing flame.
3.	Deep Blue	Deep Blue	Cobalt	—
4.	Yellow when hot	Green	Iron	—
5.	Violet in both hot and cold	Colourless	Manganese	Manganous metaborate which is obtained in reducing flame is colourless thus no colour is obtained.
6.	Brown in cold	Grey or black and opaque	Nickel	When Ni is present in excess in reducing flame bead becomes black.

Precautions :

1. If the bead is coloured even in absence of the sample, it shows that wire was not clean.

2. Apply borax bead test when sample is coloured otherwise omit.

3. Use a small particles of the sample for borax bead test otherwise dark opaque bead will form which may not be more conclusive.

4. If the bead is to be heated in oxidising flame, hold it in upper portion of non-luminous flame and if in reducing flame, hold the bead in lower portion of the same flame.

5. If the bead requires heating for long time in reducing flame before the bead shows any colour change, the colour of bead in oxidising flame would be more reliable and informative. In oxidising flame, less heating would be required.

Reactions. The reactions of these salts with bead are as follows :

(1) Copper (i) *Oxidising flame.*

$$CuO + B_2O_3 \longrightarrow Cu(BO_2)_2$$
Green - Blue
(cupric metaborate)

also $$CuO + NaBO_2 \longrightarrow NaCuBO_3 \text{ (Greenish-blue)}$$
copper sodium orthoborate

(ii) *Reducing Flame.* Two reactions may take place

(a) $$2Cu(BO_2)_2 + 2NaBO_2 + C \longrightarrow 2CuBO_2 + Na_2B_4O_7 + CO$$
(colourless)
cuprous
metaborate

(b) $$2Cu(BO_2)_2 + 4NaBO_2 + 2C \xrightarrow{\text{(Red)}} 2Cu + 2Na_2B_4O_7 + 2CO\uparrow$$

(2) **Chromium.** *Oxidising flame*

$$Cr_2(SO_4)_3 + 3B_2O_3 \longrightarrow 2Cr(BO_2)_3 + 3SO_3\uparrow$$
(Green)

This gives the same colour with reducing flame.

(3) **Cobalt.**

$$2NaBO_2 + CoCO_3 \longrightarrow Co(BO_2)_2 + Na_2CO_3$$
Blue

(4) **Iron.** *Oxidising flame.*

$$FeCl_3 + 3NaBO_2 \longrightarrow Fe(BO_2)_3 + 3NaCl$$
Yellow

In reducing flame, it forms ferrous metaborate, $Fe(BO_2)_2$

(5) **Manganese.** *Reducing flame*

$$MnO + B_2O_3 \longrightarrow Mn(BO_2)_2$$
Manganous metaborate
(colourless)

In oxidising flame, it forms manganic metaborate, $Mn(BO_2)_3$

(6) **Nickel** (a) *Oxidising flame.*

$$NiO + B_2O_3 \longrightarrow Ni(BO_2)_2$$
Nickel metaborate
(Brown)

(b) *Reducing flame.*

$$Ni(BO_2)_2 + C \longrightarrow Ni + B_2O_3 + CO\uparrow$$
Grey

The scope of this method has been extended by using Na_2CO_3 and also in some places sodium ammonium hydrogen phosphate instead of borax but their use is selective as it does not cover wide range of radicals.

The *sodium carbonate bead test* (instead of borax, use Na_2CO_3). It is especially suitable to chromium and manganese. These are oxidised to chromate and manganate ions to give their characteristic colours. Oxidising flame is only used in this test.

The *Microcosmic salt bead test* (where sodium ammonium hydrogen phosphate is taken in place of borax) is suitable for Cu, Cr and Co. These provide orthophosphates with metallic oxides which have their characteristic colours. The microcosmic salt is first dehydrated which then forms sodium metaphosphate as colourless bead.

$$Na(NH_4)HPO_4.4H_2O \longrightarrow Na(NH_4)HPO_4 + 4H_2O$$
$$Na(NH_4)HPO_4 \longrightarrow NaPO_3 + NH_3 + H_2O$$
Sodium meta-
phosphate

This reacts with metallic oxides giving coloured beads.

$$NaPO_3 + CuO \longrightarrow NaCuPO_4 \text{ (blue)}$$
$$NaPO_3 + CoO \longrightarrow NaCoPO_4 \text{ (blue)}$$
$$NaPO_3 + Cr_2O_3 \longrightarrow NaPO_3.Cr_2O_3 \text{ (Green)}$$

5·3. Charcoal Cavity Test or Charcoal Block Test (Reducing Reaction)

Principle. The metallic compound when reduced in a charcoal cavity, nature of the bead formed and its colour gives informations about the metal present.

Procedure. The compound is powdered and mixed with 3 times of its weight of Na_2CO_3 or with fusion mixture ($Na_2CO_3 + K_2CO_3$). This is placed in the cavity of the charcoal made by knife. It is moistened with few drops of water and heated strongly by means of blow pipe in the reducing flame. Some of the basic radicals show characteristic beads.

Metals	Bead and Incrustation
1. Bismuth	Brittle bead and yellow
2. Lead	Soft bead and yellow which marks the paper
3. Antimony	Brittle bead and white

[**Note :** Fusion mixture is preferred over Na_2CO_3 alone because of its lower melting point.]

5·4. Cobalt Nitrate Test (Reducing Reaction)

The principle and procedure of this test is similar to that of charcoal cavity test with slight modification *i.e.* after adding Na_2CO_3 and heating, add few drops of cobalt nitrate and heat once again strongly to get better results. The inference may be seen from following table :

Observation	Composition	Inference
1. Blue residue	$CoO.Al_2O_3$	Al (when phosphate and borate are absent)
2. Green residue	$CoO.ZnO_2$	Zinc oxide
2. Pink dirty residue	$CoO.MgO$	Magnesium oxide
3. Blue residue	$NaCo.PO_4$	Phosphate in absence of Al

Reactions :

ZnO :
$$2Co(NO_3)_2 \longrightarrow 2Co + 4NO_2 + 2O_2$$
$$ZnO + CoO \longrightarrow \underset{\text{Green}}{CoZnO_2}$$
Cobalt zincate

Al_2O_3 :
$$Al_2O_3 + CoO \longrightarrow \underset{\substack{\text{Blue Cobalt} \\ \text{aluminate}}}{Co(AlO_2)_2}$$

PO_4^{3-} :
$$Na_3PO_4 \longrightarrow NaPO_3 + Na_2O$$
$$NaPO_3 + CoO \longrightarrow \underset{\text{Blue}}{NaCoPO_4}$$

6

Semi-micro Analysis

As discussed in the introductory survey of qualitative analysis, this analysis requires less quantity of the substance. It has superiority over both, micro and macro methods as being faster and more economic. Though the micro method requires minimum quantity of specimen for the analysis but some of the reagents used in it are quite costly and the analysis as a whole sometimes becomes an expensive affair.

Apparatus. Like macro analysis, the apparatus required in this analysis are :

(*i*) Test tubes

(*ii*) Beakers

(*iii*) Flasks

(*iv*) Porcelain dish

(*v*) Dropper

Their make up is slightly different. The chart given ahead indicates most of the apparatus required for semi-micro analysis.

The major difference comes in the process of separating the precipitate from solution. We do not use any funnel to separate it as it consumes quite good time. We make use of centrifuge machine and the sample is subjected to a very rapid spin and settles down in the form of compact mass within no time. This is finally decanted carefully or supernatant liquid is taken by dropper. As centrifuge machine is used, the filtrate here is called *centrifugate*. For this purpose, both, hand driven and electrical driven machines are available. To heat and to boil the contents, we make use of water bath.

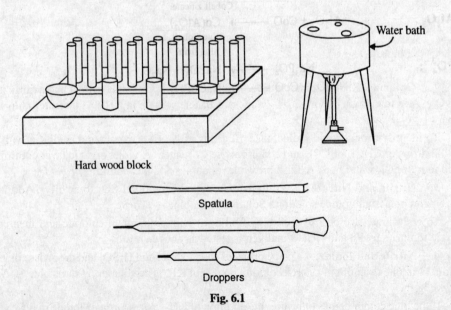

Hard wood block

Spatula

Droppers

Fig. 6.1

Analysis of Acid Radicals (Anions). The analysis of anions may be divided in three folds :

1. Preliminary examination of the mixture.

2. Confirmatory tests.

3. Combination tests.

Preliminary examination of the acid radicals of the mixture is done first by dil. HCl then concentrated H_2SO_4, exactly in the similar manner as in macro method and hence need not to discuss again. Here 30 mg of the substance with 2 ml of dil. HCl and next time the same amount with conc. H_2SO_4 is taken for it.

Like macro analysis, here too we prepare sodium carbonate extract to carry out the combination tests.

After suspecting the acid radicals, we try the different combination tests. These tests are discussed with different heads. The reactions of different combination tests can be understood by macroanalysis discussed earlier.

1. Sulphide, Sulphite, Sulphate and Thiosulphate. To 10 drops of soda extract, add 30 mg of $CdCO_3$. Shake it well and filter.

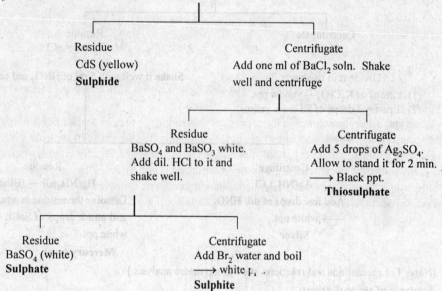

Residue
CdS (yellow)
Sulphide

Centrifugate
Add one ml of $BaCl_2$ soln. Shake
well and centrifuge

Residue
$BaSO_4$ and $BaSO_3$ white.
Add dil. HCl to it and
shake well.

Centrifugate
Add 5 drops of Ag_2SO_4.
Allow to stand it for 2 min.
$\longrightarrow$ Black ppt.
Thiosulphate

Residue
$BaSO_4$ (white)
Sulphate

Centrifugate
Add Br_2 water and boil
$\longrightarrow$ white ρ₁ .
Sulphite

2. Chloride, Bromide and Iodide. Take 10–20 drops of soda extract and acidify it with dil. HNO_3. Add few drops of $AgNO_3$. A white precipitate soluble in NH_4OH indicates chloride. Confirm it by *chromyl chloride test*.

To identify bromide and iodide, take 10 drops of Na_2CO_3 extract and acidify it with dil. HCl. Add 2 drops of $CHCl_3$ or CCl_4 and add dropwise Cl_2 water. A violet organic layer confirms *iodide*. Add more of Cl_2 water — A reddish brown layer comes and this confirms *bromide*.

3. Nitrite and Nitrate. To 5 drops of soda extract, add H_2SO_4 to acidify. Add 2 drops of starch and potassium iodide solution. Solution turns blue — *Nitrite*.

To test nitrate, we destroy nitrite by boiling it with ammonium chloride and then converting nitrate to nitrite by Zn and HCl. Finally, it is tested as above.

4. Nitrite and Iodide. To 5–10 drops of soda extract add H_2SO_4 and dropwise till it becomes acidic. Put one drop of only starch solution on it (not KI). The solution, if turns blue — *Nitrite* and *Iodide* both.

If the blue colour comes only after the addition of KI — *Nitrite alone*. Iodide may be verified by chloroform test also.

5.   **Carbonate and Sulphite.** To 20 mg of mixture (not Na_2CO_3 extract) add almost 20 mg $K_2Cr_2O_7$ and 1 ml of H_2SO_4. It is heated on water bath and the gas coming out is passed in lime water. The lime water turns milky—*Carbonate*.

The other combination tests are also performed in the same way as in macro-analysis except we take small quantity of the sample and reagents.

Analysis of Basic Radicals. The semi-micro analytical method of basic radicals also demands the preparation of the solution of the sample for analysis. The attempts to prepare the solution are done similarly, *i.e.*, first in water then dil. HCl, dil. HNO_3, conc. HCl and conc. HNO_3 finally. The solution prepared in conc. acids are diluted and then their subsequent analysis is done.

Analysis of Group I (Ag^+, Hg_2^{+2}, Pb^{+2}). Take about 2 ml of original solution and 5–10 drops of dil. HCl. Centrifuge the solution. Centrifugate is taken for second group analysis and the residue is washed by few drops of water. Heat this precipitate with 1 ml of water to boiling, quickly centrifuge and follow the flow sheet.

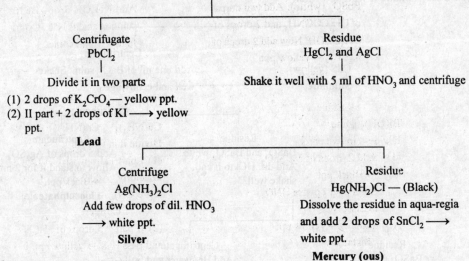

[Note. For precautions and reactions, see I group macro analysis.]

Analysis of second group

[Hg^{2+}, Pb^{2+}, Bi^{3+}, Cu^{2+}, Cd^{2+}, As^{3+}, Sb^{3+}, Sn^{2+} and Sn^{4+}]

Precipitation of second group radicals is carried out in two centrifuge tubes.

(1) Pass H_2S to distinctly acidic solution (made by dil. HCl). The solution must be hot. This is done to precipitate As^{3+} completely. Centrifuge and collect the residue.

(2) Now make this centrifugate dilute, cool and then pass the H_2S to precipitate Cd completely. Centrifuge it and leave the centrifugate for III group analysis.

Collect both the residues and wash it with hot water and shake the ppt. well with 1 ml of yellow amm. sulphide and warm the solution. Centrifuge it.

1.   The residue will contain the members of the II A group (copper group) (Cu^{2+}, Cd^{2+}, Bi^{3+}, Pb^{2+}, Hg^{2+}).

2.   The centrifugate will contain the members of the II B group (Arsenic Group) (As^{3+}, Sb^{3+}, Sn^{2+}, or Sn^{4+}).

Now residue and centrifugate are analysed separately.

Analysis of Residue (Copper Group). To residue add 10 drops of dil. HNO_3 and heat it for three minutes. Centrifuge.

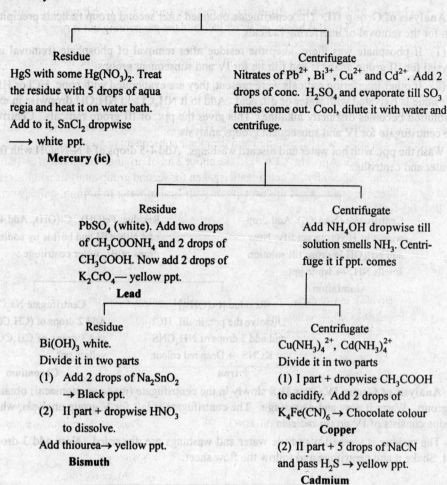

Residue
HgS with some $Hg(NO_3)_2$. Treat the residue with 5 drops of aqua regia and heat it on water bath. Add to it, $SnCl_2$ dropwise
$\longrightarrow$ white ppt.
Mercury (ic)

Centrifugate
Nitrates of Pb^{2+}, Bi^{3+}, Cu^{2+} and Cd^{2+}. Add 2 drops of conc. H_2SO_4 and evaporate till SO_3 fumes come out. Cool, dilute it with water and centrifuge.

Residue
$PbSO_4$ (white). Add two drops of CH_3COONH_4 and 2 drops of CH_3COOH. Now add 2 drops of K_2CrO_4— yellow ppt.
Lead

Centrifugate
Add NH_4OH dropwise till solution smells NH_3. Centrifuge it if ppt. comes

Residue
$Bi(OH)_3$ white.
Divide it in two parts
(1) Add 2 drops of Na_2SnO_2
$\rightarrow$ Black ppt.
(2) II part + dropwise HNO_3
to dissolve.
Add thiourea$\rightarrow$ yellow ppt.
Bismuth

Centrifugate
$Cu(NH_3)_4^{2+}$, $Cd(NH_3)_4^{2+}$
Divide it in two parts
(1) I part + dropwise CH_3COOH
to acidify. Add 2 drops of
$K_4Fe(CN)_6 \rightarrow$ Chocolate colour
Copper
(2) II part + 5 drops of NaCN
and pass $H_2S \rightarrow$ yellow ppt.
Cadmium

Analysis of Arsenic Group. The centrifugate received after the treatment with yellow amm. sulphide contains soluble thiocomplexes. It is acidified with dil. HCl. The solution is given good shake and warmed. Now add 2 drops of conc. HCl and heat to boiling. Pass H_2S in very hot solution, centrifuge :

Residue
As_2S_3 (yellow)
Arsenic

Centrifugate
Sb^{3+}, Sn^{4+}
Expel the H_2S by boiling and divide it in two parts.
(1) I part + little Al or Fe metal and boil. Take the supernatant liquid by a dropper and add it to $HgCl_2$ solution $\rightarrow$ white ppt.
Tin (Sn^{4+})
(2) II part + NH_3 dropwise till alkaline. Add 2 drops of oxalic acid and pass H_2S. Orange ppt. $\rightarrow$
Antimony

Analysis of Group III. The centrifugate obtained after second group radicals precipitation, is taken for the removal of interfering radicals.

(1) If phosphate was there, keep the residue after removal of phosphate (removal in macro analysis) for III group analysis and filtrate for IV and subsequent groups.

(2) If other interfering radicals were present, they are evaporated with conc. HCl or HNO_3 and the residue is lixivated with water or dil. acid. Add to it NH_4Cl and NH_4OH dropwise in excess till the solution becomes distinctly alkaline. This gives the ppt. of III group radicals. Centrifuge and keep centrifugate for IV and subsequent groups analysis.

Wash the ppt. with hot water and discard washings. Add 4-5 drops of 4 N-NaOH with few drops of water and centrifuge.

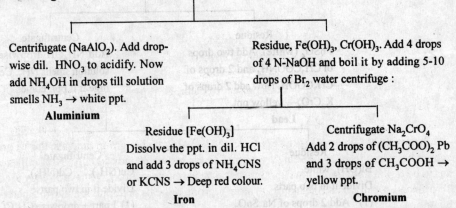

Centrifugate ($NaAlO_2$). Add drop-wise dil. HNO_3 to acidify. Now add NH_4OH in drops till solution smells $NH_3 \rightarrow$ white ppt.
Aluminium

Residue, $Fe(OH)_3$, $Cr(OH)_3$. Add 4 drops of 4 N-NaOH and boil it by adding 5-10 drops of Br_2 water centrifuge :

Residue [$Fe(OH)_3$] Dissolve the ppt. in dil. HCl and add 3 drops of NH_4CNS or KCNS $\rightarrow$ Deep red colour.
Iron

Centrifugate Na_2CrO_4 Add 2 drops of $(CH_3COO)_2$ Pb and 3 drops of $CH_3COOH \rightarrow$ yellow ppt.
Chromium

Analysis of Group IV. Pass H_2S slowly in the centrifugate (made ammoniacal) obtained from III group. Warm it now and centrifuge. The centrifugate is left for V group analysis, whereas the residue consists of IV group radicals.

The residue is washed with little water and washings are discarded. Now add 3 drops of dil. HCl. Shake it and centrifuge and follow the flow sheet.

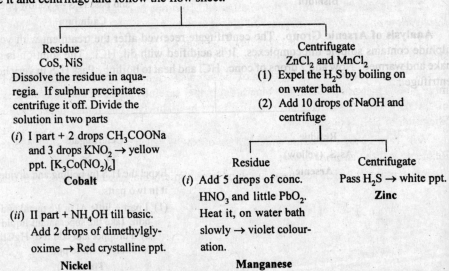

Residue CoS, NiS Dissolve the residue in aqua-regia. If sulphur precipitates centrifuge it off. Divide the solution in two parts
(*i*) I part + 2 drops CH_3COONa and 3 drops $KNO_2 \rightarrow$ yellow ppt. [$K_3Co(NO_2)_6$]
Cobalt

(*ii*) II part + NH_4OH till basic. Add 2 drops of dimethylgly-oxime $\rightarrow$ Red crystalline ppt.
Nickel

Centrifugate $ZnCl_2$ and $MnCl_2$
(1) Expel the H_2S by boiling on water bath
(2) Add 10 drops of NaOH and centrifuge

Residue
(*i*) Add 5 drops of conc. HNO_3 and little PbO_2. Heat it, on water bath slowly $\rightarrow$ violet colour-ation.
Manganese

Centrifugate Pass $H_2S \rightarrow$ white ppt.
Zinc

Analysis of the Group V. To the centrifugate obtained after IV group radicals precipitation, boiled to expel H_2S. Add 2 drops of NH_4OH followed by 10 drops of $(NH_4)_2CO_3$. Boil the solutic leave it as such for 10 minutes. Centrifuge it and leave the centrifugate for VI group analysis. T precipitate so obtained is dissolved by adding 6N acetic acid dropwise and shaking the solution. Add 2 drops of potassium chromate solution. Centrifuge and follow this flow sheet.

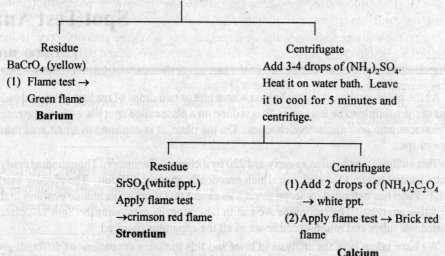

Analysis of Group VI (Mg^{2+}, Na^+, K^+, NH^{4+}). Like other group radicals, the VI group radicals cannot be precipitated by one single reagent. These are tested individually. NH_4^+ still creeps out as it has to be tested from the original mixture because in few operations, we have already used NH_4^+ ions. To proceed, we divide the centrifugate after V group radicals precipitation into 4-5 parts and test one by one.

1.   **Magnesium.** (*i*) add 2 drops ammonium hydrogen phosphate. Cool it under tap water, stir and scratch with the help of glass rod. *A white crystalline ppt. appears.*

(*ii*) In another part, add 1 drop of magneson and about 5 drops of NaOH. *A sky blue precipitate* of the magnesium hydroxide is obtained.

2.   **Sodium.** To 0.5 ml of the solution add 10 drops of magnesium uranyl acetate solution. A *light yellow ppt.* of sodium magnesium uranyl acetate, $NaMg(UO_2)_3(CH_3COO)_9.5H_2O$ is obtained.

3.   **Potassium.** To the last portion of the solution, add few drops of sodium cobaltinitrate and shake it well. After 2-3 minutes, *yellow ppt.* of $K_2NaCo(NO_2)_6$ appears.

4.   **Ammonium.** (*i*) Take 10 mg of the sample (original mixture) in a semimicro beaker, add 5 ml of NaOH to it and cover it with blue litmus paper carefully so that it may not touch even the traces of NaOH. Keep the beaker on water bath and heat slowly. The litmus paper turning *red* due to liberation of NH_3 is observed.

(*ii*) Take 20 mg of mixture and add 1 ml of NaOH. Heat and pass the evolved gas to Nesseler's Reagent. It turn *yellow* or *brown.*

Most of the chemical reactions involved in semimicro analysis are the same as in macro analysis.

7

Spot Test Analysis

(Micro analysis)

As the name implies, it is a technique where one or two drops of reagent is applied on one or two drops of the solution to be analysed. This is done on a plate called spot plate where the reagent brings out characteristic and distinctive changes. On the plate, it is confined to small area thus takes the shape of spot.

This analysis was applied as early as 1920 by a chemist Tananaev. Though ideal analysis should require special reagent for each ion which practically appears difficult. The pioneer worker of this field, F. Fiegl has extended its scope to such an extent that it occupies unique position in the modern qualitative analysis. The technique we use in laboratory is most simple. Spot test plate, dropper, assorted test tubes and micro-crucible are in all the apparatus required.

We have taken here the analysis in brief and this includes the cations of different groups. For convenience, these have also been put up here groupwise. In each testing, we take 1-2 drops of the original solution (acidic or alkaline depending upon cation to analyse) on spot plate followed by the addition of the reagent in the system given below :

I. **Group Radicals :**

  1. **Lead**

  (*i*) 1 drop of aqueous or slightly alkaline test solution.

  (*ii*) 1 drop of dithiazone → Brick red complex.

  2. **Silver**

  (*i*) 1 drop of fairly acidic solution;

  (*ii*) 1 drop of paradimethylamino benzylidine rhodamine → red or red violet colour.

  3. **Mercurous Ions (Hg_2^{2+})**

  (*i*) 1 drop of aqueous solution.

  (*ii*) 1 drop of gallic acid (prepared in alcohol) → yellow or orange colour.

II. **Group Radicals**

  4. **Mercuric Ions (Hg^{2+})**

  (*i*) 1 drop of diphenyl carbazone

  (*ii*) 1 drop of very dil. HNO_3.

  (*iii*) 1 drop of salt solution.

  → violet or blue colouration.

  5. **Bismuth**

  (*i*) 1 drop of acidic solution of substance.

  (*ii*) 1 drop of thiourea (10%).

  (*iii*) 1 drop of dil. HNO_3 → yellow colour.

  6. **Copper**

  (*i*) 1 drop of weakly acidic or ammoniacal solution.

  (*ii*) 1 drop of dithioxamide (or Rubianic acid) → greenish black or black ppt.

7. **Cadmium**

(i) 1 drop of test solution.

(ii) 1 drop of NaOH (strong).

(iii) 1 drop of allyl thiourea.

(iv) Heat slightly → yellow ppt.

8. **Arsenic**

(i) 1 drop of test solution.

(ii) 1 drop of conc. HNO_3.

(iii) 1 drop of amm. molybdate.

(iv) Heat slightly → canary yellow ppt.

[Note. PO_4^{3-} also gives this test.]

9. **Antimony (Sb^{4+})**

(i) 2 drops of test solution in test tube.

(ii) 5 drops of H_2SO_4.

(iii) 1 drop of KI.

(iv) 5 drops of C_6H_6 and shake.

(v) 1 drop of Rhodamine – B reagent → violet colour in organic layer.

10. **Tin (Sn^{2+})**

(i) 1 drop of salt solution.

(ii) 1 drop of cocatheline reagent → violet colour.

[Note. If the tin salt is tetravalent, reduce it by Zn and HCl then perform above test.]

III. Group Radicals

11. **Aluminium**

(i) 1 drop of aqueous solution on test plate.

(ii) 1 drop of aluminium Auric tricarboxylate → bright red colour.

12. **Chromium**

(i) 1 drop of aqueous solution on test plate.

(ii) 1 drop of conc. H_2SO_4.

(iii) 1 drop of freshly prepared diphenyl carbazide reagent → violet colour.

13. **Iron**

Ferrous and ferric ions are tested separately.

(a) Ferrous ions —

(i) 1 drop of test solution.

(ii) 1 drop of $K_3Fe(CN)_6$ → Blue colour.

(b) Ferric ions —

(i) 1 drop of test solution.

(ii) 1 drop of HNO_3.

(iii) 1 drop of KCNS.

→ Blood red colour.

IV. Group Radicals

14. **Zinc**

(i) 1 drop of acidic solution of salt on plate.

(*ii*) 2 drops of sodium thiosulphate.

(*iii*) 1 drop of KCN.

(*iv*) 4-5 drops of diphenyl thiocarbazone and shake → organic layer red.

15. **Manganese**

(*i*) 1 drop of test solution.

(*ii*) 1 drop of conc. NaOH.

(*iii*) 1 drop of tartaric acid.

(*iv*) 1 drop of benzidine solution.

→ Blue colour.

16. **Cobalt**

(*i*) 1 drop of aqueous test solution.

(*ii*) 1 drop of acetic acid.

(*iii*) 1 drop of α-nitroso β-naphthol reagent → Reddish brown ppt.

17. **Nickel**

(*i*) 1 drop of aqueous solution.

(*ii*) 1drop of NH_4OH

(*iii*) 1 drop of dimethyl glyoxime → pink red ppt.

V. **Group Radicals**

18. **Barium**

(*i*) 1 drop of test solution

(*ii*) 1 drop of sodium rhodizonate.

(*iii*) 1 drop of HCl after two minutes → Red brown colour.

[If the colour appears on addition of reagent but disappears on acidification, it indicates *strontium* not Barium.]

19. **Calcium**

(*i*) 3 drops of solution in test tube.

(*ii*) 1 drop of amm. sulphide. Now heat and throw away if any precipitate comes.

(*iii*) 1 drop of dihydroxy tartarate osazone.

→ yellow ppt.

VI. **Group Radicals**

20. **Magnesium**

(*i*) 1 drop of aqueous solution.

(*ii*) 1 drop of conc. NaOH solution.

(*iii*) 1 drop of magneson → Blue colour.

21. **Sodium**

(*i*) 1 drop of aqueous solution.

(*ii*) 1 drop of alcohol.

(*iii*) 1 drop of Zn or Ni uranyl acetate → reagent yellow crystalline ppt.

22. **Potassium**

(*i*) 1 drop of aqueous solution.

(*ii*) 1 drop of dipicryl amine.

→ orange red ppt.

23. **Ammonium (NH_4^+)**

(*i*) 1 drop of test solution.

(*ii*) 1 drop of dil. NaOH.

(*iii*) 1 drop of Nesseler's reagent.

→ Brown ppt.

[**Important.** *While applying spot test for a particular cation, one should, keep in mind the interference of other cations if present.*]

8

Review Questions

The students are advised before going through this chapter, to read the reactions involved in analyses of 'cations' and 'anions' and the precautions given there.

Q. 1. What is the purpose of qualitative and quantitative analysis ?

Ans. The purpose of qualitative analysis is to detect the constituents (elements or ions) present in a substance. The determination of the amounts of the individual component of a substance is the primary object of quantitative analysis.

Q. 2. What is the scope of qualitative analysis ?

Ans. Qualitative analysis should precede quantitative analysis. Qualitative analysis has to be carried out even when it is necessary to determine the percentage of a certain component present in a given substance. This is because the suitable method for quantitative determination of the particular component can be applied only after the knowledge of elements or ions present. Qualitative analysis is of enormous scientific and practical importance, being one of the most important methods for investigating substances and their changes. It also plays an important part in branches of science allied to chemistry — mineralogy, geology, physiology and microbiology, as well as in medicine, agriculture and technology.

Q. 3. What do mean by an analytical reaction and a reagent ?

Ans. The chemical change taking place is known as an analytical reaction and the substance causing it, is a reagent.

Q. 4. Explain the distinction between macro-, micro-, semimicro-, and ultramicro-methods of qualitative analysis.

Ans. In *macro analysis,* relatively large amounts of substance (0.5–gm) or solutions (20–50 ml!) are used. The reactions are carried out in ordinary test tubes (10–20 ml), beakers or flasks. Precipitates are separated from solutions by ordinary filtration.

In *micro analysis,* the amount of substance is usually 1/100 of the amount involved in macro analysis works, *i.e.* a few milligrams of a solid or some tenth of a millilitre of a solution. Highly sensitive reagents are used, which make it possible to detect the individual components even when they are present in small amounts. These reactions are carried out by *microcrystalloscopic* or by the *spot test* methods. In *microcrystalloscopic* method, the reactions are carried out on a glass *microscope slide* and presence of a given ion or element is ascertained from the type of the crystals formed. These are examined under microscope.

Semi-micro analysis occupies an intermediate position between the macro-and the micro-methods. The amount of substance is about 50 mg of solid or about 1 ml solution. It has many advantages over macro analysis.

In *ultra micro analysis,* the amount used is less than 1 mg of substance. The various analytical operations are performed under microscope.

Q. 5. What are reaction conditions ? Explain.

Ans. One of most important conditions for a reaction is a *suitable medium* which may be adjusted by addition of acid or alkali to the solution.

Another important condition is the *temperature* of the solution. If the solubility of a precipitate rises sharply with the temperature, precipitation should not be carried out in hot solution and reaction should be carried out in the cold, *i.e.* at room temperature. However, certain reactions proceed only on heating.

Other important condition is a *concentration* of the particular ion in solution.

Q. 6. What reactions are called sensitive ones ? What are the characteristics of the sensitivity of a reaction ?

Ans. If the solubility of a substance is very low and it is precipitated even at very low concentration, the reaction is said to be sensitive. Normally, the solubility of the compound formed is considerable, the sensitivity of the reaction is low and the test succeeds only if the concentration of the ion to be detected is relatively high. The flame test for volatile sodium salts is highly sensitive, as the flame of burner becomes coloured by extremely small amounts of such salts.

The sensitivity of a reaction is characterised by the interconnected quantities, the *detectable minimum* and the *dilution limit*. The *detectable minimum* is the minimum amount of a substance or ion which can be detected by a given reaction under specified conditions. As this amount is very small, it is usually expressed in microgram, denoted by μg; $1 \mu g = 10^{-6}$ gm. The detectable minimum does not fully characterise the sensitivity of reaction because not only the absolute amount but also the concentration of the substance or ion detected is significant. Therefore, *dilution limit* is generally also stated ; as the lowest concentration of the substance (or ion) at which it can be detected by the given reaction. The dilution limit is represented by the ratio 1 : G, where G is the weight of solvent per one part by weight of the substance or ion to be detected. The dilution limit is generally taken to be the concentration of the ion at which one-half of the total number of tests give positive results.

Q. 7. What is difference between the terms "specific" and "selective" when used in connection with reagents or reactions ?

Ans. Reaction (and reagents) which under the experimental conditions are indicative of one substance (or ion) only, are known as specific. Only ammonium salts, on heating with alkali, give characteristic smell of ammonia. Therefore, the alkali reaction (or reagent) is, the specific reaction (or reagent) for the NH_4^+ ion. Reactions (and reagent), which give a similar effect with only limited number of ions (or substances), are known as *selective*. Fewer the number of ions (or substances) giving a positive test, the more selective a given reaction (reagent) is. Hence we may describe reactions (or reagent), as having varying degree of selectivity : however, a reaction (or reagent) can be either specific or non-specific.

Q. 8. What is the essential principle of systematic analysis ?

Ans. If selective reactions are not available and the selectivity cannot be raised in any way, the ions cannot be detected by specific reactions *i.e.* by fractional method. In such cases, it is necessary to work out a definite sequence for the detection of the individual ions; this is *systematic analysis*. In this procedure, each given ion is detected only after all the other ions which interfere with detection *i.e.*, which also react with the reagent used have been detected and removed from the solution. It means that in systematic analysis, it is necessary to use *separation reactions* as well as *detection reactions* for individual ions. Separation reactions are in most cases based on solubility differences between analogous compounds of the respective ions. For example, the separation of Ba^{2+} from Ca^{2+} is based on the solubility difference of $BaCrO_4$ and $CaCrO_4$. Volatility differences may also be used in some instances. For example, NH_4^+ can be separated from K^+, Na^+ and Mg^{2+} by evaporation of the solution followed by ignition of the dry residue. Ammomium salts are volatilised while non-volatile Na, K and Mg salts remain in the residue.

Q. 9. What are group reagents ? What are the advantages of using the group reagents in qualitative analysis ?

Ans. In systematic analysis of a mixture, the ions are not separated individually but done in groups; use is made of the common behaviour of the ions in the same group with certain reagents. The reagent used to separate a group of ions from the others is called group reagent for that group of ions.

The use of the group reagents has considerable advantages, as the complex problem of analysis is subdivided into a number of simpler operations. If any group is entirely absent, it is then pointless to carry out tests for the individual ions of this group and a considerable amount of work, time and reagents can be saved. Thus even if there are specific reactions for all ions present, the systematic analytical procedure retains its significance, being more convenient and economic than fractional analysis of a substance of unknown composition.

Q. 10. What is the basis of classification of ions in analytical chemistry ?

Ans. The classification of ions in analytical chemistry is based on difference in solubilities of their salts or hydroxides etc. allowing their separation to different groups.

Q. 11. Explain the difference between 'anions' and 'cations' analysis.

Ans. In cation analysis, the separation of cations into groups is fairly clear because of large difference in solubility products of soluble and insoluble salts. Whereas the systematic separation of anions into groups on the basis of solubility product is neither practicable nor convenient. During identification of anions, upon acidification, many anions form unstable acids which are either lost or destroyed. Oxidising anions tend to react with reducing anions and thus lose their identities as a result of oxidation-reduction reaction. For example, the sulphite ion will reduce the chlorate ion to produce chloride and itself oxidises to sulphate thus identities of both sulphite and chlorate ions are lost. Such complications do not happen with the analysis of cationic constituents. Though certain classifications for anions have been given but they are not rigid and have no theoretical basis because the same anion may fall in more than one group.

Q. 12. Explain why dil. HCl is preferred over dil. H_2SO_4 in preliminary tests for anions.

Ans. It is because of the formation of a layer of sulphate with dilute H_2SO_4 on their particles which remain insoluble and prevent further reaction. But, if dil. HCl is used such hinderance does not arise and reaction proceeds appreciably and presence of ions is well ascertained.

Q. 13. In the confirmation of sulphide by lead acetate, the test mixture is sometimes treated with zinc and dil. HCl. Why ?

Ans. Sulphides of Fe, Mn, Zn and alkali metals are decomposed by dil. HCl or dil. H_2SO_4 with evolution of H_2S whereas sulphides of Pb, Cd, Ni, Co, Sb, and Sn (IV) are decomposed by conc. HCl and not by dil. HCl or dil. H_2SO_4. Other sulphides such as mercuric sulphide are not decomposed even by conc. HCl. Hence the presence of sulphide ion in *insoluble sulphides* is detected by reducing them with nascent hydrogen, *i.e.* Zn (or Sn) and dil. HCl.

Q. 14. What is the purpose of preparing sodium carbonate extract for identification of anions ?

Ans. The presence of cations other than Na^+, K^+ or NH_4^+ may interfere with the anion analysis. These cations, therefore, are usually removed by precipitation as insoluble carbonates, basic carbonates, or hydroxides on boiling the test mixture with excess of sodium carbonate solution. The anions are left in filtrate as sodium salts. This filtrate is called sodium carbonate extract which is used for the analysis of anions except carbonate and bicarbonate. The latter two are tested from original mixture.

Q. 15. The sodium carbonate extract contains an excess of carbonate ions. Is this excess of CO_3^{2-} harmful or not ?

Ans. The excess of CO_3^{2-} ions in soda-extract is of both, an advantage and a disadvantage. It is advantageous as CO_3^{2-} ions maintain an alkalinity, sufficient to prevent unwanted redox reactions between anions.

Q. 16. Sometimes, instead of colourless a coloured sodium carbonate extract is obtained. What do you infer from it ?

Ans. For example, if copper is present, it goes into extract as soluble complex due to which the extract becomes coloured. There are also other cations *e.g.*, arsenic, antimony which form soluble complex with Na_2CO_3 and pass into extract. In order to set free the extract from such cations, the extract is acidified and then H_2S is passed where Cu, As and Sb are precipitated as sulphides and removed leaving the extract colourless. The excess of H_2S is boiled off and then this colourless extract is used to test various anions except Cl^- and S^{2-}.

Q. 17. What is 'water extract' and how is it prepared ?

Ans. Small quantity of original test mixture is boiled with suitable quantity of distilled water (5-10 ml) in a test tube for 2-3 minutes. Cool and filter or centrifuge. The filtrate or centrifugate is called 'water extract'.

Q. 18. What for 'water extract' is used ?

Ans. It is generally used for testing of nitrate, nitrite, acetate and thiosulphate ions as their salts are usually water soluble.

Q. 19. Give reasons why PO_4^{3-}, F^- and S^{2-} are usually tested and confirmed from the original mixture and not from soda extract.

Ans. There are a few salts *e.g.* phosphates of Fe and Al ; sulphides of Zn, Ca and Sb; and a few fluorides which are not decomposed on boiling Na_2CO_3 solution. Hence, it is safer to detect PO_4^{3-}, F^- and S^{2-} not in soda-extract but in either residue left after Na_2CO_3 treatment or original mixture.

Q. 20. Explain what happens when H_2S is passed in Cl_2, Br_2 and I_2 water respectively ?

Ans. H_2S reacts with chlorine forming HCl and free sulphur.

$$H_2S + Cl_2 \longrightarrow 2HCl + S$$

With Br_2 and I_2 water, similar reaction takes place and HBr and HI are being produced respectively. The sulphur separates out.

$$H_2S + Br_2 \longrightarrow 2HBr + S$$
$$H_2S + I_2 \longrightarrow 2HI + S$$

In all these cases, H_2S is oxidised and halogen is simultaneously reduced to its hydracid.

Q. 21. What is chlorine water ? Give its characteristic properties.

Ans. It is the solution of Cl_2 gas in water. Cl_2 decomposes water to a little extent, producing HClO and HCl. In this reaction, one atom of chlorine is oxidised and another atom is reduced.

$$H_2O + Cl_2 \rightleftharpoons 2H^+ + Cl^- + ClO^-$$

Chlorine water is a strong oxidising agent. The decomposition of chlorine takes place very slowly in dark and in presence of oxidisable substances.

Q. 22. Explain why test of I^- ion is not observed in excess of cadmium ions ?

Ans. Iodide ions form CdI_2 with cadmium ions, but in presence of large excess of cadmium ions, iodide ions are removed through the formation of complex ion CdI_4^{2-}.

$$Cd^{2+} + 2I^- \longrightarrow CdI_2$$
$$CdI_2 + 2I^- \longrightarrow CdI_4^{2-}$$

Hence, we do not get a clearer test of iodide ions in presence of large excess of cadmium.

Q. 23. What is the action of metallic zinc on sulphites ?

Ans. Sulphites are decoposed by dilute acids evolving SO_2 which reacts with Zn an HCl to give H_2S gas. This may be detected by its smell or the blackening of paper moistened with lead acetate solution. The equations are :

$$Na_2SO_3 + 2HCl \longrightarrow 2NaCl + H_2O + SO_2$$
$$3Zn + 6HCl + SO_2 \longrightarrow 3ZnCl_2 + 2H_2O + H_2S\uparrow$$

on the test paper

$$H_2S + (CH_3COO)_2 Pb \longrightarrow PbS \downarrow + 2CH_3COOH$$

Q. 24. If the neutral solution of test mixture gives rose-red colour with sodium nitroprusside, what do you infer ? Is this colour affected by addition of excess $ZnSO_4$ solution, $K_4Fe(CN)_6$ or an acid?

Ans. The rose-red colour with sodium nitroprusside indicates the presence of sulphite ion and absence of thiosulphate and sulphide ions. The rose-red colour is intensified by addition of excess $ZnSO_4$ solution. A red ppt. is formed if a drop of $K_4[Fe(CN)_6]$ is added. The chemistry of this reaction is yet not clearly understood. Acids destroy the colour, therefore, acid solution must be neutrilized before performing sodium nitroprusside test for SO_3^{2-}.

Q. 25. What ions interfere in tests for CO_3^{2-} ? Describe the procedure if these ions are present.

Ans. The presence of SO_3^{2-} and $S_2O_3^{2-}$ ions interfere with detection of CO_3^{2-} because the SO_2 evolved by the action of acids on sulphites and thiosulphates is absorbed by $Ca(OH)_2$ and can form a white ppt. of $CaSO_3$. Therefore, if these ions have been found, they must be oxidised. Take the mixture and excess $K_2Cr_2O_7$ or $KMnO_4$ and then dil. H_2SO_4. Warm and then pass gas in lime water. If lime water turns milky, carbonate is present.

Q. 26. How will you test S^{2-} ions in sulphides which are (a) decomposed by acids and (b) not decomposed by acids ?

Ans. (a) Acids such as dil. H_2SO_4 and HCl decompose many sulphides with the evolution of H_2S which can be identified by the blackening of paper moistened with lead acetate. When paper is held near the mouth of test in which H_2S is evolved.

(b) Sulphides which are not decomposed by acids can be decomposed if they are mixed with zinc dust (be sure that zinc dust should not contain sulphide as an impurity) and treated with HCl. H_2S can be detected as described above.

Q. 27. Addition of $AgNO_3$ to an unknown solution gives (a) a black precipitate or (b) a ppt. which gradually turns yellow then brown and finally black. What anions are indicated ? Explain with the help of equations.

Ans. (a) Formation of black ppt. with $AgNO_3$ indicates presence of *sulphide ion*. This black ppt. of Ag_2S is insoluble in NH_4OH, but dissolves in dil. HNO_3 on warming.

(b) When $AgNO_3$ is added in excess, it gives a white ppt. of $Ag_2S_2O_3$ which gradually turns yellow then brown and finally black owing to conversion into silver sulphide. This shows the presence of $S_2O_3^{2-}$ ion.

$$2AgNO_3 + Na_2S_2O_3 \longrightarrow \underset{\text{White}}{Ag_2S_2O_3} \downarrow + 2NaNO_3$$

$$Ag_2S_2O_3 + H_2O \longrightarrow \underset{\text{Black}}{2AgS} \downarrow + H_2SO_4$$

It must be remembered that $Ag_2S_2O_3$ ppt. dissolves in excess of thiosulphate to form the complex ions $[Ag(S_2O_3)]^-$ and $[Ag_2(S_2O_3)_3]^{4-}$

Therefore, the ppt. can form only if Ag^+ ions are added in excess.

Q. 28. Which of the anions among Cl^-, Br^- and I^- is most easily oxidised, and why ?

Ans. The electrode potential of the couples

$I_2/2I^- = + 0.54$; $Br_2/2Br^- = + 1.07$; $Cl_2/2Cl^- = + 1.36$

Thus among the above ions I^- is most powerful reducing agent because oxidation potential of $I_2/2I^-$ is the lowest. Therefore, I^- will be most easily oxidised, followed by Br^- and then Cl^-. This can be supported by the fact that during the testing of I^- in presence of Br^- and Cl^- by Cl_2 water and CCl_4. I^- being stronger than the three reducing agents, is oxidised first. The reaction is represented by $2I^- + Cl_2 \rightarrow I_2 + 2Cl^-$ and is accompanied by appearance of violet colour in CCl_4 layer. On further addition of Cl_2 water, violet colour disappears (due to oxidation of I_2 to HIO_3). When more Cl_2

(oxidising agent) is added, Br^- (less powerful reducing agent than I^-) oxidises to Br_2 which produces orange colour in CCl_4 layer. This shows that I^- is oxidised more easily than Br^- and Cl^-.

Q. 29. Which anions interfere in detection of I^- and Br^- ions by the action of chlorine water and why ? Describe the procedure to be used if these ions are present.

Ans. S^{2-}, $S_2O_3^{2-}$ and SO_3^{2-} ions interfere with the detection of I^- and Br^- ions by action of Cl_2 water because they are stronger reducing agents than I^- and Br^-. Therefore, oxidation of I^- and Br^- can begin only after S^{2-} (it is most powerful reducing agent among anions as electrode potential of $S/S^{2-} = -0.51$ volt), $S_2O_3^{2-}$ and SO_3^{2-} have been completely oxidised. They may be oxidised $e.g.$, by $KMnO_4$ in acid solution. For further details, see foot note in the combination test of I^- and Br^- in acid radicals.

Q. 30. Name the two most important tests for the NO_3^- ion and write the equation.

Ans. (*i*) Ring test :

$$2NaNO_3 + conc. \ H_2SO_4 \longrightarrow Na_2SO_4 + 2HNO_3$$
$$6FeSO_4 + 3H_2SO_4 + 2HNO_3 \longrightarrow 3Fe_2(SO_4)_3 + 4H_2O + 2NO\uparrow$$
$$NO + FeSO_4 \longrightarrow [Fe(NO)]\ SO_4$$
Brown

(*ii*) **Reduction to NH_3 :**

$$3NaNO_3 + 8Al + 5NaOH + 2H_2O \longrightarrow 3NH_3\uparrow + 8NaAlO_2$$
(power)   (dilute)

Q. 31. How NO_3^- ion can be distinguished from NO_2^- ?

Ans. (*i*) NO_2^- gives brown NO_2 gas with dil. acid but NO_3^- does not.

(*ii*) NO_2^- in presence of dil. acid, gives blue or violet colour with KI + starch solution whereas NO_3^- does not.

(*iii*) When solution of nitrate is heated with twice the volume of saturated $MnCl_2$ solution in concentrated HCl, it turns dark brown owing to formation of complex $[MnCl_6]^{2-}$ ion.

$$2MnCl_2 + 12HCl + 2HNO_3 \longrightarrow 2H_2[MnCl_6] + 4H_2O + 2NO\uparrow$$

The NO_2^- ion gives yellow colour due to NO_2 gas dissolved in the liquid.

(*iv*) Since nitrogen has valency (+3) in HNO_2 can be both, oxidised and reduced. In other words, the NO_2^- can itself be reduced (*i.e.*, can also act as oxidising agent). It is oxidised by the action of $KMnO_4$ solution, acidified with dil. H_2SO_4, on heating. The colour of $KMnO_4$ disappears as a result of this reaction.

$$5NO_2^- + 2MnO_4^- + 6H^+ \longrightarrow 5NO_3^- + 2Mn^{2+} + 3H_2O$$

whereas NO_3^- being powerful oxidising agent, does not decolourise $KMnO_4$ solution.

Q. 32. What are the reduction products of nitric acid ?

Ans. The usual reduction products are either nitrogen dioxide (NO_2) or nitric oxide (NO). In some cases, reduction of nitric acid can go further with formation of free N_2 or even NH_3 (which forms the salt NH_4NO_3 with excess acid).

Q. 33. What anions interfere in the $FeCl_3$ test for acetate ion and how should the test be carried out in their presence ?

Ans. The anions CO_3^{2-}, SO_3^{2-}, PO_4^{3-}, $[Fe(CN_6)]^{4-}$, CNS^- and I^- interfere in the $FeCl_3$ test of acetate. All these ions can be removed by addition of $AgNO_3$ or Ag_2SO_4 solution to the neutral solution of acetate. They will be precipitated as silver salts which are filtered off. With the filtrate test for acetate with neutral $FeCl_3$ is carried out.

Q. 34. Why are substances powdered for analysis ?

Ans. Solid substances are powdered to make treatment with solvents easier. The powdering is necessary also because uniform mixing of individual constituents is ensured only if a heterogeneous sample is thoroughly subdivided.

Q. 35. How can you confirm that a substance is partially soluble in water ?

Ans. The partial solubility of a substance in water is ascertained by filtering off a small portion of the water from the residue and then evaporating off (up to dryness) reidue appears few drops of filtrate in a porcelain dish or on a metal plate. If residue appears, the substance is partially soluble.

Q. 36. Explain why too much substance should not be taken for analysis ?

Ans. It is advantageous to take less amount of substance for analysis because less time is needed for such analytical operations as filtration, washing and precipitation etc. Ofcourse, it is rational not to take too little either because some constituents present in small amounts may be missed.

Q. 37. Explain why excess of the acid used to dissolve a substance must be removed by evaporation before analysis ?

Ans. Evaporation of excess of acid is necessary because precipitation of most of the cations of II group is incomplete at high H^+ ion concentration.

Q. 38. Give examples of substances : (a) insoluble in HCl but soluble in HNO_3; (b) insoluble in HNO_3 but soluble in HCl.

Ans. (a) Sulphides of Cu, Pb, Bi, As, Sb, Co, Ni are insoluble in HCl but dissolve in conc. HNO_3.

(b) Water insoluble oxidising agents, e.g. SnO_2, MnO_2 and PbO_2, are insoluble in HNO_3 but dissolve in HCl.

Q. 39. State how the following substances may be dissolved : $PbSO_4$, AgCl, AgBr and AgI, $BaSO_4$ and $SrSO_4$?

Ans. All these substances are insoluble in acids. Lead sulphate dissolves in 30% ammonium acetate solution and excess of caustic alkalies.

$$PbSO_4 + 2CH_3COONH_4 \longrightarrow Pb(CH_3COO)_2 + (NH_4)_2SO_4$$

AgCl dissolves in NH_4OH forming soluble complex salt $Ag[(NH_3)_2]Cl$

$$AgCl + 2NH_4OH \longrightarrow \underset{\text{silver diammine chloride}}{[Ag(NH_3)_2]Cl} + 2H_2O$$

AgI and AgBr can be dissolved by action of a piece of metallic zinc and H_2SO_4 on heating, e.g.

$$2AgI + Zn + H_2SO_4 \longrightarrow 2Ag\downarrow + ZnSO_4 + 2HI$$

The black ppt. of metallic silver formed is dissolved in hot HNO_3 (sp. gr. 1.2).

$$\underset{\text{ppt.}}{2Ag} + 2HNO_3 \longrightarrow \underset{\text{soluble}}{2AgNO_3} + H_2\uparrow$$

The sulphates of Ba, Sr and Ca are dissolved by repeated boiling with concentrated Na_2CO solution and the carbonates of Ba, Sr and Ca so produced are dissolved in dil. CH_3COOH.

Q. 40. Explain why repeated boiling with conc. Na_2CO_3 solution is essential for dissolvin sulphates of V group cations ? Why not single boiling ?

Ans. For example,

$$BaSO_4 + Na_2CO_3 \rightleftharpoons BaCO_3 + Na_2SO_4$$

Sulphates of V group cations are less soluble than their carbonates. This reaction is reversib and can be taken to completion only by repeated treatment of the residue with conc. Na_2CO_3 sol tion. The filtrate (containing SO_4^{2-} ions formed as Na_2SO_4 in the reaction) is decanted from t residue and replaced with fresh portion of Na_2CO_3 solution. This is repeated several times. After the residue left is treated with 2N acetic acid. This dissolves the carbonates of V group which a formed.

$$BaCO_3 + 2CH_3COOH \longrightarrow (CH_3COO)_2Ba + H_2O + CO_2\uparrow$$
ppt.   soluble

Q. 41. Explain why in the precipitation of I group cations as chlorides, it is preferred to add a slightly more quantity of dil. HCl than required ?

Ans. The slight excess of dil. HCl not only causes complete precipitation of I group chlorides but this also prevents the formation of white precipitates of BiOCl and SbOCl along with I group chlorides.

$$BiOCl + 2HCl \rightleftharpoons BiCl_3 + H_2O$$
ppt.   soluble

$$SbOCl + 2HCl \rightleftharpoons SbCl_3 + H_2O$$
ppt.   soluble

Q. 42. Explain why in I group precipitation.

(a) **large excess of dil. HCl is avoided.**

(b) **Conc. HCl can not be used as group reagent.**

Ans. [a and b]. The large amount of dil. HCl or conc. HCl is avoided becaue the chloride ions in high concentration increase the solubility of precipitated I group chlorides through the formation of soluble complex ions.

$$AgCl + 2Cl^- \longrightarrow AgCl_3^-$$
$$PbCl_2 + 2Cl^- \longrightarrow PbCl_4^{2-}$$

Q. 43. Sometimes, during I group analysis, the residue left after ammonia treatment is white instead of black. What inference will you draw ?

Ans. It shows the absence of mercurous ion. The white residue may be either lead oxychloride formed by ammonia acting upon some undissolved lead chloride or excess of silver Chloride undissolved by ammonia.

$$2PbCl_2 + 2NH_4OH \longrightarrow Pb_2OCl_2 + 2NH_4Cl + H_2O$$
white ppt.

Q. 44. Name the two solutions that give

(a) **a black ppt. on mixing.**

(b) **a red ppt. on mixing, soluble in excess of one of them.**

(c) **a white ppt. on mixing, soluble in ammonia.**

(d) **a yellow ppt. on mixing, which dissolves on heating but reappears on cooling in a golden** ellow plate.

Ans. (a) A colourless solution of lead nitrate when mixed with a colourless solution of sulphuretted ydrogen, give black ppt. of PbS.

$$Pb(NO_3)_2 + H_2S \longrightarrow PbS\downarrow + 2HNO_3$$
Black

(b) A colourless solution of KI when added to a colourless solution of mercuric chloride, a red ot. of mercuric iodide is formed. The red ppt. dissolves in excess of KI solution with the formation f a soluble complex potassium mercuri-iodide.

$$2KI + HgCl_2 \longrightarrow 2KCl + HgI_2\downarrow$$
red

$$2KI + HgI_2\downarrow \longrightarrow K_2[HgI_4]$$
soluble

(c) A colourless solution of silver nitrate when added to a colourless solution of a chloride, a rdy white ppt. of AgCl is formed. The precipitated AgCl dissolves in ammonia, forming a soluble mplex, diammine silver chloride

$$AgNO_3 + NaCl \longrightarrow \underset{\text{curdy white}}{AgCl\downarrow} + NaNO_3$$

$$AgCl\downarrow + 2NH_4OH \longrightarrow \underset{\text{soluble}}{[Ag(NH_3)_2]\,Cl} + 2H_2O$$

(d) A colourless solution of KI when added to colourless solution of lead nitrate, a yellow ppt. of lead iodide is formed. On heating the ppt. with water, it dissolves but reappears on cooling in golden spangles.

$$Pb(NO_3)_2 + 2KI \longrightarrow PbI_2 + 2KNO_3$$

Q. 45. Discuss action of ammonia on

(a) **$HgCl_2$**

(b) **alkaline solution of mercuric iodide in KI and**

(c) **silver chloride.**

Ans. (a) See "the chemical reactions involved in I group analysis".

(b)
$$2KI + HgI_2 \longrightarrow \underset{\substack{\text{Colourless soluble} \\ \text{Pot. mercuric iodide}}}{K_2[HgI_4]}$$

$$2K_2[HgI_4] + NH_3 + 3NaOH \longrightarrow 4KI + 3NaI + 2H_2 \; +O\underset{Hg}{\overset{Hg}{\diagdown\;\diagup}}\!\!NH_2I\downarrow$$

Brown ppt.

Oxydimercuri ammonium
iodide (Million's base)

(c) When NH_3 gas is passed over solid AgCl, it is absorbed by AgCl with formation of $AgCl.3NH_3$, $2AgCl.3NH_3$, and other compounds. But when solution of NH_3 or NH_4OH is added to AgCl, a soluble complex, silver diammine chloride is formed.

$$AgCl + 2NH_4OH \longrightarrow \underset{\text{soluble}}{[Ag(NH_3)_2]\,Cl} + 2H_2O$$

Q. 46. Explain the following :

(a) **$AgNO_3$ is added to a solution of iodine in water.**

(b) **Excess of $AgNO_3$ is added to a solution of Cl_2 in water.**

(c) **$AgNO_3$ is added to chloro penta-aqua-chromium sulphate.**

Ans. (a) I_2 is sparingly soluble in water and the solution contains hydroiodic acid (HI) and hypoiodous acid (HOI) to small extent due to slight hydrolysis of iodine.

$$I_2 + H_2O \rightleftharpoons HI + HOI$$

When $AgNO_3$ is added to solution of I_2 in water, the iodide ion formed from HI is removed as yellow ppt. of AgI

$$Ag^+ + NO_3^- + H^+ + I^- \longrightarrow AgI\downarrow + HNO_3$$

(b) Cl_2^- water contains a little HCl and HOCl due to hydrolysis :

$$Cl_2 + H_2O \rightleftharpoons HCl + HOCl^-$$

On adding excess of $AgNO_3$ solution to it, the HCl present in little amount reacts with $AgNO_3$ producing a curdy white ppt. of AgCl. Then further hydrolysis occurs, as the reaction is reversible and gradually the whole of Cl_2 is hydrolysed to HCl and HOCl and AgCl is formed by reaction of HCl with $AgNO_3$. In the meantime, $AgNO_3$ reacts with HOCl with the formation of silver hypochlorite (AgClO) which immediately undergoes disproportionation with the precipitation of AgCl and formation of $AgClO_3$ in solution.

$$HOCl + AgNO_3 \longrightarrow AgOCl + HNO_3$$

$$3AgOCl \longrightarrow 2AgCl\downarrow + AgClO_3$$

Ultimately, all Cl_2 is removed partly as curdy white ppt. of $AgCl$ and partly as $AgClO_3$.

(c) When $AgNO_3$ is added to chloropenta-aqua-chromium sulphate, $[Cr(H_2O)_5Cl]SO_4$, no ppt. of $AgCl$ is formed because in the solution, there are no free chloride ions.

Q. 47. Describe the action of the following on a solution of cupric sulphate, giving equations.

(a) KI, (b) KCN, (c) ammonium thiocyanate, (d) NH_4OH and (e) excess of NaOH.

Ans. (a) At first :

$$CuSO_4 + 2KI \longrightarrow CuI_2\downarrow + K_2SO_4$$
$$\text{green}$$

But CuI_2 being very unstable, soon decomposes into white ppt. of Cu_2I_2.

$$2CuI_2 \longrightarrow Cu_2I_2\downarrow + I_2$$
$$\text{white}$$

Cu_2I_2 dissolves in excess of KI solution giving a clear solution, coloured brown by free I_2

$$Cu_2I_2 + 2KI \longrightarrow 2K[CuI_2]$$
$$\text{Soluble}$$
$$\text{Pot. cuprous}$$
$$\text{iodide}$$

(b) At first, unstable $Cu(CN)_2$ is produced which at once changes with the loss of cyanogen to white ppt. of cuprous cyanide. This white ppt. dissolves in excess of KCN forming soluble complex $K_3[Cu(CN)_4]$. On boiling cyanogen gas is evolved. Therefore, this is a method for getting a supply of cyanogen gas in laboratory.

$$CuSO_4 + 2KCN \longrightarrow Cu(CN)_2\downarrow + K_2SO_4$$
$$\text{yellow,}$$
$$\text{unstable}$$

$$2Cu(CN)_2 \longrightarrow Cu_2(CN)_2\downarrow + (CN)_2\uparrow$$
$$\text{white}$$

$$Cu_2(CN)_2 + 6KCN \longrightarrow 2K_3[Cu(CN)_4]$$
$$\text{soluble}$$

(c) $$CuSO_4 + 2NH_4SCN \longrightarrow Cu(SCN)_2\downarrow + (NH_4)_2SO_4$$
$$\text{Black}$$
$$\text{cupric}$$
$$\text{thiocyanate}$$

This black ppt. gradually changes into white ppt. of cuprous thiocyanate.

$$Cu(SCN)_2 \rightleftharpoons CuSCN\downarrow + (SCN)^-$$
$$\text{white}$$

(d) At first :

$$2CuSO_4 + 2NH_4OH \longrightarrow CuSO_4Cu(OH)_2\downarrow + (NH_4)_2SO_4$$
$$\text{Pale blue}$$
$$\text{Basic copper}$$
$$\text{sulphate}$$

It dissolves in excess of amm. hydroxide to form deep blue solution of tetra-ammine copper (II) sulphate

$$CuSO_4.Cu(OH)_2 + (NH_4)_2SO_4 + 6NH_4OH \longrightarrow 2[Cu(NH_3)_4]SO_4 + 8H_2O$$
$$\text{Soluble}$$
$$\text{deep blue}$$

(e) A blue ppt. of $Cu(OH)_2$ is formed which is insoluble in moderate solution of NaOH and converted on boiling into black cupric oxide.

$$CuSO_4 + 2NaOH \longrightarrow Cu(OH)_2\downarrow + Na_2SO_4$$
$$\text{Blue ppt.}$$

$$Cu(OH)_2 \xrightarrow{\text{Boiling}} CuO\downarrow + H_2O$$
$$\text{Black}$$

If excess of alkali is not used, the ppt with NaOH has the composition, $CuSO_4.Cu(OH)_2$.

Q. 48. What is aqua regia ? Explain its functioning.

Ans. It is the mixture of 3 parts by volume of conc. HNO_3 and one part by volume of conc. HCl. Conc. HNO_3 acts upon conc. HCl producing water, nitrosyl chloride and free chlorine.

$$HNO_3 + 3HCl \longrightarrow H_2O + NOCl\uparrow + Cl_2\uparrow$$

In this reaction, valency of nitrogen has been reduced from +5 to +3 and two atoms of chlorine have been oxidised to form chlorine molecule. Therefore, it behaves very much like chlorine water but it is more efficient because more chlorine can be formed in small volume of solution and NOCl is also good oxidising agent.

Q. 49. In II group analysis, the black residue on treatment with 1 : 1 water and HNO_3 changes to white. Does it confirm the absence of mercuric ion ?

Ans. The treatment of HgS with 1 : 1 HNO_3 may result in the formation of white $Hg(NO_3)_2$. $2HgS$. So the lack of black ppt. may not confirm the absence of mercuric ion.

Q. 50. If the black HgS after being dissolved in aqua regia is not boiled, does $SnCl_2$ bring characteristic test of Hg^{2+} ?

Ans. No, it does not give test for Hg^{2+} ions. Aqua regia liberates chlorine which, if not removed, will oxidise Sn^{2+} to Sn^{4+}. So the solution should be boiled well to expel Cl_2, because only Sn^{2+} ions, not Sn^{4+} ions, give characteristic test of Hg^{2+},

$$SnCl_2 + Cl_2 \longrightarrow SnCl_4$$

Q. 51. How does aqua regia act in dissolving black HgS ?

Ans. In dissolving HgS, aqua regia acts as an oxidising agent and also complexing agent. The NO_3^- ion in presence of hydronium ion (H_3O^+) oxidise the S^{2-} ion to free sulphur and the Cl^- ions with Hg^{2+} ions form soluble complex ion, $HgCl_4^{2-}$. These reactions are shown by the following equations :

$$3HgS + 2NO_3^- + 8H_3O^+ \longrightarrow 3S + 2NO + 12H_2O + 3Hg^{2+}$$
$$3Hg^{2+} + 12Cl^- \longrightarrow 3HgCl_4^{2-}$$

$$\overline{3HgS + 2NO_3^- + 8H_3O^+ + 12Cl^- \longrightarrow 3HgCl_4^{2-} + 2NO + 3S + 12H_2O}$$

The presence of Hg^{2+} is confirmed by adding stannous chloride solution to the solution containing $HgCl_4^{2-}$ ions. A white ppt. of Hg_2Cl_2 is obtained in the reaction.

$$2HgCl_4^{2-} + SnCl_4^{2-} \longrightarrow Hg_2Cl_2\downarrow + SnCl_6^{2-} + 4Cl^-$$

Q. 52. Explain why the following acids are not suitable for acidifying the solution before precipitation of II group by H_2S :

(a) H_2SO_4 (b) HNO_3 (C) CH_3COOH.

Ans. Before H_2S is passed, it is preferable to acidify the solution with HCl. If we use H_2SO_4, we would introduce SO_4^{2-} ions which precipitate Ba^{2+}, Sr^{2+}, Ca^{2+} and Pb^{2+} as acid insoluble sulphate and will complicate the analysis. HNO_3 is unsuitable because of its oxidising action which may oxidise S^{2-} ions to free sulphur concentrations. Finally, CH_3COOH is a weak acid and would not give ion concentration to prevent precipitation of zinc sulphides, etc.

Q. 53. What should be the H^+ ion concentration for precipitation of II group cations b

H_2S ? Explain what errors can arise during the analysis if the acidity of the solution during precipitation was (a) too high; (b) too low.

Ans. See the precaution No. 3 in II group analysis in Chapter 3.

Q. 54. Explain why a hot solution is used at first during II group precipitation by H_2S. Why is it subsequently diluted with cold water ?

Ans. The precipitation of II group should on performed in a hot solution to avoid the formation of colloidal sulphide solution. However, H_2S (like all other gases) becomes less soluble at higher temperature and as a result, the separation of II group cations is incomplete. Therefore, at the end of the precipitation, the filtrate is diluted with an equal volume of cold water and again saturated with H_2S. This not only cools the solution but also lowers the H^+ ion concentration thus making the precipitation of Cd^{2+} complete.

Q. 55. Explain why the precipitated II group sulphides must be filtered off without delay ?

Ans. When precipitated II group sulphides are left to stand in contact with solution, "delayed precipitation" of zinc sulphide takes place (this is the gradual precipitation of an impurity on standing). Therefore, filtration should not be delayed too much, in order to avoid "loss" of Zn^{2+}.

Q. 56. Why does H_2S precipitate the corresponding sulphide from complex cadmium cyanide but not from the analogous copper compound ?

Ans. See the reactions involved in copper and cadmium separation in chapter 3.

Q. 57. If the original solution is yellow but changes to green on passing H_2S, what does it indicate ?

Ans. This shows the presence of chromate ion (CrO_4^{2-}). The conversion of yellow to green means the conversion of CrO_4^{2-} to Cr^{3+}. In such a case, pass H_2S through the solution for sometime in order to convert whole of chromate ions to Cr^{3+} so that chromate as Cr^{3+} is precipitated in III group.

$$CrO_4^{2-} + 4H_2S \longrightarrow Cr^{3+} + 4H_2O + 4S\downarrow$$

Q. 58. If a solution of Bi^{3+} is diluted with water, a white ppt. of BiOCl is formed but when H_2S is passed, the ppt. changes to black. Would the same be true for a suspension of $Bi(OH)_3$? The solubility product of Bi_2S_3 is 5×10^{-23} and that of $Bi(OH)_3$ and H_2S are respectively 1.2×10^{-36} and 1.08×10^{-23}.

Ans. No.

Q. 59. In the Cu and Cd separation by KCN, why is it advised to use in excess KCN solution ?

Ans. In absence of an excess of KCN, however, an appreciable ionization of cuprocyanide complex ion takes place :

$$[Cu(CN)_4]^{3-} \rightleftharpoons Cu^+ + 4CN^-$$

The above ionization increases as the solution is diluted. From the diluted solutions the compounds $K_2[Cu_2(CN)_4]$, $K[Cu_2(CN)_3]$ and finally $Cu_2(CN)_2$ are obtained which are of less complex nature. All these compounds, even in solid state, are decomposed by H_2S with precipitation of black Cu_2S. Consequently, in order to prevent the precipitation of copper by H_2S, sufficient KCN must be added i.e., more than enough to form the complex salt $K_3[Cu(CN)_4]$.

Q. 60. (a) Pass H_2S through an acidified solution of $SnCl_4$ containing excess of $NH_4F \longrightarrow$ o dull yellow ppt. of SnS_2. Why ?

(b) Pass H_2S through an acidified solution of $SnCl_4$ containing NH_4F and a large excess of oric acid $\longrightarrow$ dull yellow ppt. of SnS_2. Why ?

Ans. (a) With excess NH_4F, the acidified $SnCl_4$ solution forms a complex compound ammonium annifluoride, $(NH_4)_2[SnF_6]$ in which tin is present as anionic complex ion. Hence in solution, we ve no free Sn^{4+} ions to give the dull yellow ppt. of SnS_2 with H_2S.

$$SnCl_4 + 4NH_4F \longrightarrow SnF_4 + 4NH_4Cl$$
$$SnF_4 + 2NH_4F \longrightarrow (NH_4)_2[SnF_6]$$

(b) Boric acid produces ammonium borofluoride and no stanni fluoride (SnF_4) because boron has greater affinity with fluorine than tin. Hence in solution, free Sn^{4+} ions produce dull yellow ppt. of SnS_2 with H_2S.

$$H_2BO_3 + 3NH_4F \longrightarrow BF_3 + 3NH_4OH$$
$$BF_3 + NH_4F \longrightarrow NH_4BF_4$$
$$\text{Amm. borofluoride}$$

$$SnCl_4 + 2H_2S \longrightarrow SnS_2 + 4HCl$$

Q. 61. What would you notice when you add carefully KI solution to the following ? Give equations.

(a) Pb $(NO_3)_2$ solution, (b) acidified $CuSO_4$ solution, (c) $HgCl_2$ solution, and (d) acidified KNO_2 solution.

Ans. (a) Yellow ppt. which dissolves in water on heating and reappears on cooling.

$$Pb(NO_3)_2 + 2KI \longrightarrow PbI_2\downarrow + 2KNO_3$$
$$\text{yellow}$$

(b) White ppt. of Cu_2I_2. The solution becomes brown due to liberated I_2.

$$2CuSO_4 + 4KI \longrightarrow Cu_2I_2\downarrow + 2K_2SO_4 + I_2$$
$$\text{white}$$

(c) Red ppt. of HgI_2 is first formed. It dissolves in excess of KI, giving a colourless solution of potassium mercuric iodide.

$$HgCl_2 + 2KI \longrightarrow HgI_2\downarrow + 2KCl$$
$$\text{red}$$

$$HgI_2 + 2KI \longrightarrow K_2[HgI_4]$$
$$\text{colourless}$$

(d)
$$HCl + KNO_2 \longrightarrow HNO_2 + KCl$$
$$HCl + KI \longrightarrow HI + KCl$$
$$2HNO_2 + 2HI \longrightarrow I_2\uparrow + 2H_2O + 2NO\uparrow$$

Q. 62. What are thiosalts ?

Ans. Thiosalts are the salts of corresponding thioacids such as thiostannic (H_2SnS_3) and thioarsenious (H_3AsS_3), etc.

Q. 63. What are thioacids ? Give examples.

Ans. These acids are similar in composition to the corresponding oxyacids but the oxygen atoms are replaced by atoms of an analogous element, sulphur; for example :

arsenious acid	$H_3AsO_3 \longrightarrow$	thioarsenious acid H_3AsS_3
arsenic acid	$H_3AsO_4 \longrightarrow$	thioarsenic acid H_3AsS_4
antimonous acid	$H_3SbO_3 \longrightarrow$	thioantimonous acid H_3SbS_3
antimonic acid	$H_3SbO_4 \longrightarrow$	thioantimonic acid H_3SbS_4
stannic acid	$H_2SnO_3 \longrightarrow$	thiostannic acid H_2SnS_3

Q. 64. How are thiosalts formed ?

Ans. The reactions of thiosalts formation are similar to the reaction in which salts of oxyacid are formed from acidic and basic oxides. For example, the reaction

$$SnS_2 + Na_2S \longrightarrow Na_2SnS_3$$
$$\text{sod. thiostannate}$$

is similar to the reaction.

$$SnO_2 + Na_2O \longrightarrow Na_2SnO_3$$
sod. stannate

This comparison shows that, *in contrast to sulphides of Groups I, II-A, III, IV and VI, the sulphides of II-B group are distinctly acidic.*

Q. 65. What explanation do you offer for the acidic nature of sulphides of Group II-B ?

Ans. The acidic nature of II-B group sulphides is associated with the position of the corresponding elements in periodic table. In periodic table Sn is in IV-A group and As and Sb in V-A group. The elements located here exhibit non-metallic properties which are manifested in acidic nature of their oxides and sulphides. In accordance with general rule, the acidic nature should be most pronounced in group V-A elements, As and Sb, especially in their highest oxidation states, *i.e.*, in the compounds As_2S_5 and Sb_2S_5. The acidic nature of Sn is less pronounced and appears only in compounds of highest oxidation state, such as in SnS_2. The sulphides of bivalent tin, SnS is a basic sulphide and, therefore, does not form thiosalts with Na_2S and $(NH_4)_2S$ but forms with amm. polysulphide.

Q. 66. What are polysulphides ? Does ammonium polysulphide exactly corresponds to the formula $(NH_4)_2S_2$. How is it prepared ?

Ans. Polysulphides are salts corresponding to various polysulphides of hydrogen — H_2S_2, H_2S_3 etc. The reagent used in practice, *i.e.*, ammonium polysulphide does not exactly correspond to the formula $(NH_4)_2S_2$ but is a mixture of various polysulphides from $(NH_4)_2S_2$ to $(NH_4)_2S_9$. It is prepared by dissolving sulphur in ammonium sulphide, $(NH_4)_2S$ and always contains an excess of the latter.

Q. 67. How can precipitation of As^{5+} by H_2S be accelerated ?

Ans. See the chemical reaction involved in precipitation of As^{5+} by H_2S, in chapter 3.

Q. 68. What is the action of water on trivalent and pentavalent antimony salts ? Write the equations for reactions.

Ans. In hydrochloric acid solution, Sb^{2+} and Sb^{5+} probably exist as $[SbCl_4]^-$ or $[SbCl_6]^{3-}$ and $SbCl_6]^-$ respectively which hydrolyse upon dilution with water to give white ppts. of antimony oxychloride and of basic salt SbO_2Cl respectively. The equation of the reactions are :

Sb^{3+} : $[SbCl_4]^- + H_2O \rightleftharpoons SbOCl + 3Cl^- + 2H^+$
white

or

Sb^{5+} : $[SbCl_6]^{3-} + H_2O \rightleftharpoons SbOCl + 5Cl^- + 2H^+$

        $[SbCl_6]^{3-} + 2H_2O \rightleftharpoons SbO_2Cl\downarrow + 5Cl^- + 4H^+$
white
Basic salt

These white precipitates dissolve on heating with excess HCl.

Q. 69. Explain why metallic iron reduces Sn^{4+} to Sn^{2+} but not to metallic tin ?

Ans. Let us allow both type of changes

$$\begin{array}{c} \overset{\displaystyle -0.14\ V\ \rightarrow}{} \\ Fe + Sn^{4+} \longrightarrow Sn^{2+} + Fe^{2+} \\ \underset{\displaystyle -0.44\ V \longrightarrow}{} \end{array}$$

$$\begin{array}{c} Fe + Sn^{4+} \longrightarrow Sn + 2Fe^{2+} \\ -0.009\ V \\ -0.44\ V \end{array}$$

In first type of change, we get 0.58 V, whereas in second type of change only 0.449 V. It is, therefore, first change is preferable.

Q. 70. **Write equations showing how these reactions are used in qualitative scheme (a) stannous salt conversion of to stannic one and vice versa, (b) conversion of a trivalent Sb compound to a pentavalent Sb compound and vice versa, (c) trivalent As can be converted to pentavalent As and vice versa.**

Ans. (a) (i) Sn^{2+} to Sn^{4+} :

$$2HgCl_2 + SnCl_2 \longrightarrow Hg_2Cl_2 \downarrow + SnCl_4$$
$$\text{white}$$

In excess of Sn^{2+} ions :

$$Hg_2Cl_2 + SnCl_2 \longrightarrow 2Hg \downarrow + SnCl_4$$
$$\text{grey}$$

(ii) Sn^{4+} to Sn^{2+} :

In presence of Zn and HCl, $SnCl_4$ changes to $SnCl_2$

$$Zn + 2HCl \longrightarrow ZnCl_2 + 2H^+$$
$$SnCl_4 + 2H^+ \longrightarrow SnCl_2 + 2HCl$$

or

$$Fe + SnCl_4 \longrightarrow SnCl_2 + FeCl_2$$

(b) (i) Sb^{3+} to Sb^{5+} :

Sb^{3+} probably exists as $SbCl_4^-$ in HCl solution Into it, pass H_2S. Then dissolve yellow ppt. of Sb_2S_3 in yellow amm. sulphide whereby it dissolves to give amm. thioantimonate in which Sb is pentavalent.

$$2SbCl_4^- + 3H_2S + 6OH^- \longrightarrow Sb_2S_3 \downarrow + 6H_2O^+ + 8Cl^-$$
$$Sb_2S_3 + 3(NH_4)_2S_2 \longrightarrow 2(NH_4)_3 Sb^{5+}S_4 + 3S \downarrow$$

(ii) Sb^{5+} to Sb^{3+} :

Sb^{5+} in HCl exists as $SbCl_6^-$ which with H_2S gives ppt. of Sb_2S_6. The ppt. dissolves in conc. HCl producing $SbCl_3$ in which Sb is trivalent.

$$2SbCl_6^- + 5H_2S + 10H_2O \longrightarrow Sb_2S_5 \downarrow + 10H_3O^+ + 12Cl^-$$
$$Sb_2S_5 + 6HCl \longrightarrow 2SbCl_3 + 2H_2S + 3S$$

Also HI reduces Sb^{5+} to Sb^{3+} in acid solution with separation of iodine.

$$Sb^{5+} + 2I^- \longrightarrow Sb^{3+} + I_2$$

(c) (i) As^{3+} to As^{5+} :

In neutral medium, arsenious acid reduces I_2 and itself is oxidised to arsenic acid in which As is pentavalent

$$H_3As^{3+}O_3 + I_2 + H_2O \rightleftharpoons 2HI + H_3As^{5+}O_4$$

(ii) As^{5+} to As^{3+} :

HI reduces arsenate to arsenite in acidic medium.

$$Na_3As^{5+}O_4 + 2HI \longrightarrow Na_3As^{3+}O_3 + I_2 + H_2O$$

Q. 71. **Complete and balance the following equations :**

(a) $HgS + KClO_3 + HCl \longrightarrow$

(b) $As_2S_5 + NH_4OH \longrightarrow$

(c) $H_3AsO_4 + Zn + H^+ \longrightarrow$

(d) $SnCl_6^- + Fe \longrightarrow$

Ans. (a) $8KClO_3 + 24HCl + 9HgS \longrightarrow 9HgCl_2 + 9S \downarrow + 8KCl + 6ClO_2 + 12H_2O$

(b) $As_2S_5 + 6NH_4OH \longrightarrow 3H_2O + (NH_4)_3.AsS_4 + (NH_4)_3.AsSO_3$

(c) $\quad H_3AsO_4 + 8Zn + 8H^+ \longrightarrow 8Zn^{2+} + 4H_2O + AsH_3\uparrow$

(d) $\qquad SnCl_6^{2-} + Fe \longrightarrow Fe^{2+} + SnCl_6^{4-}$

$\qquad$ chlorostannate ion $\qquad\qquad\qquad$ chlorostannite ion

Q. 72. Why is iron precipitated as ferric hydroxide and not as ferrous hydroxide in III group ?

Ans. At room temperature, the solubility products of $Fe(OH)_2$ and $Fe(OH)_3$ are respectively 2×10^{-14} and 6×10^{-38}. The purpose of adding NH_4Cl before NH_4OH addition is to decrease the OH^- ion concentration to such an extent that only III group hydroxides are precipitated and not the higher group hydroxides (because solubility products of III group hydroxides are very much lower than that of higher group hydroxides). Under this condition solubility product of $Fe(OH)_3$ will be exceeded and not of $Fe(OH)_2$ as solubility product of $Fe(OH)_3$ is much lower than that of $Fe(OH)_2$. Therefore, $Fe(OH)_3$ will precipitate completely but not $Fe(OH)_2$.

Q. 73. Explain why the following conditions are necessary for complete precipitation of $Cr(OH)_3$: (a) solution must be boiling and (b) Excess of NH_4OH must be avoided.

Ans. [a and b]. The greyish-green ppt. of $Cr(OH)_3$ is partly soluble in excess of NH_4OH, in the cold forming a soluble violet or pink complex chromammine.

$$Cr(OH)_3 + 6NH_4OH \longrightarrow [Cr(NH_3)_6] (OH)_3 + 6H_2O$$

But in boiling solution this soluble complex is not formed and thus $Cr(OH)_3$ precipitates completely.

Q. 74. Would you expect any hydroxide which does not form a complex ammine and is soluble in an excess of ammonia solution ?

(b) Can a hydroxide be readily soluble in NaOH solution and practically insoluble in an excess of NH_4OH ?

Ans. [a and b]. Aluminium hydroxide.

Q. 75. Name the hydroxides which form ammines soluble in excess of NH_4OH. Give equations.

Ans. $Cr(OH)_3$ and $Zn(OH)_2$.

$$Cr(OH)_3 + 6NH_4OH \longrightarrow [Cr(NH_3)_6] (OH)_3 + 6H_2O$$
$$\text{Chromammine}$$

$$Zn(OH)_2 + 2NH_4OH \longrightarrow [Zn(NH_3)_2] (OH)_2 + 2H_2O$$
$$\text{Zinc ammine hydroxide}$$

Q. 76. When the mixture contains an iodide, what should be done before adding group reagent to precipitate III group cations ?

Ans. If an iodide is present in mixture, boil the 2nd group filtrate, free from H_2S, with conc. HNO_3 and continue the boiling until all iodine is given off. Otherwise the iodine will be precipitated as soon as ammonia is added to the solution.

Q. 77. The conditions necessary to precipitate III group cations and to prevent the precipitation of $Mg(OH)_2$, are apparently contradictory. How are these conditions satisfied ?

Ans. The complete precipitation of III group cations is assured by keeping the solution alkaline made by NH_4OH. On the other hand, OH^- ion concentration must be maintained sufficiently low to prevent the precipitation of magnesium hydroxide along with III group hydroxides. The addition of NH_4OH, buffered with large excess of NH_4Cl, accomplishes both purposes. The large excess of NH_4^+ ions formed from NH_4Cl represses the OH^- ion concentration (common ion effect) to the extent necessary to prevent precipitation of higher group hydroxides and at the same time to precipitate III group hydroxides completely.

Q. 78. Why do we prefer NH_4OH to other alkalies in III group precipitation ?

Ans. NH_4OH is a weak volatile base. In III group, we want a weak base, otherwise hydroxides of higher groups will also be precipitated along with those of III group hydroxides.

Q. 79. For common ion effect would it be permissible to add amm. sulphate in place of amm. chloride in III group precipitation ?

Ans. No. Amm. sulphate will give, no doubt, sufficient ammonium ion to suppress the OH^- ion concentration, but it cannot be used in place of amm. chloride because sulphate ion will cause precipitation of V group cations as sulphate, in third group, which is not wanted.

Q. 80. In third group, would it be permissible to use NaCl + NaOH buffer instead of $NH_4Cl + NH_4OH$?

Ans. The ionization of NaOH is very high whereas NH_4OH ionizes very little. Hence the presence of Na^+ ions in NaCl + NaOH buffer is unable to suppress OH^- ions concentration necessary to prevent the precipitation of higher group hydroxides in III group. Also Cr and Al hydroxides dissolve in excess of NaOH, their complete precipitation will not occur.

$$Al(OH)_3 + NaOH \longrightarrow NaAlO_2 + 2H_2O$$
Soluble

$$Cr(OH)_3 + NaOH \longrightarrow NaCrO_2 + 2H_2O$$
Green soln.
Sod. chromite

Therefore, for above reasons (NaCl + NaOH) buffer cannot be used in place of $(NH_4Cl + NH_4OH)$.

Q. 81. A solution, containing $NaAlO_2$ and Na_2CrO_4, when boiled with NH_4Cl, gives white ppt. but not green. Why ?

Ans. $NaAlO_2$ is decomposed on boiling with amm. salt, giving white ppt. $Al(OH)_3$ whereas sodium chromate (Na_2CrO_4) remains unaffected. That is why green ppt. of $Cr(OH)_3$ does not produce.

$$NaAlO_2 + NH_4Cl + H_2O \longrightarrow Al(OH)_3\downarrow + NaCl + NH_3\uparrow$$

Q. 82. Why metallic zinc and iron, both precipitate metallic copper from a solution of its ions ?

Ans. Both are more electropositive than copper.

Q. 83. What happens when aluminium metal is placed in a solution of copper sulphate ?

Ans. Since Al occurs much above copper in electrochemical series, aluminium displaces copper when it is placed in a solution of copper sulphate Copper is thrown down as red powdery ppt.

$$3CuSO_4 + 2Al \longrightarrow Al_2(SO_4)_3 + 3Cu$$
Red

Q. 84. Name metals more electropositive than zinc.

Ans. Li, K, Ba, Ca, Na, Mg, Al, Mn

Q. 85. Why in III group filtrate, NH_4Cl is added before making it ammoniacal to precipitate IV group cations with H_2S ?

Ans. $$NH_4Cl \rightleftharpoons NH_4^+ + Cl^- \qquad \text{(very strong ionization)}$$
$$NH_4OH \rightleftharpoons NH_4^+ + OH^- \qquad \text{(very weak ionization)}$$

Due to common ion effect, NH_4^+ ions produced from NH_4Cl ionization, will repress the OH^- ion concentration to such an extent that higher group hydroxide do not precipitate along with IV group sulphides.

Q. 86. Explain why in IV group precipitation the solution is alkalined with NH_4OH before passing H_2S ?

Ans. NH_4OH is a weak base and H_2S is weak acid. They ionize as follows :

$$NH_4OH \rightleftharpoons NH_4^+ + OH^- \qquad \qquad ...(i)$$
$$H_2S \rightleftharpoons 2H^+ + S^{2-} \qquad \text{(over all ionisation)} ...(ii)$$

The OH^- ions from NH_4OH combine with ions from H_2S to produce practically unionized H_2O molecules :

$$OH^- + H^+ \rightleftharpoons H_2O$$

This will cause the decrease in H^+ ion concentration. The ionization constant of H_2S is

$$K = \frac{C_{H^+} \times C_{S^{2-}}}{C_{H_2S}}$$

So to maintain K constant, the concentration of S^{2-} ions should increase and H_2S should decrease i.e., equilibrium (ii) will shift to the right side. This results in the high concentration of S^{2-} ions which is needed for precipitation of IV group cations as sulphide because IV group sulphides have comparatively high solubility products.

Q. 87. Cobalt sulphide is precipitated even from an acid solution of a cobalt salt by H_2S in presence of sodium acetate. How do you account for it ?

Ans. When CH_3COONa is added to an acid solution of a cobalt salt, the mineral acid is replaced by acetic acid; acetic acid is feebly ionized and in presence of excess of sod. acetate this feeble ionization is also further set back due to common ion effect. So the acidity of solution is practically destroyed and hence by passing H_2S through such a solution, cobalt sulphide is precipitated.

$$HCl + CH_3COONa \longrightarrow NaCl + CH_3COOH$$
$$CoCl_2 + H_2S \rightleftharpoons CoS\downarrow + 2HCl$$

The produced HCl is also used up by reaction with CH_3COONa and so the precipitation of CoS takes place completely.

Q. 88. Explain the following :

(a) **When NH_4OH is added to $ZnCl_2$ in presence of NH_4Cl, no ppt. is produced.**

(b) **When NH_4OH is added to $ZnCl_2$ solution in absence of NH_4Cl, white ppt. is produced.**

(c) **Zinc amm. phosphate is soluble in both acid and ammonia.**

Ans. *(a)* The ppt. of $Zn(OH)_2$ is not formed. It is because of the fact that due to common ion effect NH_4Cl considerably suppresses the ionization of NH_4OH. Due to low concentration of OH^- ions in solution, no precipitation of $Zn(OH)_2$ takes place. But, zinc ammine chloride is formed in solution.

$$ZnCl_2 + 4NH_4OH \longrightarrow [Zn(NH_3)_4]Cl_2 + 3H_2O$$
$$\text{soluble}$$

(b) In absence of NH_4Cl, ammonium hydroxide gives at first a white ppt. of $Zn(OH)_2$ which is readily soluble in excess of NH_4OH

$$ZnCl_2 + 4NH_4OH \longrightarrow Zn(OH)_2\downarrow + 2H_4Cl$$
$$Zn(OH)_2 + 2NH_4OH \longrightarrow [Zn(NH_3)_4](OH)_2 + 4NH_2O$$
$$\text{soluble}$$

(c) Dissolution of zinc amm. phosphate in all mineral acids is due to formation of water soluble compounds $ZnCl_2$, NH_4Cl and H_3PO_4.

$$ZnNH_4PO_4 + 3HCl \longrightarrow ZnCl_2 + NH_4Cl + H_3PO_4$$

Zinc amm. phosphate dissolves in NH_4OH due to formation of zinc ammine complex and amm. phosphate (both are water soluble).

$$3ZnNH_4PO_4 + 12NH_4OH \longrightarrow [Zn(NH_3)_4]_3 (PO_4)_2 + (NH_4)_3 PO_4 + 12H_2O$$

Q. 89. Explain why a copper foil dissolves in an aqueous solution of zinc salt containing an excess of KCN with separation of zinc, while a piece of zinc dissolves in aqueous solution of copper sulphate with separation of metallic copper.

Ans. Copper has greater tendency to form complex than zinc. Copper is a transitional element whereas Zn is not. The solution of zinc salt with excess of KCN contains the complex $K_2[Zn(CN)_4]$

which is an imperfect complex. On adding a copper foil to this solution of $K_2[Zn(CN)_4]$, copper passes into solution forming a complex $K_3[Cu(CN)_4]$ which is perfect complex (*i.e.* stable complex), and zinc is thrown out of solution.

$$3K_2[Zn(CN)_4] + 2Cu \longrightarrow 2K_3[Cu(CN)_4] + 2(CN)_2 + 3Zn\downarrow$$

On the other hand, zinc is placed above copper in electrochemical series and so on addition of zinc to copper sulphate solution, copper is thrown out of solution and zinc passes into solution.

$$CuSO_4 + Zn \longrightarrow ZnSO_4 + Cu$$

Q. 90. How many methods do you know for separation of Co and Ni ?

Ans. (*i*) $NaHCO_3$-method, (*ii*) KCN-method and (*iii*) KNO_2-method.

Q. 91. Explain KNO_2-method of Ni and Co separation.

Ans. A neutral solution containing Ni and Co is made acidic by CH_3COOH and is then treated with KNO_2 solution. A yellow ppt. of pot. cobaltinitrite is obtained.

$$CoCl_2 + 2KNO_2 \longrightarrow Co(NO_2)_2 + 2KCl$$
$$Co(NO_2)_2 + 2KNO_2 + 2CH_3COOH \longrightarrow Co(NO_2)_3 + NO + H_2O + 2CH_3COOK$$
$$CoCl_2 + 7KNO_2 + 2CH_3COOH \longrightarrow K_3[Co(NO_2)_6] + 2KCl + NO + H_2O + 2CH_3COOK$$
$$\text{yellow ppt.}$$

If the original solution containing Co and Ni is dilute, the yellow ppt. appears only after standing for several hours or more rapidly on shaking (or rubbing the inner sides of test tube with a glass rod) Whereas nickel does not form such complex. Therefore, cobalt is precipitated and then nickel is tested in filtrate. In this way, cobalt is separated from nickel.

Q. 92. V group cations form a number of sparingly soluble salts with various ions e.g. sulphates, phosphates, oxalates and carbonates. In the form of which salt, should Group V cations be separated from IV group filtrate ? Explain.

Ans. The answer to this question is determined by a number of considerations. First, the salt must be sufficiently insoluble *i.e.*, they should have lowest possible solubility products. V group cations can not be completely precipitated in aqueous solution in the form of sulphates as one of them ($CaSO_4$) has fairly large solubility product (5.6×10^{-6}). Another disadvantage is that sulphates being salts of strong acid, are practically insoluble in acids and it would, therefore, be relatively difficult to dissolve them after precipitation.

It is also difficult to separate group V in the form of phosphates or oxalates as PO_4^{3-} and C_2O_4 ions introduced into the solution would complicate the subsequent analysis.

The most suitable procedure is to separate group V cations as carbonates. The solubility products of these salts are very small (of the order of 10^{-9}) and, therefore, practically complete precipitation of all the group V cations is possible. Moreover, it is very simple to dissolve the precipitate for further analysis of V group because, in contrast to sulphates, group V carbonates are readily soluble in acids, being salts of weak acids. Finally, the excess of the precipitant, CO_3^{2-} ions is easily removed when solution is acidified because the carbonic acid (H_2CO_3) formed decomposes into CO_2 and H_2. From this, it follows that *the most important analytical property of group V cations, used for separating from group IV, is the practical insolubility of $CaCO_3$, $SrCO_3$ and $BaCO_3$ in water.*

Q. 93. Only ammonium carbonate can be used as the group reagent. Why not Na_2CO_3 K_2CO_3 ?

Ans. Only ammonium carbonate can be used as the group reagent for group V because the use of Na_2CO_3 or K_2CO_3 would introduce Na^+ or K^+ ions, to be tested ahead. The introduction of NH_4^+ ion does not lead to error because ammonium is detected in the original mixture.

Q. 94. Would NH_4HCO_3 be suitable group reagent for V group cations precipitation ?

Ans. No, V group bicarbonates, are water soluble.

Q. 95. Describe the conditions influencing the completeness of precipitation of group V cations as carbonate.

Ans. (1) **pH** : One of the most important conditions for complete precipitation of sparingly soluble salts of weak acids, such as carbonates of group V cations, is a suitable pH. Amm. carbonate hydrolyses almost completely in solution :

$$(NH_4)_2CO_3 + H_2O \rightleftharpoons NH_4HCO_3 + NH_4OH$$

Because of this, amm. carbonate solution is really a mixture of approximately equivalent amounts of NH_4OH and NH_4HCO_3 i.e., it is an ammonium buffer mixture of pH ≈ 9.2. At this pH all V-group carbonates are precipitated almost completely, while K^+, Na^+, NH_4^+ cations which form soluble carbonates, remain in solution. Although basic magnesium carbonate, $(MgOH)_2CO_3$ and $Mg(OH)_2$ are sparingly soluble, they are not precipitated at pH ≈ 9.2. The precipitation of $Mg(OH)_2$ begins at pH>9.3 and is complete at pH = 11.3. The conditions for precipitation of $(MgOH)_2CO_3$ are similar. *Thus group V cations are separated completely when the precipitation is performed at pH ≈ 9.2.; at pH<9.2 the precipitation of V group cations is incomplete, while at pH > 9.3 Mg^{2+} is partially precipitated.*

It might appear that the pH required for precipitation of Mg^{2+} would arise due to the presence of excess $(NH_4)_2CO_3$. In reality, this is true only if the buffer capacity of the solution is sufficient, i.e., if the solution is capable of consistently maintaining a constant pH.

The following considerations show that the buffer capacity of solution must be taken into account in this case. Ammonium carbonate solution is a mixture of equivalent amounts of NH_4OH and NH_4HCO_3. Therefore, the reaction which takes place when group V cations are precipitated by amm. carbonate is essentially represented by the equation.

$$MCl_2 + NH_4HCO_3 + NH_4OH \longrightarrow MCO_3\downarrow + 2NH_4Cl + H_2O \ (M = Ca, Sr \ or \ Ba)$$

or in ionic form

$$M^{2+} + HCO_3^- + NH_4OH = MCO_3\downarrow + NH_4^+ + H_2O$$

The equation shows that NH_4OH is used up in the reaction while NH_4^+ ions accumulate in the solution. Similar consumption of ammonia and accumulation of NH_4^+ ions must evidently occur if amm. carbonate reacts with any acid which may be present in solution :

$$2H^+ + HCO_3^- + NH_4OH \longrightarrow CO_2\uparrow + 2H_2O + NH_4^+$$

The pH of ammonium buffer mixture is given by the formula

$$pH = 14 - pK_{NH_4OH} + \log \frac{[NH_4OH]}{[NH_4^+]} = 9.25 + \log \frac{[NH_4OH]}{[NH_4^+]}$$

As NH_4OH is used up while NH_4^+ ions accumulate in solution, the ratio of their concentrations decreases and becomes less than unity. Accordingly, the solution pH falls to below 9.25. If this fall of pH is at all considerable i.e. buffer capacity of solution is inadequate, the precipitation of group V cations is incomplete. However, the buffer capacity of a mixture increases with increasing concentration of its components. Therefore, it is not enough to introduce a buffer mixture (amm. carbonate solution) with the required pH; it is also necessary to ensure that its buffer capacity is sufficient. In practice, before precipitation of group V cations by amm. carbonate, ammonia is added to the solution until its odour appears, followed by a little NH_4Cl. Ammonia neutralizes any free acid but solution becomes so alkaline that Mg^{2+} may be precipitated. Addition of NH_4Cl lowers the pH to the required value of ~9. At this pH Mg^{2+} can not precipitate. At the same time, the buffer capacity of the solution is raised by the addition of ammonia and ammonium chloride to such an extent that the consumption of NH_4OH ion and accumulation of NH_4^+ ions can not produce any appreciable changes solution pH.

(ii) **Temperature.** It is another important condition. The reason is that solid ammonium carbonate partially decomposes on keeping, liberating water and forming ammonium carbamate.

$$(NH_4)_2CO_3 \rightleftharpoons NH_2COONH_4 + H_2O$$

Therefore, commercial amm. carbonate usually contains amm. carbamate, which lowers consider-ably the CO_3^{2-} ion concentration in its solution. However, this effect is easily prevented if precipita-tion of group V is done in a solution heated to about 60-80°C because at higher temperature the equilibrium of above reaction is shifted to the right, i.e., towards conversion of amm. carbonate into amm. carbamate. The precipitation in hot condition is also useful because it accelerates the conver-sion of originally amorphous carbonate ppt. into a crystalline form, which is easier to filter off and which has less tendency to absorb extra ions.

It follows that *group V cations should be precipitated by the group reagent in the presence of ammonia and amm. chloride (at pH = 9.2) in a solution heated to 60—80°C.*

Q. 96. What is the purpose of making solution ammoniacal before V group precipitation ?

Ans. $(NH_4)_2CO_3$ in aqueous solution is almost completely hydrolysed.

$$(NH_4)_2CO_3 + H_2O \rightleftharpoons NH_4HCO_3 + NH_4OH$$

In presence of NH_4OH, the equilibrium shifts to the left side and, therefore, concentration of CO_3^{2-} does not fall and V group carbonates get completely precipitated.

Q. 97. What is the harm in adding a large excess of amm. chloride in V group preci-pitation ?

Ans. The carbonates of V group are soluble in excess of NH_4Cl because ammonium chloride has an acidic reaction as a result of hydrolysis.

$$NH_4Cl + H_2O \rightleftharpoons NH_4OH + HCl \qquad \text{(acidic nature)}$$
$$MCO_3 + 2HCl \longrightarrow MCl_2 + CO_2 + H_2O \quad (M = Ca, Sr \text{ or } Ca)$$

Therefore, in presence of excess NH_3Cl, precipitation of V group carbonates will not be complete.

Q. 98. It is always advised to avoid too much quantity of NH_4Cl in III group precipitation. Why ?

Ans. Too much quantity of NH_4Cl in III group precipitation is avoided because it will create difficulty in complete precipitation of V group carbonates as these become soluble in excess of NH_4Cl.

Q. 99. Why should the solution after adding amm. carbonate, not be boiled ?

Ans. The solution must not be boiled as this decomposes amm. carbonate into ammonia and carbon dioxide.

$$(NH_4)_2CO_3 \longrightarrow 2NH_3 + CO_2 + H_2O$$

Thus complete precipitation will not occur. The boiling also enhances the solvent action of ammonium ions in reversing the carbonate precipitation (i.e., it may result in some redissolution of carbonates) :

$$MCO_3 + 2NH_4Cl \longrightarrow MCl_2 + 2NH_3 + CO_2 + H_2O \quad (M = Ca, Sr \text{ or } Ba$$

Q. 100. After dissolving V group ppt. in minimum amount of acetic acid, why is it preferred to use $CH_3COONa + K_2Cr_2O_7$ solution over K_2CrO_4 solution, in order to separate Ba^{2+} as $BaCrO$ from Ca^{2+} and Sr^{2+} ?

Ans. Solubility products of $BaCrO_4$ and $SrCrO_4$ are respectively $2\cdot4 \times 10^{-10}$, $3\cdot6 \times 10^{-5}$ and that of $CaCrO_4$ is even higher.

With K_2CrO_4 :

Sometimes, when Sr^{2+} is present in high concentration or K_2CrO_4 is added in excess, the solubility product of $SrCrO_4$ is also reached. With the result, $SrCrO_4$ is also precipitated along with $BaCrO_4$. The solubility product of $CaCrO_4$ is quite high and thus it is not reached i.e. Ca^{2+} remains in the solution.

With $CH_3COONa + K_2Cr_2O_7$ solution :

The precipitation of $SrCrO_4$, in this case, is prevented irrespective of high concentration of Sr^{2+} or CrO_7^{2-} or both. The explanation is as follows :

$$CH_3COONa + H_2O \rightleftharpoons CH_3COOH + NaOH$$

or
$$CH_3COO^- + H_2O \rightleftharpoons CH_3COOH + OH^-$$

OH^- ions thus produced, convert CrO_7^{2-} to CrO_4^{2-}

$$Cr_2O_7^{2-} + 2OH^- \longrightarrow 2CrO_4^{2-} + H_2O$$

Since in this case CrO_4^{2-} ions are obtained indirectly, the possibility of high concentration of chromate ions is minimised and, therefore, $SrCrO_4$ does not get any chance to precipitate along with $BaCrO_4$. Excess CH_3COONa should be taken so that a part of it remains unused, i.e., so that an acetate buffer mixture is formed. This maintains pH practically constant of about 5 which is necessary for complete precipitation of $BaCrO_4$.

Q. 101. Explain which carbonates of V group are dissolved in minimum amount of acetic acid and not in mineral acid before separating Ba^{2-} as $BaCrO_4$.

Ans. Mineral acids are not used as $BaCrO_4$, CaC_2O_4 are soluble in them but in acetic acid, being very weak, they are insoluble.

Q. 102. Explain why $BaCrO_4$ and CaC_2O_4 are insoluble in acetic acid but soluble in dilute mineral acid ?

Ans. When dil. mineral acid is added to suspension of $BaCrO_4$, the following equilibrium will exist.

$$BaCrO_4 \rightleftharpoons Ba^{2+} + CrO_4^{-2} \qquad \qquad ...(i)$$
$$2H^+ + 2CrO_4^{2-} \rightleftharpoons 2HCrO_4^- \rightleftharpoons Cr_2O_7^{2-} + H_2O \qquad ...(ii)$$

Thus in presence of excess of H^+ ions (given by mineral acids as they are strongly ionizing) the CrO_4^{2-} ions are removed by reaction (i), thereby shifting equilibrium in equation (ii) more and more to the right side. Consequently, $BaCrO_4$ passes into solution in the form of $BaCr_2O_7$ as all dichromates are water soluble. Whereas acetic acid gives only very small H^+ ion concentration and, therefore, has little effect an CrO_4^{2-} ions. Thus $BaCrO_4$ does not go into solution in presence of acetic acid.

When mineral acid is added to a suspension of CaC_2O_4 in water, the following equilibria exist :

$$CaC_2O_4 \rightleftharpoons Ca^{2+} + C_2O_4^{2-} \qquad \qquad ...(iii)$$
$$\therefore \qquad K_{sp} = [Ca^{2+}][C_2O_4^{2-}]$$
$$H^+ + C_2O_4^{2-} \rightleftharpoons HC_2O_4^- \qquad \qquad ...(iv)$$
$$H^+ + HC_2O_4^- \rightleftharpoons H_2C_2O_4 \qquad \qquad ...(v)$$

Thus concentration of oxalate ions is reduced, from its initial value, by combining with H^+ ions (from mineral acid) to form $H_2C_2O_4$ according to equilibria (iv) and (v), and thereby shifting equilibrium (iii) to the right side, i.e., more CaC_2O_4 passes into solution. In this way, if sufficient H^+ ions are present, the process will continue until the whole of CaC_2O_4 has gone into solution. Whereas acetic acid is feebly ionizing and thus H^+ ions given by it are not enough to reduce the oxalate ion concentration according to (iv) and (v). Therefore, equilibrium (i) remains practically undisturbed, i.e., CaC_2O_4 does not pass into solution.

Q. 103. Explain the difference between the behaviour of $BaCrO_4$ with HCl and with CH_3COOH.

Ans. See Question No. 102.

Q. 104. Why is barium chromate and no dichromate formed by action of $K_2Cr_2O_7$ on barium salt ? Why is CH_3COONa added in this reaction ?

Ans. $K_2Cr_2O_7$ forms a yellow ppt. of $BaCrO_4$ with Ba^{2+}, and not $BaCr_2O_7$ as might be expected. The reason is as follows : A solution of $K_2Cr_2O_7$ contains in addition to $Cr_2O_7^{2-}$, a small amount of CrO_4^{2-} by the interaction of $Cr_2O_7^{2-}$, with water.

$$H_2O \rightleftharpoons H^+ + OH^-$$
$$\underline{Cr_2O_7^{2-} + OH^- \rightleftharpoons 2CrO_4^{2-} + H^+}$$

or
$$Cr_2O_7^{2-} + H_2O \rightleftharpoons 2CrO_4^{2-} + 2H^+ \qquad \qquad ...(i)$$

However, the CrO_4^{2-} ion concentration is high enough to exceed the solubility product of $BaCrO_4$ before the solubility product of $BaCr_2O_7$ is reached. Therefore, $BaCrO_4$ is the compound precipitated :

$$2CrO_4^{2-} + 2Ba^{2+} \longrightarrow 2BaCrO_4 \downarrow \qquad\qquad ...(ii)$$

Adding equations (i) and (ii), we have the overall equation for the reaction

$$Cr_2O_7^{2-} + 2Ba^{2+} + H_2O \rightleftharpoons 2BaCrO_4 + 2H^+$$

The above reaction reverses if H^+ ions are produced in excess (this will happen when barium salt of strong acid is taken e.g., $BaCl_2$ or $Ba(NO_3)_2$). However, complete precipitation Ba^{2+} can be achieved if CH_3COONa is added as well as $K_2Cr_2O_7$; the excess of H^+ ions is then replaced by the weak acetic acid, in which $BaCrO_4$ is insoluble.

$$CH_3COO^- + H^+ \longrightarrow CH_3COOH$$

Q. 105. State the guiding rule to dissolve the ppt. by a reagent.

Ans. "For a ppt. to dissolve, one of the ions, it yields in aqueous solution must form a compound of low degree of dissociation with one of the ions of the reagent dissolving the ppt." This is the guiding rule for dissolving the ppt. by a reagent. In other words, *if we want to dissolve a salt which is sparingly soluble in water, we must act on it with an acid which is stronger than the acid formed in the reaction.*

Q. 106. What are general characteristics of group VI cations ?

Ans. Group VI contains K^+, Na^+, Mg^{2+} and NH_4^+. Most of their salts are soluble in water. The solubility of their sulphides, hydroxides, carbonates and chlorides in water is especially important for analysis, as this distinguishes group VI cations from all other groups. Owing to the solubility of these salts, group VI cations are not precipitated by group reagents, such as HCl, H_2S and $(NH_4)_2CO_3$, and remain in solution when cations of other groups are separated as insoluble salts. There is no group reagent which precipitates all four cations of group VI. As cations of the other groups interfere with the detection of K^+, Na^+ and Mg^{2+} ions, in systematic analysis these are detected only after all other cations have been completely removed from solution.

Sodium and potassium salts of strong acid are not hydrolysed and their solutions have pH = 7. The potassium and sodium salts of weak acids are hydrolysed to a considerable extent and their solutions are alkaline (pH > 7). Solutions of ammonium salts of strong acids have an acid reaction (pH < 7).

The K^+, Na^+ and Mg^{2+} cations have completed outermost shells of eight electrons, similar to the outermost shell of the inert gases. The Mg^{2+} ion differs from other group VI cations ; it belongs to second group of periodic table and is intermediate between analytical groups V and VI, so it can be included in either. The Mg^{2+} cation resembles group V cations, in that its basic carbonate is sparingly soluble in water. However, it is soluble in ammonium salts, and since Group V is precipitated by ammonium carbonate in presence of NH_4Cl, the Mg^{2+} ion remains in solution with Group VI and is not precipitated along with group V. It is, therefore, more conveniently studied with group VI. All the group VI cations are colourless in aqueous solution.

Q. 107. Write down the equation for reaction of Mg^{2+} with disodium hydrogen phosphate in presence of NH_4OH. Why is NH_4Cl added in this test ?

Ans.
$$MgCl_2 + Na_2HPO_4 \longrightarrow MgHPO_4 + 2NaCl$$
$$MgHPO_4 + NH_4OH \longrightarrow MgNH_4PO_4 \downarrow + H_2O$$

Adding : $MgCl_2 + Na_2HPO_4 + NH_4OH \longrightarrow MgNH_4PO_4 \downarrow + 2NaCl + H_2O$

$$\text{white}$$

or ionically : $Mg^{2+} + HPO_4^{2-} + NH_4OH \longrightarrow MgNH_4PO_4 \downarrow + H_2O$

The ammonium salt is added to prevent the precipitation of $Mg(OH)_2$ by the action of NH_4OH.

9

Volumetric Analysis
(Titrimetric Analysis)

Introduction

Volumetry is a subject which involves estimation of quantities of chemical species (element, compound, radical or ion) indirectly by measuring the volume of the solution of that particular species in a suitable solvent. It is the mass of the chemical species which takes part in the reaction but we measure the volume of the solution for calculation of the reacting mass of that very species. *The masses of the species react and form under volumetric conditions always in the ratio of their so-called 'chemical equivalent weights'*. This is enumerated as *'Law of Equivalence'* which is applied in all volumetric estimations. The technique involved in it is called titration.

The main task in titration is the standardisation of a given solution with the help of known standard solution where a chemical species of the nature of primary standard has been used. Keeping in view the precautions taken in the preparation of standard solution, we weigh out accurately almost desired weight of the substance to dissolve in a solvent water. The solution is made homogeneous by proper shaking. The standard solution may be pipetted and filled in rinsed *burette* as directed according to the convenience of the exercise. This standard solution is now titrated against the unknown solution using suitable indicator. Sometimes even indicator is avoided where one of the titrants possesses sharp colour in solution. Titration is simple mixing of two solutions to react in a beaker or conical flask. At the time of completion of the reaction, a visible colour may be achieved which is regarded as the practical end point. It shows that reaction has been theoretically complete just before addition of one of the titrants by a fraction of drop. The volumes of titrants are sufficient to calculate the quantity of the mass of either the species dissolved in the solution if mass of the one species is already known from Law of Equivalence ($N_1V_1 = N_2V_2$).

Primary and secondary standard substances. The substance which fulfills the following requisites is called *'primary standard substance or primary standard'*. It must

1. be easily available in highly pure state.
2. be easily soluble in the desired solvent.
3. not suffer decomposition in presence of solvent.
4. be stable in and unaffected by the air.
5. have large equivalent weight so that weighing errors are minimised.
6. Its composition must not change on standing.

Commonly used primary standards are : Anhydrous Na_2CO_3, Oxalic acid, succinic acid, $K_2Cr_2O_7$, ferrous ammonium sulphate, NaCl and $AgNO_3$ etc.

Though the hydrated salts are not good primary standards but those which are not efflorescent such as $CuSO_4.5H_2O$, $Na_2B_4O_7.10H_2O$ and $H_2C_2O_4.2H_2O$ work satisfactory. The standard solution of the primary standard substance is called *'primary standard solution'*. This is prepared by dissolving an accurately weighed out sample of the dry reagent in known volume of the solvent.

Those substances which do not fulfil the above requirements and their standard solutions can not be prepared by direct weighing, are called *'Secondary standard substances or secondary standards'*. The common secondary standards are alkali hydroxides and inorganic acids etc. An approximate weight of the secondary standard is dissolved in a known volume and the exact strength of the solution is found by titrating it against standard solution of the suitable reagent. This process is called 'Standardisation'.

Equivalent weight. It is defined as the number of parts by weight, a chemical species combined with or displaced by 1.008 parts of hydrogen, 8 parts of oxygen or 35.5 parts by weight of chlorine.

$$\text{Eq. wt. of an acid} = \frac{\text{Mol. wt.}}{\text{basicity}}$$

$$\text{Eq. wt. of a base} = \frac{\text{Mol. wt.}}{\text{acidity}}$$

$$\text{Eq. wt. of an electrolyte} = \text{Eq. wt. of cation} + \text{Eq. wt. of anion}$$

Equivalent weight is not always constant. It may change from reaction to reaction Such variations are often observed in the reaction involving oxidation reduction e.g. Eq. wts. of copper sulphate, ferrous sulphate and of sodium thiosulphate are half of their moleulcar weights in usual reactions but in redox reactions both, mol. wt. and eq. wt. are often equal.

The equivalent wt. expressed in gm is called gm equivalent wt. of the substance.

Strength of the solution. It is expressed in following ways :

(i) *Percentage.* Number of gms of a substance dissolved in 100 gm of its solution. It is generally used in dilute solutions of electrolytes in water. 10% NaCl means that 100 gm of its solution contains 10 gm of NaCl.

(ii) *Molality.* It is number of gm-moles of substance dissolved in 1000 gm of solvent.

(iii) *Molarity.* It is number of gm-moles of the substance dissolved per litre of the solution. If one gm-mole of a substance is dissolved in a solvent and the solution is made up to the one litre, such a solution is called *Molar Solution.*

If w gm of a substance is dissolved in V ml of the solution and mol. wt. of the substance is M,

$$\text{Molarity} = \frac{w}{M} \times \frac{1000}{V}$$

$$= \frac{\text{gm / litre}}{\text{mol. wt.}}$$

(iv) *Normality.* It is number of gm equivalents of the substance dissolved per litre of the solution. If one gm. equivalent of an acid, base or salt is dissolved in water and the solution is made up to one litre, such a solution is called '*Normal Solution*'.

If w gm of a substance is dissolved in V ml of the solution and equivalent wt. of the substance is E,

$$\text{Normality} = \frac{w}{E} \times \frac{1000}{V}$$

$$= \frac{\text{gm / litre}}{\text{Eq. wt.}}$$

(v) *Formality.* Number of gram formula weights of the substance dissolved in a litre of solution.

$$\text{Formality} = \frac{\text{gm / litre}}{\text{Formula wt.}}$$

Standard solution. A solution of known strength is called *standard solution*. It is prepared either by direct weighing of the substance or after titrating it with another standard solution i.e., standardisation with the help of another standard solution.

Calculation of the normality of prepared solution :

(i) If mass of the substance taken is   = w gm

(ii) Equivalent weight of the substance   = E

In 250 ml measuring flask the normality of solution

$$= \frac{1000\,w}{250 \times E}$$

$$= \frac{4 \times \text{wt. of the substance taken}}{\text{Eq. wt.}}$$

Calculation of the required weight for preparing the standard solution

Suppose equivalent wt. of the substance is E and we have to prepare a solution of N normality in V ml measuring flask.

∴ Substance in gm per litre $= N \times E$

∴ Substance in gm in V ml $= \dfrac{N \times E \times V}{1000}$

Titration. It is the process for standardisation of a solution. Two solutions are generally used in every titration where strength of one is generally known. The solution to be titrated is taken in a conical flask by using a pipette of 10, 20 or 25 ml and the other solution is taken in the burette. Indicator is added at desired stage to the solution taken in the conical flask and burette solution is generally poured into it. A sharp colour change occurs at the completion of the reaction. This point is called 'End point'.

There are so many types of titrations but the most common are :

1. Neutralization titrations

2. Redox titrations which also include iodine titrations

3. Precipitation titrations

4. Complexometric titrations.

Acidimetry and Alkalimetry

The determination of concentration of acid by standard alkali is known as acidimetry and the reverse process is alkalimetry. Though it has a wide range in the volumetric analysis but few experiments are discussed below.

EXERCISE 1

To determine the strength of NaOH and Na_2CO_3 present in a solution and find their percentage composition.

Theory. The mixture is titrated with a standard solution of HCl or H_2SO_4 by the selective use of indicators. With HCl, the reactions proceed in the following ways :

$$NaOH + HCl \longrightarrow NaCl + H_2O$$
$$Na_2CO_3 + HCl \longrightarrow NaCl + NaHCO_3$$

This $NaHCO_3$ further reacts with HCl

$$NaHCO_3 + HCl \longrightarrow NaCl + H_2O + CO_2$$

The phenolphthalein is added as first indicator which loses its colour of alkaline medium when NaOH and half Na_2CO_3 are neutralised. In fact, it does not give colour in $NaHCO_3$ solution.

Now use of methyl orange as second indicator is made which shows the complete neutralisation.

Procedure. Take 25 ml supplied solution and add 2 drops of phenolphthalein indicator. Now titrate it with standard solution of HCl (prepare if not supplied) until the additional drop of HCl fades the pink colour of solution. Note this reading. Now add two drops of methyl orange and add slowly HCl again until the last drop of HCl gives red colour. Note this second reading. Repeat it for the correct readings.

Note : The second reading of burette includes previous volume as well, used with phenolphthalein indicator.

Observation. 1. Volume of HCl used with phenolphthalein indicator $= V_1$ ml

2. Volume of HCl used with methyl orange indicator *i.e.*, vol. obtained after subtracting first

reading from second reading of burette = V_2 ml

Calculation. The volume used in first reading shows the complete neutralisation of NaOH and half neutralization of Na_2CO_3 as half of it has been converted to $NaHCO_3$ (still it is alkaline) hence

$$V_1 \text{ ml} \equiv NaOH + 1/2 Na_2CO_3$$

and
$$V_2 \text{ ml} \equiv 1/2 \ Na_2CO_3 \quad \text{(by latter reaction shown in theory of it)}$$

1. Hence complete Na_2CO_3 required $\quad = 2 V_2$ ml
2. NaOH required only $\quad = (V_1 - V_2)$ ml

If the normality of HCl be N_1, the normality of NaOH solution, N_2 will be

$$N_2 = \frac{N_1(V_1 - V_2)}{25}$$

$$\text{Strength of NaOH} = \text{Eq. wt. of NaOH} \times \text{normality}$$

$$= 40 \times \frac{N_1(V_1 - V_2)}{25} \text{gm/litre}$$

$$\text{Amount of NaOH in 100 ml} = \frac{40 \times N_1(V_1 - V_2)}{25 \times 10} \text{ or say } x_1$$

Similarly, it may be worked out for Na_2CO_3.

$$\text{The strength of } Na_2CO_3 = 53 \times \frac{2V_2 \times N_1}{25} \text{gm/litre}$$

$$\text{Amount of } Na_2CO_3 \text{ in 100 ml} = 53 \times \frac{2V_2 \times N_1}{25 \times 10} \text{ or say } x_2$$

1. Percentage of NaOH in mixture : $\dfrac{x_1}{x_1 + x_2} \times 100$

2. Percentage of Na_2CO_3 in mixture : $\dfrac{x_1}{x_1 + x_2} \times 100$

EXERCISE 2

To determine the strength of Na_2CO_3 and $NaHCO_3$ present in a solution and also find their percentage composition.

Theory. The principle involved in it, is just similar to that of discussed previously. Here also Na_2CO_3 will react in two steps and the same two indicators will be used. The chemical reactions taking place are as follows :

$$Na_2CO_3 + HCl \longrightarrow NaCl + NaHCO_3$$
$$NaHCO_3 + HCl \longrightarrow NaCl + H_2O + CO_2$$

Procedure. Pipette out 25 ml of the supplied solution and add two drops of phenolphthalein indicator. Add gradually a standard solution of HCl till the last drop of it fades the pink colour of solution. Note this end point carefully. Now, add two drops of methyl orange indicator and titrate with HCl and see the light pink as end point. Note also this reading. (This reading will also include previous volume used with first indicator). Repeat experiment to ensure the accuracy of your readings.

Observations. Let

1. Volume of HCl used when phenolphthalein used $\quad = V_1$ ml
2. Volume of HCl used in all $\quad = V_2$ ml

Calculation. V_1 ml. is the volume for half neutralization of Na_2CO_3 as $NaHCO_3$ formed to reacts alkaline. Therefore,

total volume used to neutralise Na_2CO_3 $\quad = 2V_1$ ml

total volume of the acid used for complete neutralisation of Na_2CO_3 and $NaHCO_3$ both $\quad = V_2$ ml

Hence the volume of HCl required for $NaHCO_3$ $= (V_2 - 2V_1)$ ml

If the normality of HCl be N_1 and that of Na_2CO_3 be N_2,

$$N_2 = \frac{N_1 \times 2V_1}{25}$$

$$\text{Strength} = \text{Eq. wt.} \times \text{normality}$$

$$\text{Strength of } Na_2CO_3 = \frac{53 \times N_1 \times 2V_1}{25} \text{ gm/litre}$$

$$\text{Amount of } Na_2CO_3 \text{ in 100 ml} = \frac{53 \times N_1 \times 2V_1}{25 \times 10} \text{ or say } x_1$$

$$\text{Similarly, the normality of } NaHCO_3 = \frac{N_1 \times (V_2 - 2V_1)}{25}$$

$$\text{Strength of } NaHCO_3 = \frac{84 \times N_1(V_2 - 2V_1)}{25} \text{ gm/litre}$$

$$\text{Amount of } NaHCO_3 \text{ in 100 ml} = \frac{84 \times N_1 \times (V_2 - 2V_1)}{25 \times 10} \text{ or say } x_2$$

$$\text{Percentage of } Na_2CO_3 = \frac{x_1}{x_1 + x_2} \times 100$$

$$\text{Percentage of } NaHCO_3 = \frac{x_2}{x_1 + x_2} \times 100$$

EXERCISE 3

To determine the percentage of ammonia in an ammonium salt.

Theory. When an ammonium salt (except carbonate and bicarbonate) is decomposed by boiling with an excess of sodium hydroxide solution, ammonia is liberated according to the equation :

$$NH_4Cl + NaOH \longrightarrow NaCl + H_2O + NH_3 \uparrow$$

If a known quantity of NaOH (in excess of that required to decompose the ammonium salt) is used, and, after the decomposition is complete, the unused NaOH is determined, the difference between these quantities gives the amount of NaOH equivalent to the ammonium salt taken.

Procedure. Weigh accurately about 1 to 2 gm of the amm. salt, and transfer it in to a 250 ml beaker and dissolve in about 75 ml of water. Add 50 ml (measured) N-NaOH solution *i.e.*, amount which is more than sufficient to decompose the ammonium salt and boil gently to expel the ammonia. At intervals of about five minutes, test for ammonia in the escaping steam with a piece of moist red litmus paper—the test paper being held outside the beaker. Boil for about half an hour. When the decomposition is complete, *i.e.*, when no ammonia is detected in the vapours, cool the solution and transfer it to a 250 ml measuring flask. Dilute the solution up to the mark and shake it thoroughly.

Fill the burette with N/10 HCl. Pipette out 25 ml of the above solution into a conical flask. Add 2-3 drops of methyl orange as an indicator. Now add slowly HCl solution from burette shaking the flask constantly till a pink colour is just obtained. Note the burette reading and repeat the titration until two concordant readings are obtained.

Calculation

Mass of the salt taken $= w$ gm

Vol. of N-NaOH added $= 50$ ml

Vol. made up after boiling the solution $= 250$ ml

Vol. of the solution taken for titration $= 25$ ml

Vol. of N/10 HCl required to titrate the unused alkali $= V$ ml

∴ Vol. of N-NaOH left after neutralisation of amm. salt in 25 ml solution

$$\equiv V \text{ ml of } N/10 \text{ HCl}$$
$$\equiv V \text{ ml of } N/10 \text{ NaOH}$$
$$\equiv V/10 \text{ ml of } N\text{-NaOH}$$

Thus vol. of N-NaOH left unreacted in 250 ml of solution

$$= \frac{250 \times V}{25 \times 10} \text{ or } V \text{ ml}$$

$\therefore$ Vol. of N-NaOH reacted with amm. salt

$$= (50 - V) \text{ ml}$$

Now 1 gm. equivalent NaOH $\equiv$ 1 gm equivalent NH_3

or   1000 ml N-NaOH $\equiv$ 17 gm NH_3

$\therefore$ $(50 - V)$ ml N-NaOH $\equiv \dfrac{17(50 - V)}{1000}$ gm NH_3

Hence amount of NH_3 in 1 gm amm. salt

$$= \frac{17(50 - V)}{w \times 1000} \text{ gm}$$

$\therefore$ Percentage (W/W) of NH_3 in amm. salt

$$= \frac{17(50 - V)\,100}{w \times 1000}$$

$$= \frac{17(50 - V)}{10 \, w}.$$

Note. 1. If the ammonium salts contain free acid (as an impurity), the amount of acid present must be determined by a separate titration.

2. The ammonium radical in salts like ferrous ammonium sulphate can not be determined by this method, as part of sodium hydroxide is used up in the precipitation of hydroxide of the heavy metal.

Oxidation-Reduction (Redox.) Titrations

Redox-titrations include all those varieties of titrations where one reactant is oxidised and the other is reduced. The indicator is so chosen as to give the colour in either of the conditions.

Oxidation is defined as the process of loss of one or more electrons and reduction, the gain of one or more electrons by atoms or ions. Both, oxidation and reduction are complementary to one another and take place simultaneously, *i.e.* one cannot take place without the other.

The reagent undergoing reduction is called *oxidising agent* (*oxidant*), *i.e.*, an oxidising agent gains electrons and is reduced to lower valency state. The reagent which undergoes oxidation is called *reducing agent* (*reductant*) *i.e.* a reducing agent loses electrons and is oxidised to higher-valency state. The reducing agents are generally sodium oxalate, ferrous sulphate, ferrous ammonium sulphate, oxalic acid etc. The commonly employed oxidants are $KMnO_4$, $K_2Cr_2O_7$ (both acidified), I_2 solutions etc.

In redox titrations, the estimations are generally done by the following methods :

[A] Potassium permanganate method

[B] Potassium dichromate method

[C] Iodine method

The first two methods will be discussed in this chapter and the third one is discussed in the next chapter.

Calculation of equivalent weights of oxidants and reductants.

The equivalent weight of a substances involved in oxidation-reduction (redox) titrations is calculated by any one of the following methods :

1.   Available oxygen method,

2.   Oxidation number method and

3.   Ion-electron method.

Out of these, oxidation number method is convenient. Few examples follow.

(1) *Equivalent weight of KMnO₄ in acidic medium.* The reduction of $KMnO_4$ in acidic medium is represented by the following equation :

$$2KMnO_4 + 3H_2SO_4 \longrightarrow K_2SO_4 + 2MnSO_4 + 3H_2O + 5O$$

The oxidation of Mn in $KMnO_4$ = +7

in $MnSO_4$ = +2

This shows the gain is of 5 electrons. Therefore,

$$\text{The Eq. wt. of } KMnO_4 = \frac{M. wt.}{5}$$

$$= \frac{158}{5} = 31.6$$

Eq. wt. of $KMnO_4$ in acidic medium is **31.6**.

(2) *Eq. wt. of Ferrous sulphate.* It is oxidised according to following equation.

$$2FeSO_4 + H_2SO_4 + [O] \longrightarrow Fe_2(SO_4)_3 + H_2O$$

Since ferrous is converted to ferric sulphate, the loss of only one electron per ferrous ion is there. It is, therefore,

$$\text{Eq. wt. of } FeSO_4 = \frac{M. wt.}{1} = \mathbf{152}$$

Eq. wt. of $FeSO_4.7H_2O$ = 278

(3) *Eq. wt. of ferrous ammonium sulphate,* $(NH_4)_2SO_4.FeSO_4.6H_2O$

It is a double salt and is, therefore, quite resistant to oxidation and so is perferred over ferrous sulphate. It reacts in the same manner as does ferrous sulphate. Its equivalent weight is equal to that of its molecular weight *i.e.* 392.

(4) *Eq. wt. of oxalic Acid.*

It is oxidised in the following way

$$\begin{matrix} COOH \\ | \\ COOH \end{matrix} . 2H_2O + O \longrightarrow 2CO_2 + 3H_2O$$

The oxidation state of C_2 in $H_2C_2O_4$ = +6

"   "   "   C   in CO_2 = + 4

So for C_2 this will be   = + 8

The loss of electrons in 2 carbons is 2. Therefore,

$$\text{Eq. wt. of oxalic acid} = \frac{M. wt.}{2}$$

$$= \frac{126}{2}$$

$$= 63$$

(5) *Eq. wt. of Potassium dichromate.* Potassium dichromate in acidic medium reacts with a reducing agent as follows :

$$K_2Cr_2O_7 + 6FeSO_4(NH_4)_2SO_4 + 7H_2SO_4$$

$$\longrightarrow 3Fe_2(SO_4)_3 + K_2SO_4 + Cr_2(SO_4)_3 + 6(NH_4)_2SO_4 + 7H_2O$$

Here Cr_2^{+12} changed to Cr_2^{+6}.

$$\text{Eq. wt.} = \frac{\text{M. wt.}}{6}$$
$$= 49.03$$

EXERCISE 4

To determine the strength of supplied oxalic acid solution by titrating it against approx. N/30 KMnO$_4$ solution.

Theory. The $KMnO_4$, in presence of dil H_2SO_4 oxidises the oxalic acid to CO_2 and H_2O :

$$2KMnO_4 + 3H_2SO_4 + 5H_2C_2O_4 \longrightarrow K_2SO_4 + 2MnSO_4 + 10CO_2 + 8H_2O$$

or $$2[MnO_4]^- + 5[C_2O_4]^{2-} + 16H^+ \longrightarrow Mn^{2+} + 8H_2O + 10CO_2$$

Since the above reaction proceeds very slowly at room temperature, the solution in the titration flask is heated to about 60—80°C before adding $KMnO_4$ from burette. $KMnO_4$ acts as a self indicator.

Procedure. It involves three stages.

(a) *Preparation of standard (approx. N/30) oxalic acid solution.* Equivalent wt. of $H_2C_2O_4.2H_2O$ is 63 hence weigh accurately about $63/120 = 0.525$ gm of oxalic acid crystals and dissolve in water in 250 ml measuring flask and then make up the volume by adding distilled water.

If w is mass of oxalic acid dissolved in 250 ml of distilled water, the normality of oxalic acid solution $= 4w/63$.

(b) *Standardisation of KMnO$_4$ solution with the help of known oxalic acid solution.* Rinse and fill the burette with $KMnO_4$ solution and note the initial reading. Pipette out 25 ml of standard oxalic acid solution in a clean titrating flask and add to it 25 ml of dil. H_2SO_4. Heat the contents gently to about 60—80°C (*i.e.*, just unbearable to touch by hand). Now run in $KMnO_4$ from burette very slowly (about 10-15 ml per minute) and shake the flask after each addition. Avoid addition of large amounts of $KMnO_4$ at a time otherwise a brown ppt. of hydrated manganese dioxide is formed. During titration if the temperature of flask decreases heat it again. Continue the addition of $KMnO_4$ solution until a light pink colour just appears and persists for about 30 seconds. It is the end point. Note the burette reading. Repeat the titration to get at least two concordant readings.

(c) *Determination of the strength of supplied oxalic acid solution (unknown solution).*

Fill the burette with the same $KMnO_4$ solution. Pipette out 25 ml of supplied oxalic acid solution in a conical flask, add 25 ml of dil. H_2SO_4 and heat to about 60—80°C and then proceed with the titration just as above.

Observations

It is recorded here in tabular form as follows :

(A) Titration with known solution of oxalic acid

S.No.	Vol. of oxalic acid	Burette reading		Vol. of KMnO$_4$ used
		Initial	Final	
1.	25 ml	0·0 ml	.... ml	.... ml
2.	25 ml	0·0 ml	.... ml	.... ml
3.	25 ml	0·0 ml	.... ml	.... ml
				Say V$_1$ ml

(B) Titration with unknown solution of oxalic acid.

S.No.	Vol. of Oxalic acid solution	Burette reading		Vol. of KMnO$_4$ used
		Initial	Final	
1.	25 ml.	0·0 ml	.... ml	.... ml
2.	25 ml	0·0 ml	.... ml	.... ml
3.	25 ml	0·0 ml	.... ml	.... ml
				Say V$_2$ ml

Calculations : V$_1$ is the volume of KMnO$_4$ solution used with 25 ml of standard oxalic acid solution.

∵   Vol. × normality of KMnO$_4$ solution = vol. × normality of standard oxalic acid solution

∴   Normality of KMnO$_4$ solution $= \dfrac{25}{V_1} \times \dfrac{4w}{63}$

Equivalent wt. of oxalic acid is 63.

If V$_2$ is the vol. of KMnO$_4$ solution used with 25 ml of supplied oxalic acid solution

∵   Vol. × normality of supplied oxalic acid solution

$$= \text{Vol.} \times \text{normality of KMnO}_4 \text{ solution}$$

∴   Normality of supplied acid solution $= \dfrac{V_2 \times 25 \times 4 \times w}{25 \times V_1 \times 63}$

∴   Strength of supplied oxalic acid solution in gm/litre

$$= \frac{V_2 \times 25 \times 4 \times w \times 63}{25 \times V_1 \times 63}$$

Strength of supplied oxalic acid solution $= \dfrac{4 \times w \times V_2}{V_1} \text{ gm lit}^{-1}$

Result. The strength of given oxalic acid solution is .... gm lit^{-1}.

Precautions

1.   KMnO$_4$ solution attacks rubber hence it should always be taken in a glass stoppered burette.

2.   If sulphuric acid is not added in sufficient quantity, a brown ppt. of hydrated manganese dioxide is formed and end point is not properly detected and also the readings are not reliable.

3.   Due to intense colour of KMnO$_4$ solution, its lower meniscus is not properly read hence note the burette reading at the upper meniscus.

4.   At the end point, the pink tinge must persist for 30 seconds. Beyond this time, the pink colour may disappear due to reduction brought about by certain substances present in air.

5.   Take chemically pure KMnO$_4$ because ordinary KMnO$_4$ contains oxides of nitrogen which reduce KMnO$_4$.

6.   At the end point, we get a light pink colour. Add a drop of oxalic acid to it, if pink colour disappears, the end point is correct.

EXERCISE 5

To dertermine the strength in gm/litre of a given solution of ferrous ammonium sulphate (Mohr's salt) being provided with approx. N/30 KMnO$_4$ solution.

Theory. The acidified KMnO$_4$ being an oxidising agent converts ferrous ions to ferric ions according to the following equation :

$$2KMnO_4 + 3H_2SO_4 \longrightarrow K_2SO_4 + 2MnSO_4 + 3H_2O + 5O$$

$$10[FeSO_4(NH_4)_2SO_4.6H_2O] + 5O + 5H_2SO_4$$
$$\longrightarrow 5Fe_2(SO_4)_3 + 10(NH_4)_2SO_4 + 65H_2O$$

or

$$2MnO_4^- + 16H^+ + 10e \longrightarrow 2Mn^{2+} + 8H_2O$$
$$10Fe^{2+} \longrightarrow 10Fe^{3+} + 10e$$

$$2MnO_4^- + 16H^+ + 10Fe^{2+} \longrightarrow 2Mn^{2+} + 10Fe^{3+} + 8H_2O$$

This titration is done in cold because on heating, ferrous sulphate present in the Mohr's salt gets oxidised by air.

Procedure. It involves three steps :

(a) *Preparation of standard N/30 ferrous ammonium sulphate solution* (*known solution*).

Since equivalent wt. of ferrous ammonium sulphate is 392, hence weigh accurately about $\dfrac{392}{120}$ = 3.266 gm of the ferrous amm. sulphate for 250 ml. solution of N/30 strength. Transfer it to a 250 ml measuring flask, add first 5-10 ml of dil. H_2SO_4 and dissolve it. Add distilled water to make up the volume to 250 ml.

(b) *Standardisation of KMnO$_4$ solution by known ferrous ammonium sulphate solution*

(i) Rinse and fill the glass stoppered burette with $KMnO_4$ solution. Read the upper level of solution.

(ii) Pipette out 25 ml of known ferrous ammonium sulphate in a clean conical flask and add about 25 ml dil. H_2SO_4.

(**Note.** *Do not heat the contents of titration flask*).

(iii) Add the solution of $KMnO_4$ from burette very slowly and shake the flask after each addition until a *light pink colour* just appears. Note the reading of the burette. The correctness of end point is judged by adding a drop of ferr. amm. sulphate solution to the litration mixture. If the pink colour disappears, the end point is correct.

(iv) Repeat the titration until two concordant readings are obtained.

(c) *Determination of strength of given ferrous ammonium sulphate* (*unknown*) *solution.*

(i) Rinse and fill the burette with standardised $KMnO_4$ solution.

(ii) Pipette out 25 ml of unknown ferrous amm. sulphate solution in a titration flask and proceed for titration exactly in the same way as you have done with known ferrous amm. sulphate.

Observations

(A) Titration with known solution of ferrous ammonium sulphate (F.A.S.)

S.No.	Vol. of F.A.S.	Burette reading		Vol. of KMnO₄ used
		Initial	Final	
1.	25 ml	0·0 ml	.... ml	.... ml
2.	25 ml	0·0 ml	.... ml	.... ml
3.	25 ml	0·0 ml	.... ml	.... ml
				Say V$_1$

(B) Titration with given solution of F.A.S.

S.No.	Vol. of F.A.S.	Burette reading		Vol. of KMnO₄ used
		Initial	Final	
1.	25 ml	0·0 ml	.... ml	.... ml
2.	25 ml	0·0 ml	.... ml	.... ml
3.	25 ml	0·0 ml	.... ml	.... ml
				Say V$_2$

Calculations. Let

w = mass of ferrous amm. sulphate dissolved in 250 ml solution.

V_1 = Vol. of $KMnO_4$ used with 25 ml. known solution.

V_2 = Vol. of $KMnO_4$ used with 25 ml. of given solution.

$$\text{Normality of known soln.} = \frac{4 \times w}{392}$$

∵ Vol. × normality of $KMnO_4$ soln. = Vol. × normality of known soln.

∴ $$\text{Normality of } KMnO_4 \text{ soln.} = \frac{25 \times 4 \times w}{392 \times V_1}$$

∵ Vol. × normality of known soln. = Vol. × normality of $KMnO_4$ known soln.

$$\text{Normality of known soln.} = \frac{V_2 \times 25 \times 4 \times w}{25 \times 392 \times V_1}$$

Eq. wt. of ferrous amm. sulphate is 392.

$$\text{Strength} = \text{Normality} \times \text{Eq. wt.}$$

∴ Strength of given ferrous amm. sulphate solution in gm/litre

$$= \frac{4 \times w \times V_2}{V_1}$$

Result. The strength of given ferrous ammonium sulphate soution is .... gm lit^{-1}.

EXERCISE 6

To estimate ferrous (Fe^{+2}) and ferric (Fe^{+3}) ions given in the mixture.

Theory. First of all, the ferrous ions are estimated by $KMnO_4$ solution. After this, we reduce ferric ions in the same amount of fresh solution to ferrous state and this time titration against $KMnO_4$ estimates total ions present in the solution.

Procedure. Pipette out 10 ml of supplied solution and add nearly 10 ml of dil. H_2SO_4 in it and titrate it against standard $KMnO_4$ taken in burette. The apperance of light pink colour shows end point.

Now, take again 10 ml of supplied solution and boil it with zinc and sulphuric acid for 10 minutes so as to convert all ferric ions to ferrous state. Now this is titrated as above with $KMnO_4$. Note the end point of this also. This shows the ferrous ions originally present and ferrous ions obtained by reduction of ferric ions.

Observation. (*i*) Volume of $KMnO_4$ used first time = V_s ml

(*ii*) Volume of $KMnO_4$ used with reduced solution = V ml

Calculation. V_s volume used to oxidise ferrous ions already present, to ferric ions in 10 ml solution.

Let N_2 be normality of $KMnO_4$ (say 1/10) and N_1 normality of ferrous ions.

∴ $$N_1 = \frac{1/10 \times V_s}{10}$$

Now strength of ferrous ions in gm/litre

$$= \text{Eq. wt.} \times \text{Normality}$$

$$= 56 \times \frac{1}{10} \times \frac{V_s}{10}$$

$$= \frac{56 \times V_s}{100}$$

Similarly for Fe^{+++} ions :

Volume of $KMnO_4$ required for ferric ions only

$$= (V - V_s) \text{ ml.}$$

Strength of ferric ions in gm/litre $= 56 \times \dfrac{1}{10} \times \dfrac{(V - V_s)}{10}$

$$= \dfrac{56 \times (V - V_s)}{100} \text{ gm/lit.}$$

By knowing the total volume, the complete determination of the ions is possible. To determine the percentage composition, calculate these values in 100 ml.

EXERCISE 7

To determine the percentage of MnO₂ and available oxgyen in pyrolusite.

Theory. The ore pyrolusite which is mainly of the composition MnO_2, is reduced by means of oxalic acid taken in excess and the excess of the acid is titrated against standard solution of $KMnO_4$.

Procedure. Weigh the exact quantity of the powdered ore. Say it is 1.5 gm. To it, add 50 ml of normal solution of oxalic acid and 4-5 ml of conc. H_2SO_4. Heat the contents gently on water bath until all the black particles disappear. Filter this solution, throw the residue and dilute the filtrate to 250 ml. Titrate 25 ml of it against the standard solution of $KMnO_4$ taken in burette. The appearance of pink colour is the end point. Note this reading carefully.

[The use of $K_2Cr_2O_7$ is suggested to obtain better results and here internal indicator diphenyl amine is used which gives the blue colour at end point.]

Observation

Let :

1. Amount of pyrolusite ore taken $= 1.5$ gm
2. The volume of $KMnO_4$ used $= V$ ml
3. Strength of $KMnO_4$ $= \dfrac{N}{10}$

Calculation. We have taken 25 ml of reaction mixture which has been already diluted to 250 ml and titrated against $N/10$ $KMnO_4$. 25 ml consumes V ml of $N/10 KMnO_4$, that means we will require $10 \times V$ ml of $N/10$ $KMnO_4$ (or V ml. of $N\text{-}KMnO_4$) to neutralise 250 ml completely. As the oxalic acid is one normal, the volume of N-oxalic acid which has not reacted will be V ml.

The volume of the N-oxalic acid reacted with MnO_2

$$= (50 - V) \text{ ml}$$

∴ $(50 - V)$ ml of $N\text{-}H_2C_2O_4 \equiv (50 - V)$ ml of $N\text{-}MnO_2$

Now according to reaction

$$MnO_2 + H_2SO_4 \longrightarrow MnSO_4 + H_2O + \underset{\substack{\text{available} \\ \text{oxygen}}}{O}$$

Eq. wt. of $MnO_2 = \dfrac{1}{2} \times 86 \cdot 93$

$$= \dfrac{86 \cdot 93 \times (50 - V)}{2 \times 10} \text{ gm say } x \text{ gm}$$

This is the actual amount of MnO_2 in 1.5 gm of ore.

∴ Percentage of MnO_2 in the ore $= \dfrac{x \times 100}{1 \cdot 5}$

Now amount of available oxygen $= \dfrac{16 \times x}{86 \cdot 93} \text{ gm}$

∴ Percentage of available oxygen in the ore

$$= \dfrac{16 \times x \times 100}{86 \cdot 93 \times 1 \cdot 5}$$

EXERCISE 8

To determine calcium in a given calcium chloride solution.

Theory. The calcium is precipitated as calcium oxalate. The washed precipitate is decomposed by dil. H_2SO_4 to free oxalic acid which is then titrated with standard $KMnO_4$ solution.

Procedure

Precipitation of calcium oxalate. Pipette out 25 ml of supplied $CaCl_2$ solution (1N) into a 500 ml beaker and cover it with a watch-glass. Add a few drops (2-3 drops) of methyl red and then dil. NH_4OH until the solution is just neutral (*i.e.* colour is yellow). Add 5 ml dil. HCl and dilute the solution to about 150 ml and heat to boiling, always keeping the beaker covered with the watch glass. Prepare 6% solution of amm. oxalate and heat it to boiling in a separate beaker. When both solutions, of $CaCl_2$ and of amm. oxalate are in boiling condition, add the latter to the former, immediately after removing it from the flame, with constant stirring.

A white precipitate of CaC_2O_4 is obtained. Now add dil. NH_4OH drop by drop until neutral or slightly alkaline (colour is yellow). Stir continuously to avoid "*bumping*" and boil for a few minutes. This procedure gives a coarse-grained precipitate which is easily filtered off and washed. Allow the ppt. to settle for sometime. Check whether or not precipitation is complete by adding a few more drops of amm. oxalate solution to the supernatant liquid.

Decant off the supernatant liquid through gravimetric filter paper (better quality ordinary filter paper may also be used). Add hot water containing little amm. oxalate to the precipitate in the beaker (ppt. is appreciably soluble in pure water but practically insoluble in dilute amm. oxalate solution). Stir and then allow to stand for 2 minutes. Again decant off the clear liquid through the same filter paper. Repeat this process thrice. Finally, transfer the precipitate to the filter paper (make use of "*policeman*"). Wash the beaker for several times and pour all the washings on the filter paper. Wash the precipitate on filter paper with hot water (containing a little NH_4OH) until it becomes free from oxalate and chloride ions (test with $AgNO_3$ solution). Discard the washings.

Now place the beaker under the funnel, pierce the apex of filter paper with a *pointed* glass rod and wash the precipitate into the beaker with hot water. Pour about 25 ml of hot $2N$-H_2SO_4 into the filter paper in small portions at a time, taking care that acid comes in contact with every part of filter paper and that some of it is poured behind the double fold of the filter paper in case of any precipitate has lodged there. When every possible particle of ppt. has been removed from the filter paper, wash the paper with hot water. If the solution in the beaker is turbid (which is rarely formed), add further quantity of dil. H_2SO_4 to get a clear solution. Transfer in a 250 ml measuring flask and make up the volume.

Standardisation of $KMnO_4$ solution. Prepare approx. $N/10$ $KMnO_4$ solution and titrate it against standard oxalic acid solution. For details see exercise 1 (Chapter 12).

Titration of calcium oxalate solution against standardised $KMnO_4$ solution.

Pipette out 25 ml of the calcium oxalate solution into a conical flask, heat to about 60°—80°C, and titrate against standard $KMnO_4$ solution added from burette. Shake the flask after each addition. Avoid addition of large amount of $KMnO_4$ at one time otherwise a brown precipitate will be formed. At the end point, light pink colour appears. Repeat the titration to get at least two concordant reading.

Calculation

$$\text{Normality of } KMnO_4 \text{ solution} = xN$$
$$\text{Vol. of } CaC_2O_4 \text{ solution taken} = 25 \text{ ml}$$
$$\text{Vol. of } KMnO_4 \text{ used} = V \text{ ml}$$
$$V \text{ ml of } xN \text{ } KMnO_4 \equiv V_1 \text{ ml of N-}KMnO_4$$
$$\therefore \qquad V_2 = xV \text{ ml}$$
$$\text{Thus vol. of N-}KMnO_4 \text{ would be} = x \cdot V \text{ ml}$$

$$1 \text{ gm eq. } KMnO_4 \equiv 1 \text{ gm eq. } H_2C_2O_4$$
$$\equiv 1 \text{ gm eq. } CaC_2O_4$$
$$\equiv \frac{Ca}{2} \quad i.e., (20 \cdot 04)$$

or $\qquad 1000 \text{ ml of } N\text{-}KMnO_4 \equiv 20.04 \text{ gm Ca}$

or $\qquad 1 \quad " \quad " \quad " \quad \equiv 0.02004 \text{ gm Ca}$

$\therefore \qquad xV \quad " \quad \equiv 0.02004 \times x \text{ V gm Ca}$

Thus amount of calcium in the supplied volume of $CaCl_2$ soln.

$$= \frac{250 \times 0 \cdot 02004 \times x \times V}{25} \text{ gm}$$
$$= 10 \times 0.02004 \times x \text{ V gm}$$
$$= 0.02004 \times x \times V \text{ gm}$$

Also $\qquad 1 \text{ ml of } N\text{-}KMnO_4 = 0.0551 \text{ gm anhydrous } CaCl_2.$

Hence amount of anhydrous $CaCl_2$ present in the supplied volume of solution.

$$= 10 \times 0.0551 \times x \times V \text{ gm}.$$
$$= 0.55 \times x \times V \text{ gm}.$$

POTASSIUM DICHROMATE METHODS

In presence of an acid (dil. HCl or dil H_2SO_4) $K_2Cr_2O_7$, acts as an oxidising agent and each molecule of it gives up three atoms of oxygen available for oxidation of reducing agent as shown below :

$$K_2Cr_2O_7 + 4H_2SO_4 \longrightarrow K_2SO_4 + Cr_2(SO_4)_3 + 4H_2O + 3O$$
$$K_2Cr_2O_7 + 8HCl \longrightarrow 2KCl + 2CrCl_3 + 4H_2O + 3O$$
$$[Cr_2O_7]^{2-} + 8H^+ \longrightarrow 2Cr^{3+} + 4H_2O + 3O$$

That is in presence of H^+ ions, each Cr atom by gaining 3 electrons (provided by reducing agent) is reduced to trivalent Cr^{3+} ion (i.e., valency of Cr changes from +6 to + 3). Hence in acidic medium, the half reaction is represented as

$$[Cr_2O_7]^{2-} + 14H^+ + 6e \rightleftharpoons 2Cr^{3+} + 7H_2O$$

Indicators and end point detection

(i) **External indicator.** About 0.1% solution of potassium ferricyanide acts as external indicator. Use freshly prepared pot. ferricyanide indicator because the old solution of it becomes contaminated with pot. ferrocyanide.

By means of a glass rod, place a series of drops of the indicator on a white tile or a plate and near end point, take a drop of titration mixture and put it on one of these drops on tile and see the colour. If ferrous ions (Fe^{2+}) are present, a strong blue colour will be developed due to formation of a blue complex according to the following equation :

$$2K_3Fe(CN)_6 + 3FeSO_4 \longrightarrow Fe_3[Fe(CN)_6]_2 + 3K_2SO_4$$
ferrous ferricyanide
(Dark blue complex)

Repeating the process, the blue colour becomes fainter. When a drop of solution first fails to produce blue colour on the indicator's drop, this will be the end point i.e. all Fe^{2+} ions are oxidied to Fe^{3+} ones. Usually at the end point, the colour of the indicator's drop becomes light brownish yellow due to reaction of indicator with Fe^{3+} ions to produce brown coloured ferric ferricyanide complex.

$$Fe_2(SO_4)_3 + 2K_3[Fe(CN)_6] \longrightarrow 3K_2SO_4 + 2Fe[Fe(CN)_6]$$
Brown

(ii) **Internal indicators.** The most comonly used internal indicators are :

(a) 1% solution of diphenyl amine in chemically pure conc. H_2SO_4 (sp gravity 1.84).

The end point is obtained as follows :

Take 25 ml of solution of ferrous ions, add about 25 ml dil. H_2SO_4, 6-8 drops of indicator and then 5 ml syrup of phosphoric acid. Now add $K_2Cr_2O_7$ solution from burette until the last drop causes a change from green (due to formed chromium sulphate) to blue-violet which persists even on shaking. This is the end point.

(b) N-phenyl anthranilic acid (0.005 M solution) is also used as an indicator. Take 25 ml of solution of reducing agent in a conical flask, acidify with about 25 ml H_2SO_4 (dilute) and then add few drops of the indicator solution. Titrate the solution against $K_2Cr_2O_7$ and when yellowish green solution changes sharply to purple red, indicates the end point.

Oxidised form of this indicator is purple-red and reduced form is colourless.

EXERCISE 9

> To determine the strength of a given solution of ferrous ammonium sulphate by titrating it against $K_2Cr_2O_7$ solution and using pot. ferricyanide as an external indicator.

Theory. Acidic $K_2Cr_2O_7$ is a strong oxidising agent. When it is added to ferrous ammonium sulphate solution containing dil. H_2SO_4 or dil HCl, only $FeSO_4$ is oxidised and $(NH_4)_2SO_4$ remains unchanged hence not shown in the equation. The reactions taking place are as follows :

$$K_2Cr_2O_7 + 4H_2SO_4 \longrightarrow K_2SO_4 + Cr_2(SO_4)_3 + 4H_2O + 3O$$
$$6FeSO_4 + 3H_2SO_4 + 3O \longrightarrow 3Fe_2(SO_4)_3 + 3H_2O$$

The complete reaction is :

$$K_2Cr_2O_7 + 6FeSO_4 + 7H_2SO_4 \longrightarrow 3Fe_2(SO_4)_3 + K_2SO_4 + Cr_2(SO_4)_3 + 7H_2O$$

or $\qquad (Cr_2O_7)^{2-} + 14H^+ + 6Fe^{2+} \longrightarrow 6Fe^{3+} + 2Cr^{3+} + 7H_2O$

At the end point, the indicator fails to produce blue colour when treated with a drop of titration mixture.

Procedure : This involves following steps.

(a) *Preparation of approx. N/30 ferrous amm. sulphate solution* (*known solution*).

Equivalent wt. of it is 392·12 hence weigh accurately about 392.12/120 = 3·267 gm and transfer to a 250 ml measuring flask. Dissolve it in distilled water, add 5 ml of conc. H_2SO_4 and then make up the volume. Shake well. Normality of this solution will be :

$$= \frac{4 \times w}{392 \cdot 12}$$

where w is mass of ferr. amm. sulphate dissolved in 250 ml of water.

(b) *Standardisation of $K_2Cr_2O_7$ solution with known ferrous ammonium sulphate solution.*

Rinse and fill burette with $K_2Cr_2O_7$ solution. Note the initial reading. Pipette out 25 ml of standard ferrous amm. sulphate solution in a clean conical flask and add to it about equal volume of $N-H_2SO_4$. Put a series of drops of indicator (pot. ferricyanide solution) with the help of glass rod, on white tile or a plate. Run in solution from the burette and when nearly 15 ml. is added, take out from the flask, a drop of titration mixture and mix it with one of the drops of indicator. A strong blue colour will be produced. Continue the titrations and repeat the process at different intervals and when the colour of indicator-drop does not give blue colour, the end point is reached. Note the reading of the burette. Repeat the titration to get the exact end point.

(c) *Titration with unknown ferr. amm. sulphate solution.*

For this, proceed exactly in the same way as above.

Observations

This is recorded in tabular form.

(A) Titration with known solution of F.A.S.

S.No.	Vol. of F.A.S. used	Burette reading		Vol. of $K_2Cr_2O_7$ used
		Initial	Final	
1.	25 ml	0·0 ml	.... ml	.... ml
2.	25 ml	0·0 ml	.... ml	.... ml
3.	25 ml	0·0 ml	.... ml	.... ml
				Say V_1 ml

(B) Titration with unknown solution of F.A.S.

S.No.	Vol. of F.A.S.	Burette reading		Vol. of $K_2Cr_2O_7$ used
		Initial	Final	
1.	25 ml	0·0 ml	.... ml	.... ml
2.	25 ml	0·0 ml	.... ml	.... ml
3.	25 ml	0·0 ml	.... ml	.... ml
				Say V_2 ml

Calculations

Let $\qquad V_1$ = Volume of $K_2Cr_2O_7$ used against known solution.

$\qquad\qquad V_2$ = Volume of $K_2Cr_2O_7$ used against unknown solution.

$\qquad\qquad N_1$ = Normality of $K_2Cr_2O_7$ solution.

According to normality equation

∵   Vol × normality of known soln. = Vol. × normality of $K_2Cr_2O_7$ solution.

∴ $$25 \times \frac{4 \times w}{392 \cdot 12} = V_1 \times N_1$$

Hence, $$N_1 = \frac{25}{V_1} \times \frac{4 \times w}{392 \cdot 12}$$

In the titration against unknown solution.

$$25 \times N = V_2 \times N_1$$

where N is the normality of unknown solution.

$$N = \frac{V_2}{25} \times \frac{25}{V_1} \times \frac{4 \times w}{392 \cdot 12}$$

Strength of unknown solution. $$= 392 \cdot 12 \times \frac{V_2}{V_1} \times \frac{4 \times w}{392 \cdot 12}$$

$$= \frac{4 \times w \times V_2}{V_1} \text{ gm/litre}$$

Result. The strength of unknown ferrous ammonium sulphate solution is .... gm lit^{-1}.

EXERCISE 10

To determine the strength of a given solution of ferrous ammonium sulphate (F.A.S.) by titrating it against potassium dichromate solution using N-phenyl anthranilic acid as an internal indicator. To standardise $K_2Cr_2O_7$ solution, prepare N/30 ferrous amm. sulpate solution.

Theory. Same as in exercise 9.

Procedure

(a) *Preparation of N/30 F.A.S. solution.* Same as in exercise 9.

(b) *Standardisation of $K_2Cr_2O_7$ solution.* Rinse and fill the burette with $K_2Cr_2O_7$ solution

Pipette out 25 ml of prepared F.A.S. solution and add to it equal volume of dilute H_2SO_4 and 6 to 8 drops of N-phenyl anthranilic acid indicator. Add slowly dichromate solution from the burette to it until the colour of the titration mixture just turns to violet red. This is end point of the titration. Note it carefully. Repeat the experiment until two concordant readings are obtained. This reading will be V_1.

(c) *Titration with supplied F.A.S. solution.*

For this, proceed exactly in the same away as above. This reading will be V_2.

Observations and Calculations

Write observations and calculations as in exercise 9.

Result. The strength of supplied F.A.S. solution will be

$$= \frac{4 \times w \times V_2}{V_1} \text{gm lit}^{-1}$$

EXERCISE 11

> *To determine the strength in gms/litre of a given solution of $K_2Cr_2O_7$ provided with ferrous amm. sulphate solution and N-phenyl anthranilic acid solution as an internal indicator.*

Theory. The same as in previous exercise. At the end point the green solution changes to light purple-red.

Procedure

(a) *Preparation of $K_2Cr_2O_7$ (approx. N/30) solution.*

Equivalent wt. of $K_2Cr_2O_7$ is $49 \cdot 03$, hence weigh accurately about $49 \cdot 03/120 = 0 \cdot 41$ gm of $K_2Cr_2O_7$, transfer to a 250 ml measuring flask, dissolve in distilled water and then make up to 250 ml. Shake well.

Normality of known $K_2Cr_2O_7$ solution $= \dfrac{4 \times w}{49 \cdot 03}$

where w = mass of $K_2Cr_2O_7$ dissolved in 250 ml water.

(b) *Standardisation of ferrous amm. sulphate solution with known $K_2Cr_2O_7$ solution.*

Rinse and fill the burette with prepared $K_2Cr_2O_7$ solution. Note the initial reading. pipette out 25 ml of ferrous amm. sulphate soln. in a conical flask, add 25 ml of $2N-H_2SO_4$ and then 6-8 drops of N-phenyl anthranilic acid indicator. Add $K_2Cr_2O_7$ solution from burette very slowly and shake the flask after each addition until the solution becomes green (due to formation of chromium sulphate). Now add the dichromate solution drop by drop slowly until the addition of one drop of $K_2Cr_2O_7$ solution suddenly changes the green colour to light purple-red. This is end point. The change of colour is very sharp when end point reaches. Repeat the titration using each time the same drops of indicator until two concordant readings are obtained :

(c) *Titration with unknown $K_2Cr_2O_7$ solution.*

Proceed exactly as above.

Observations. As in previous exercise.

Calculation

Let w = mass of $K_2Cr_2O_7$ dissolved in 250 ml water

V_1 = Vol. of known $K_2Cr_2O_7$ solution used with 25 ml solution of ferrous ammonium sulphate

V_2 = Vol. of given $K_2Cr_2O_7$ solution used with 25 ml solution of ferrous ammonium sulphate

Normality of $K_2Cr_2O_7$ solution $= \dfrac{4 \times w}{49 \cdot 03}$

$\therefore$ Vol. × normality of ferrous amm. sulphate solution

$$= \text{Vol.} \times \text{normality of known } K_2Cr_2O_7 \text{ solution}$$

$\therefore$ Normality of Ferr. amm. sulphate solution

$$= \frac{V_1 \times 4 \times w}{25 \times 49 \cdot 03}$$

Similarly:

$V_2 \times$ normality of given $K_2Cr_2O_7$ solution

$$= 25 \times \text{normality of ferr. amm. sulphate}$$

$V_2 \times$ normality of given $K_2Cr_2O_7$ solution

$$= \frac{25 \times 4 \times w \times V_1}{25 \times 49 \cdot 03}$$

or normality of given $K_2Cr_2O_7$ solution $= \dfrac{4 \times w \times V_1}{V_2 \times 49 \cdot 03}$

for determination of strength multiply it by eq. wt. (i.e., 49.03)

$\therefore$ Strength in gm/litre of given $K_2Cr_2O_7$

$$= \frac{4 \times w \times V_1}{V_2}$$

i.e., Strength in gm/litre of given $K_2Cr_2O_7$ solution.

$$= \frac{4 \times \text{wt. of } K_2Cr_2O_7 \text{ dissolved in 250 ml wt.} \times \text{Vol. of known } K_2Cr_2O_7}{\text{Vol. of given } K_2Cr_2O_7 \text{ solution}}$$

[**Note.** When the end point has been obtained, add a drop of ferrous ammonium sulphate to the titration mixture. If the purple-red tinge changes to green, the end point is correct. In this way, the correctness of the end point can be judged.]

Iodine Titration

Iodimetric and Iodometric Titrations

The redox-titration using iodine directly or indirectly as an oxidising agent are called **Iodine Titrations.** These are of two types :

1. Iodimetric titrations. Iodimetric titrations are defined as those iodine titrations in which a standard iodine solution is used as an oxidant and iodine is directly titrated against a reducing agent. Iodimetric procedures are used for the determination of reducing agents like thiosulphates, sulphites, arsenites and stannous chloride etc. by titrating them against standard solution of iodine taken in a burette. Some cases of oxidation-reduction reactions are given as under :

(i) $\qquad 2Na_2S_2O_3 + I_2 \longrightarrow Na_2S_4O_6 + 2NaI$

(ii) $\qquad Na_2SO_3 + I_2 + H_2O \longrightarrow Na_2SO_4 + 2HI$

(iii) $\qquad Na_3AsO_3 + I_2 + H_2O \longrightarrow Na_3AsO_4 + 2HI$

2. Iodometric titrations. Iodometric titrations are defined as those iodine titrations in which some oxidising agent liberates iodine from an iodine solution and then liberated iodine is titrated with a standard solution of a reducing agent added from a burette. In such titrations, a neutral or an acidic solution of oxidising agent is employed. The amount of iodine liberated from an iodide, (i.e. KI) is equivalent to the quantity of the oxidising agent present. Iodometric titrations are used for the determination of $CuSO_4$, $K_2Cr_2O_7$, $KMnO_4$, ferric ions, antimonate ions, H_2O_2, MnO_2, bromine and chlorine etc. The equations for some of the reactions are as follows :

(i) $\qquad 2CuSO_4 + 4KI \longrightarrow Cu_2I_2 + 2K_2SO_4 + I_2$

(ii) $$2KMnO_2 + 3H_2SO_4 \longrightarrow K_2SO_4 + 2MnSO_4 + 3H_2O + 5O$$
$$10KI + 5H_2SO_4 + 5O \longrightarrow 5K_2SO_4 + 5H_2O + 5I_2$$

(iii) $$K_2Cr_2O_7 + 4H_2SO_4 \longrightarrow K_2SO_4 + Cr_2(SO_4)_3 + 4H_2O + 3O$$
$$6KI + 3H_2SO_4 + 3O \longrightarrow 3K_2SO_4 + 3H_2O + 3I_2$$

In the above reactions, the liberatred iodine is titrated with a standard sodium thiosulphate.

$$2Na_2S_2O_3 + I_2 \longrightarrow Na_2S_4O_6 + 2NaI$$

Equivalent weights of :

(i) *Sodium thiosulphate* (hypo) :

The oxidation state of S_2 in $Na_2S_2O_3$ $= +4$

"   "   "   " S_2 in $Na_2S_4O_6$ $= +5$

Therefore,

Equivalent wt. of $Na_2S_2O_3$ $= $ M.W. $= 158 \cdot 1062$

Equivalent wt. of $Na_2S_2O_3 \cdot 5H_2O$ $= $ M.W. $= 248 \cdot 1832$

(ii) *Sodium arsenite, Na_3AsO_3* :

$$2Na_3AsO_3 + 2I_2 + 2H_2O \longrightarrow 2Na_3AsO_4 + 4HI$$

$\therefore$ Equivalent wt. of Na_3AsO_3 $= \dfrac{M.W.}{2}$
$$= 95 \cdot 9441$$

(iii) *Arsenious oxide, As_2O_3* :

Since As_2O_3 is not completely soluble in water, it is dissolved in NaOH and then the solution is titrated with a standard solution of iodine. The equivalent weight of As_2O_3 is calculated as follows :

$$As_2O_3 + 6NaOH \longrightarrow 2Na_3AsO_3 + 3H_2O$$
$$2Na_3AsO_3 + 2I_2 + 2H_2O \longrightarrow 2Na_3AsO_4 + 4HI$$

or $$As_2O_3 + 2I_2 + 2H_2O \longrightarrow As_2O_5 + 4HI$$

Equivalent wt. of As_2O_3 $= \dfrac{M}{4}$
$$= 49 \cdot 4595$$

2.   Equivalent weight of

(i) *Copper sulphate, $CuSO_4 . 5H_2O$* :

$$2CuSO_4 + 4KI \longrightarrow 2CuI + I_2 + 2K_2SO_4$$

Equivalent wt. of hydrated copper sulphate
$$= \text{M.W. or } 249 \cdot 6786$$

(ii) *Potassium dichromate $K_2Cr_2O_7$*

$$K_2Cr_2O_7 + 4H_2SO_4 \longrightarrow K_2SO_4 + Cr_2(SO_4)_3 + 4H_2O + 3O$$
$$6KI + 3H_2SO_4 + 3O \longrightarrow 3K_2SO_4 + 3H_2O + 3I_2$$

$\therefore$ Equivalent wt. of $K_2Cr_2O_7$ $= \dfrac{M}{6} = 49 \cdot 0319$

(iii) *Iodine* :

$$I_2 + H_2O \longrightarrow 2HI + O$$

$\therefore$ Equivalent wt. of iodine $= \dfrac{2 \times 127}{2}$
$$= 127.$$

(iv) *Potassium permanganate* :

$$2KMnO_4 + 8H_2SO_4 + 10KI \longrightarrow 6K_2SO_4 + 2MnSO_4 + 8H_2O + 5I_2$$

As discussed earlier, equivalent wt. of $KMnO_4$

$$= \dfrac{M.W.}{5} = 31.6$$

Preparation of Standard Solutions

Standard solution of sodium thiosulphate (hypo) approx. N/30.

Since hydrated sodium thiosulphate crystals are efflorescent, these lose their water of crystallisation on keeping in air and decompose in solution due to bacterial action. The hydrated sodium thiosulphate is not primary standard and its standard solution can not be prepared directly by weighing.

Since equivalent wt. of $Na_2S_2O_3.5H_2O$ is 248·1832, for 250 c.c. of N/30 solution, we have to dissolve $\frac{248 \cdot 18}{120} = 2.068$ gm. The exact strength of this solution is found by standardising it with a standard solution of an oxidising agent *e.g.*, $K_2Cr_2O_7$ and $CuSO_4.5H_2O$ etc.

Standard solution of iodine N/30 approx.

Since iodine is volatile at ordinary temperature, its standard solution can not be prepared directly by weighing. Iodine is not completely soluble in water hence it is dissolved in water containing KI with which iodine forms a soluble brown complex KI_3 which liberates iodine readily.

$$KI + I_2 \rightleftharpoons KI_3$$

Since eq. wt. of iodine is 127, for 250 ml of N/120 solution, we will require 1·59 gm of iodine. Weigh approximately 1.6 gm of iodine and transfer it to a beaker containing about 3 gm solid KI and some water and shake well. When all the iodine has dissolved, transfer the solution to a 250 ml measuring flask and make up the volume. The exact strength is found out by standardising it with standard solution of sodium thiosulphate (hypo).

Indicator and End point Detection

Iodine produces, with starch solution an intensely blue iodostarch complex. In all iodine titrations, freshly prepared 1% solution of starch is used as an indicator.

In *iodimetric titrations*, the starch solution is added to the solution of reducing agent taken in the conical flask and then the iodine solution is added from the burette. The *end point* is indicated by the appearance of blue colour.

In *iodometric titrations*, the iodide (*e.g.* KI) solution and the oxidising agent are taken into a conical flask. The standard solution of reducing agent *e.g.*, sodium thiosulphate is run in from the burette. When the brown colour is faint yellowish, the starch is added. It produces a deep blue complex. Now, the thiosulphate solution is added drop by drop with proper shaking. The end point will be indicated when blue colour just disappears.

Precaution

(*a*) Iodine solution attacks rubber ; hence it should always be taken in a glass stoppered burette

(*b*) In iodometric titrations, the starch solution must be added just before the end point i reached, *i.e.* when the colour is faint yellow. It should never be added in the beginning of titratio because it will give a permanent deep blue complex which does not easily disappear even after th addition of a large quantity of thiosulphate and thus we can not identify the end point correctly.

EXERCISE 12

> *To determine the strength in gm/litre of a given copper sulphate solution being provided with an approx. N/30 sodium thiosulphate (hypo) solution.*

Theory. The strength of copper sulphate is determine by iodometric method. When KI is adde to a solution of copper sulphate a white cuprous iodide, Cu_2I_2 is precipitated and an equivalent amou of iodine is liberated. The free iodine is titrated with standard solution of hypo, using starch as indicator. As soon as all the liberated iodine has been reduced to iodide (NaI), the blue colour iodo-starch complex will disappear and the colour of precipitate in conical flask will be white that of cuprous iodide. This indicates end point. The chemical reactions taking place are :

$$2CuSO_4 + 4KI \longrightarrow Cu_2I_2\downarrow + 2K_2SO_4 + I_2$$
$$\text{white ppt.}$$

$$2Na_2S_2O_3 + I_2 \longrightarrow Na_2S_4O_6 + 2NaI$$

sodium
tetrathionate

[The titration reactions take place in the neutral medium hence any mineral acid present in the solution should be neutralized by adding sodium carbonate solution and excess of sodium carbonate must be neutralized by adding dil. acid. This should be done before starting the titration.]

Procedure.

(a) *Preparation of standard approx. N/30 CuSO$_4$.5H$_2$O solution (known solution).*

Since equivalent wt. of CuSO$_4$.5H$_2$O is 249·678, for 250 ml of N/30 solution, we require 249·678/120 = 2·08 gm of copper sulphate crystals. Weigh accurately about 2.1 gm of copper sulphate and transfer it to a 250 ml measuring flask. Dissolve it in distilled water then add a little acetic acid (5 ml) to check the hydrolysis and make up the volume. Shake well.

If the weight of CuSO$_4$.5H$_2$O is w gm, the correct normality of copper sulphate solution will be

$$\frac{4 \times w}{249 \cdot 678}.$$

(b) *Standardisation of hypo solution (approx. N/30) with the help of known copper sulphate solution.*

Fill the burette with hypo solution and note the initial reading. Pipette out 25 ml of known copper sulphate solution in a clean conical flask. Now add 1 gm solid KI (or 1 ml of 100% KI solution), mix well and cover the mouth of conical flask by watch glass or filter paper and allow the mixture to stand for 2-5 minutes in the dark. The solution becomes brown in colour due to liberated iodine. Now titrate the liberated iodine with the hypo solution added from burette. The brown colour of iodine fades slowly and when only a very faint yellow colour (light straw colour) remains, add 1 ml of starch solution. This immediately forms a deep blue iodo-starch complex. Now, add further hypo solution drop by drop, shaking well in whirling motion and titrate till the blue colour just disappears. If the colour does not return within 10 seconds, this will indicate end-point. Note this burette reading. Repeat the titration until two concordant readings are obtained.

(c) *Determination of the strength of given copper sulphate solution (unknown solution) :* Fill the burette with the same hypo solution. Pipette out 25 ml of unknown copper sulphate solution in a clean conical flask and titrate in the same way as you have done with known copper sulphate solution.

Observations. These can be recorded in tabular form :

(A) Titration with known solution of copper sulphate.

S.No.	Vol. of Copper sulphate taken	Burette reading		Vol. of hypo used
		Initial	Final	
1.	25 ml	0·0 ml	.... ml	.... ml
2.	25 ml	0·0 ml	.... ml	.... ml
3.	25 ml	0·0 ml	.... ml	.... ml
				V_1 = ... ml

(B) Titration with given solution of copper sulphate.

S.No.	Vol. of Copper Sulphate taken	Burette reading		Vol. of hypo used
		Initial	Final	
1.	25 ml	0·0 ml	.... ml	.... ml
2.	25 ml	0·0 ml	.... ml	.... ml
3.	25 ml	0·0 ml	.... ml	.... ml
				V_2 = ... ml

Calculations. Suppose the volume of hypo solution used in the titration with known solution is V_1 ml.

Volume × normality of hypo soln.   = Volume × normality of known copper sulphate solution

$$V_1 \times N_1 = 25 \times \frac{4 \times w}{249 \cdot 678}$$

or

$$N_1 = \frac{25 \times 4 \times w}{V_1 \times 249 \cdot 678}$$

Therefore, normality of hypo solution $= \frac{25 \times 4 \times w}{V_1 \times 249 \cdot 678}$

We also know :

Volume × normality of unknown copper sulphate solution

= Volume × normality of hypo solution

$$25 \times N_2 = V_2 \times \frac{25 \times 4 \times w}{V_1 \times 249 \cdot 678}$$

or

$$N_2 = \frac{25 \times V_2 \times 4 \times w}{V_1 \times 25 \times 249 \cdot 678}$$

Since the equivalent wt. of $CuSO_4.5H_2O$ is 249.678

∴ $\text{Strength} = \frac{25 \times V_2 \times 4 \times w \times 249 \cdot 678}{V_1 \times 25 \times 249 \cdot 678}$

Strength in gm./litre $= \frac{4 \times w \times V_2}{V_1}$.

Result. The strength of unknown copper sulphate solution is .... gm lit^{-1}

EXERCISE 13

To determine the strength in gm/litre of a given $K_2Cr_2O_7$ solution being provided with approx. N/30 hypo solution.

Theory. When $K_2Cr_2O_7$ reacts with KI in presence of acid (dil. H_2SO_4 or dil. HCl), the dichromate is reduced to green chromic salt and liberates an equivalent amount of iodine which is titrated with a standard hypo solution using starch solution as an indicator. This principle is employed here to determine the strength of $K_2Cr_2O_7$ *iodometrically.* When the end point is reached, the blue colour of iodo starch complex suddenly disappears and a pure green solution due to chromic salt remains.

$$K_2Cr_2O_7 + 7H_2SO_4 + 6KI \longrightarrow 4K_2SO_4 + Cr_2(SO_4)_3 + 7H_2O + 3I_2$$
$$2Na_2S_2O_3 + I_2 \longrightarrow Na_2S_4O_6 + 2NaI$$

Hence $K_2Cr_2O_7 \equiv 3I_2 \equiv 6Na_2S_2O_3$

Procedure. It involves three stages :

(a) *Preparation of standard, approx. N/30 $K_2Cr_2O_7$ solution (known solution).*

Since equivalent wt. of $K_2Cr_2O_7$ is 49·0319, for 250 ml of N/30 solution, weigh accurately abou 49·0319/120 = 0.408 gm $K_2Cr_2O_7$ crystals and transfer it to 250 ml measuring flask. Dissolve it distilled water and then make up the volume with distilled water. Shake well.

Normality of known $K_2Cr_2O_7$ solution $= \frac{4 \times w}{49 \cdot 0319}$

where w = mass of $K_2Cr_2O_7$ dissolved in 250 ml water.

(b) *Standardisation of hypo solution with the help of known $K_2Cr_2O_7$ solution.*

Fill the burette with hypo solution and note the initial reading. Pipette out 25 ml of know $K_2Cr_2O_7$ solution into a clean titrating flask, add about 1 gm solid KI (or 2 ml of 100% KI solution) a then 10 ml of dil H_2SO_4 and shake well. Cover the mouth of the conical flask by watch glass and all it to stand for 5 minutes in dark. The mixture becomes brown due to liberation of iodine. Now titrate the liberated iodine with hypo solution from burette, shaking well after each addition. The brown colour fades as the hypo solution is added. When only a faint yellowish green colour remains, add 2 ml of freshly prepared 1% solution of starch. Continue the titration with hypo solution, swirling constantly until the blue colour suddenly disappears and a pale green solution is left. This is end point. Note the burette reading (final). Repeat the titration to get at least two concordant readings.

(c) *Determination of the strength of given $K_2Cr_2O_7$ solution (unknown solution).*

Fill the burette with the same hypo solution. Pipette out 25 ml of unknown $K_2Cr_2O_7$ solution and add about 1 gm solid KI (or 2 ml. of 100% KI solution) and then 10 ml of dil H_2SO_4 and proceed for the titration as described earlier.

Observations. These can be recorded in tabular form :

(A) Titration with known solution of potassium dichromate

S.No.	Vol. of $K_2Cr_2O_7$ taken	Burette reading		Vol. of hypo soln. used
		Initial	Final	
1.	25 ml	0·0 ml	.... ml	.... ml
2.	25 ml	0·0 ml	.... ml	.... ml
3.	25 ml	0·0 ml	.... ml	.... ml
				$V_1 =$ ..... ml

(B) Titration with a given solution of $K_2Cr_2O_7$

S.No.	Vol. of $K_2Cr_2O_7$ taken	Burette reading		Vol. of hypo soln. used
		Initial	Final	
1.	25 ml	0·0 ml	.... ml	.... ml
2.	25 ml	0·0 ml	.... ml	.... ml
3.	25 ml	0·0 ml	.... ml	.... ml
				$V_2 =$ .... ml

Calculations. Suppose V_1 is the volume of hypo solution required for 25 ml of known $K_2Cr_2O_7$ solution.

$\because$ Vol. × normality of hypo solution = Vol. × normality of known $K_2Cr_2O_7$ solution

$$V_1 \times \text{normality of hypo solution} = 25 \times \frac{4 \times w}{49 \cdot 0319}$$

$\therefore$ Normality of hypo solution $= \dfrac{25 \times 4 \times w}{V_1 \times 49 \cdot 0319}$

V_2 is the volume of standardised hypo solution used for 25 ml of unknown $K_2Cr_2O_7$ solution.

$\because$ Vol. × normality of unknown $K_2Cr_2O_7$ solution

                    = Vol. × normality of hypo solution

25 × normality of known $K_2Cr_2O_7$ solution

$$= V_2 \times \frac{25 \times 4 \times w}{V_1 \times 49 \cdot 0319}$$

or   Normality of unknown $K_2Cr_2O_7$ solution

$$= \frac{V_2 \times 25 \times 4 \times w}{25 \times V_1 \times 49 \cdot 0319}$$

Since equivalent wt. of $K_2Cr_2O_7$ is 49·0319

$\therefore$ Strength of unknown $K_2Cr_2O_7$ in gm/litre

$$= \frac{V_2 \times 25 \times 4 \times w \times 49 \cdot 0319}{25 \times V_1 \times 49 \cdot 0319}$$

$$= \frac{4 \times w \times V_2}{V_1}$$

Result. The strength of given $K_2Cr_2O_7$ solution is .... gm lit^{-1}.

EXERCISE 14

To determine the strength in gm/litre of a given potassium permanganate solution being provided with approx. N/20 hypo solution.

Theory. It is obtained *iodometrically*. When $KMnO_4$ in acidic medium reacts with KI an equivalent amount of iodine is set free. The liberated iodine turns the solution brown as iodine dissolves in KI forming brown KI_3 complex which gives out iodine quickly. The liberated iodine is titrated against a standard solution of hypo.

$$2KMnO_4 + 8H_2SO_4 + 10KI \longrightarrow 6K_2SO_4 + 2MnSO_4 + 8H_2O + 5I_2$$

$$2Na_2S_2O_3 + I_2 \longrightarrow Na_2S_4O_6 + 2NaI$$

Hence $2KMnO_4 \equiv 5I_2 \equiv 10Na_2S_2O_3$

Procedure. It involves three stages :

(a) *Preparation of standard solution approx. N/20 of potassium permanganate.* Dissolve about 3.95 gm of copper sulphate in a 250 ml measuring flask and make up the volume.

Normality of prepared $KMnO_4$ solution $= \dfrac{4 \times w}{31 \cdot 6}$

where w = mass of $KMnO_4$ dissolved in 250 ml

(b) *Standardisation of hypo solution approx. N/20 with the help of prepared standard $KMnO_4$ solution.* For this, we proceed in the same way as described in copper sulphate determination. (Exercise 1).

(c) *Determination of the strength of given $KMnO_4$ solution.* Fill the burette with hypo solution. Note the initial reading. Pipette out 25 ml of the given $KMnO_4$ solution into a clean titration flask, add 10 ml of H_2SO_4 and then 1 gm solid KI (or 2 ml of 100% KI solution). Shake and cover the mouth of titration flask with watch glass and wait for 5 minutes. Titrate the mixture by adding the hypo solution from burette and shake well after each addition. Stop the addition of hypo when the colour of the solution becomes light yellow. Now add 2 ml of starch solution and continue the addition of hypo solution carefully until the blue colour disappears. This is end-point. Note the burette reading (final). Repeat the titration until two concordant readings are obtained.

Observations. These can be recorded in tabular form :

(A) Titration with known solution of $KMnO_4$

S.No.	Vol. of $KMnO_4$ taken	Burette reading		Vol. of hypo used
		Initial	Final	
1.	25 ml	0·0 ml	.... ml	.... ml
2.	25 ml	0·0 ml	.... ml	.... ml
3.	25 ml	0·0 ml	.... ml	.... ml
				$V_1 =$ ..... ml

(B) Titration with a given solution of $KMnO_4$.

S.No.	Vol. of $KMnO_4$	Burette reading		Vol. of hypo used
	taken	Initial	Final	
1.	25 ml	0·0 ml	.... ml	.... ml
2.	25 ml	0·0 ml	.... ml	.... ml
3.	25 ml	0·0 ml	.... ml	.... ml
				$V_2 =$ .... ml

Calculations. Let V_1 is the volume of hypo used against 25 ml of known solution of $KMnO_4$.

$$\text{The normality of hypo solution } = \frac{25 \times 4 \times w}{V_1 \times 31 \cdot 6}$$

Suppose V_2 is the vol. of hypo solution used with 25 ml. of the given $KMnO_4$ solution.

$$25 \times \text{normality of } KMnO_4 \text{ solution } = V_2 \times \text{normality of hypo solution.}$$

$$\text{Normality of } KMnO_4 \text{ solution } = \frac{4 \times w \times 25 \times V_2}{25 \times 249 \cdot 678 \times V_1}$$

Since equivalent wt. of $KMnO_4$ is 31.6

$$\text{The normality in gm/litre of } KMnO_4 = \frac{4 \times w \times 31 \cdot 6 \times V_2}{31 \cdot 6 \times V_1}$$

i.e., Strength in gm/litre of $KMnO_4$ solution

$$= \frac{4 \times w \times V_2}{V_1}$$

Result. The strength of unknown $KMnO_4$ solution is .... gm lit^{-1}.

EXERCISE 15

> To determine the strength in gm/litre of a given arsenious oxide solution being provided with approx. N/30 iodine solution.

Theory. Arsenious oxide, As_2O_3 is estimated *iodimetrically*. In the neutral medium, As_2O_3 is quantitatively oxidised to As_2O_5 by iodine.

$$As_2O_3 + 2I_2 + 2H_2O \rightleftharpoons As_2O_5 + 4HI$$

As the above reaction is reversible, some As_2O_5 reduces back to As_2O_3 *i.e.* HI checks the complete oxidation of As_2O_3. Hence to accomplish the complete oxidation of As_2O_3 by I_2, the formed HI should be continuously removed *i.e.*,solution is maintained almost neutral. This condition is obtained by adding $NaHCO_3$. Sodium bicarbonate does not react with iodine. As NaOH and Na_2CO_3 react with iodine, they cannot be added to neutralise HI.

$$NaHCO_3 + HI \longrightarrow NaI + H_2O + CO_2$$

The solution of As_2O_3 is prepared in NaOH as it does not completely dissolve in water. The excess of NaOH is removed by adding dil. acid (H_2SO_4 or HCl) and using phenolphthalein to check the excess addition of dil. acid.

$$As_2O_3 + 6NaOH \longrightarrow 2Na_3AsO_3 + 3H_2O$$
Sod. arsenite

As_2O_3 is converted into soluble sod. arsenite. When this solution is titrated against standard iodine solution using excess of $NaHCO_3$ and starch as indicator, As_2O_3 is quantitatively determined.

$$2Na_3AsO_3 + 2I_2 + 2H_2O \rightleftharpoons 2Na_3AsO_4 + 4HI$$

Procedure

(a) *Preparation of standard approx. N/30 solution of arsenite oxide* (*known solution*) : Since the equivalent wt. of As_2O_3 is 49·4595, hence for 250 ml of N/30 solution, we require 49·4595/120 = 0.412 gm of As_2O_3. Weigh out accurately about 0.41 gm of As_2O_3. Dissolve it in 3-5 ml of dil NaOH (10%) in a beaker (warm if necessary). Dilute the solution by about 25 ml. of distilled water, add 2 drops of phenolphthalein and then add dil. H_2SO_4 or dil. HCl dropwise with shaking untill the pink colour just disappear. Pour the contents of beaker and its washings into a 250 ml. measuring flask and then make up the volume. Shake well.

Let 'w' be the mass of As_2O_3 dissolved in 250 ml water.

$$\text{Then, normality of solution} = \frac{4 \times w}{49 \cdot 4595}.$$

(b) *Standardisation of iodine solution with the help of known As_2O_3 solution* : Take iodine solution in a glass stoppered burette and note the initial reading. Take out 25 ml of known As_2O_3 solution in a clean conical flask, add 2 gm of solid $NaHCO_3$ (20 ml of its saturated solution) and then 2 ml of freshly prepared 1% starch solution. Mix well and titrate with iodine solution, swirling the titrating flask after each addition of iodine until a permanent light blue colour is just obtained. This is the end point. Note the burette reading (final). Repeat the titration until two concordant values are obtained.

Suppose V_1 is the volume of iodine used with 25 ml of known As_2O_3 solution. Hence the normality of I_2 solution $= \dfrac{25 \times 4 \times w}{49 \cdot 4595 \times V_1}$.

(c) *Determination of strength in gm/litre of a given As_2O_3 solution (unknown solution) using standardised iodine solution.*

Fill the burette with the same iodine solution. Note the intial reading. Take out 25 ml of un-known As_2O_3 solution in a titrating flask, add 2 gms of solid $NaHCO_3$ (20 ml. of its saturated solution), 2 ml of starch solution, and then complete the titration as above.

Calculation. Let V_2 be the volume of standardised iodine solution used with 25 ml of unknown As_2O_3 solution.

$25 \times$ normality of unknown As_2O_3 solution

$$= V_2 \times \text{normality of } I_2 \text{ solution.}$$

Normality of unknown As_2O_3 solution

$$= \frac{V_2 \times 25 \times 4 \times w}{25 \times 49 \cdot 4595 \times V_1}$$

Since Equivalent wt. of As_2O_3 = 49·4595

Hence strength of unknown As_2O_3 solution in gm/litre

$$= \frac{4 \times w \times V_2}{V_1}.$$

EXERCISE 16

To estimate percentage of 'available chlorine' in bleaching powder.

Theory. When a dilute mineral acid (HCl, H_2SO_4 or CH_3COOH) reacts with bleaching powder sets free the chlorine which is known as *available chlorine*. The liberated chlorine from bleaching powder liberates equivalent amount of I_2 from KI which is titrated against standard hypo solution.

$$Ca(OCl)Cl + 2HCl \longrightarrow CaCl_2 + H_2O + Cl_2$$
$$\text{(available chlorine)}$$

$$Cl_2 + 2KI \longrightarrow 2KCl + I_2$$
$$I_2 + 2Na_2S_2O_3 \longrightarrow Na_2S_4O_6 + 2NaI$$

1 gm equivalent of hypo $\equiv$ 1 gm equivalent of chlorine.

Procedure. (*a*) *Preparation of solution of bleaching powder.* Weigh accurately about 5 gm of bleaching powder. Transfer it to a porcelain dish, make its paste and transfer it quantitavely (*i.e.*, without loss) with the help of water into 500 ml measuring flask and then make up the volume with the distilled water. Shake well. A turbid solution of bleaching powder is thus obtained.

(*b*) *Preparation and standardisation of hypo solution.* For this, see the previous copper sulphate exercise.

(*c*) *Titration of bleaching powder solution against standard hypo solution (iodometrically).* Rinse and fill the burette with standardised hypo solution and note the initial reading. Pipette out 50 ml of turbid solution into a 250 ml titrating flask, add 2 gm of solid KI and then 15 ml of 4N-H_2SO_4. Mix well, cover the mouth of flask with watch glass and allow it to stand in dark for 5 minutes. Titrate the liberated iodine by adding hypo from burette. When only a light yellow colour remains, add 2 ml of 1% starch solution (freshly prepared) and then continue the addition of hypo solution drop by drop with shaking until the blue colour just disappears. This is the end point. Note the burette reading. Repeat the titration to get at least two concordant values.

Calculations

Let w = wt. of bleaching powder dissolved in 50 ml of water.

V = volume of standard (say x.N) hypo solution used with 50 ml bleaching powder solution.

x = Normality of hypo solution.

Now 1 gm equivalent of available chlorine = 1 gm equivalent of hypo solution.

i.e., Vol. × normality of available chlorine = Volume × normality of hypo solution

∴ Normality of available chlorine $= \dfrac{V \times xN}{50}$

Since equivalent wt. of chlorine is = 35·46

∴ Strength of available chlorine in gm/litre $= \dfrac{V \times 35 \cdot 46 \times x}{50}$

or amount of available chlorine in 500 ml. of prepared bleaching powder solution.

$$= \frac{V \times 35 \cdot 46 \times x}{2 \times 50} \text{ gms.}$$

Percentage of available chlorine in the bleaching powder (*i.e.*, in 100 gm)

$$= \frac{V \times 35 \cdot 46 \times 100 \times x}{2 \times 50 \times w}$$

$$= \frac{35 \cdot 46 \times V \times x}{w}$$

Titrations Involving Precipitation Reactions
ARGENTOMETRIC TITRATIONS

The titrations which are based on the formation of insoluble precipitates (when the solutions of two reacting substances are brought into contact with each other) are called **Precipitation Titrations**. For example, when a solution of silver nitrate is added to a solution of sodium chloride, a white precipitate of silver chloride is formed.

$$AgNO_3 + NaCl \longrightarrow AgCl \downarrow + NaNO_3$$

Similarly, when $AgNO_3$ is added to ammonium thiocyanate a white precipitate of silver thiocyanate is formed.

$$AgNO_3 + NH_4CNS \longrightarrow AgCNS \downarrow + NH_4NO_3$$

Both of these reactions may be used for the volumetric estimation of silver if completion of the precipitation is judged with the help of a suitable indicator. Such volumetric estimations involving silver nitrate are called **Argentometric titrations**.

These are of two types :

1. **Titration of AgNO$_3$ against NaCl :**

This titration is done by two methods :

(a) By the use of potassium chromate as an indicator (**Mohr's method**).

(b) By use of **Adsorption indicator** (using Fluorescein).

2. **Titration of AgNO$_3$ against a thiocyanate solution :**

In this titration, potassium or ammonium thiocyanate is titrated against silver nitrate using ferric alum or ferric nitrate as indicator. The method is known as **Volhard's method**.

Determination of equivalent Weights

As it is evident from the equations,

$$AgNO_3 + NaCl \longrightarrow AgCl + NaNO_3$$
$$169.9 \quad 58.45$$

$$AgNO_3 + NH_4CNS \longrightarrow AgCNS + NH_4NO_3$$
$$169.9 \quad 76$$

$$AgNO_3 + KCNS \longrightarrow AgCNS + KNO_3$$
$$169.9 \quad 97.2$$

One mole of AgNO$_3$ reacts with one mole each of NaCl, NH$_4$CNS or KCNS. Since equivalent weights of silver nitrate is equal to its molecular weight (because one atom of Ag is equivalent to one atom of hydrogen) the equivalent weight of NaCl, NH$_4$CNS or KCNS will be equal to their molecular weights i.e.,

$$AgNO_3 \equiv NaCl \equiv NH_4CNS \equiv KCNS$$
$$169.9 \quad 58.45 \quad\quad 76 \quad\quad\quad 97.2$$

Preparation of Standard Solutions

(i) *Preparation of standard silver nitrate solution approx. N/30.* Since equivalent weight of silver nitrate is 169·9 hence for N/20 solution, 169.9/30 = 5.663 gm. of AgNO$_3$ has to be dissolved per litre.

To prepare 250 ml of silver nitrate of approx. strength N/30, weigh about 1.42 gm of A.R. silver nitrate and dissolve it in distilled water.

(ii) *Preparation of standard sodium chloride approx. N/30 solution.*

For one litre solution, weigh $\dfrac{58\cdot45}{30}$ = 1.95 gm sodium chloride.

To prepare 250 ml of sodium chloride solution, weigh about 0.49 gm A.R. sodium chloride and dissolve it in distilled water.

(iii) *Preparation of standard approx. N/30 ammonium thiocyanate solution :*

Weigh $\dfrac{76}{30}$ = 2.5 gm ammonium thiocyanate and dissolve it in one litre of distilled water.

(iv) *Preparation of standard approx. N/30 potassium thiocyanate solution :*

Weigh about $\dfrac{97\cdot2}{30}$ = 3.24 gm potassium thiocyanate and dissolve it in one litre water.

Note. Solution of ammonium and potassium thiocyanates have to be standardised by titrating against silver nitrate solution because standard solutions of these salts cannot be prepared directly by weighing as these salts are hygroscopic.

Preparation of Indicator Solutions

(a) Potassium Chromate Solution. Dissolve 5 gm. A.R. potassium chromate in 100 ml of water.

(b) Fluorescein indicator. Dissolve 0.2 gm of fluorescein in 100 ml of 70% alcohol.

(c) Ferric alum indicator. Prepare a saturated solution of A.R. ferric ammonium sulphate to which a few drops of 6 N nitric acid has been added.

(d) Dichlorofluorescein solution. Dissolve 0·1 gm dichlorofluorescein in 100 ml of 60-70 % alcohol.

EXERCISE 17

> *To determine the strength in gm/litre of a given silver nitrate solution being provided with an approx. N/30 chloride solution by Mohr's method.*

Theory. When silver nitrate solution is added to a solution of sodium chloride containing a few drops of potassium chromate, white silver chloride is precipitated initially. As soon as all the chloride ions have been precipitated out, even a drop of silver nitrate added in excess, gives a red precipitate of silver chromate. This indicates the end point.

The solubility product of silver chloride is lower than that of silver chromate. Hence so long the chloride ions are available, the less soluble silver chloride is precipitated ; the Ag^+ ions are not sufficient for silver chromate to be precipitated. As soon as the chloride ions have been precipitated out, even a slight excess of Ag^+ ions produces insoluble silver chromate which is red in colour.

$$2Ag^+ + CrO_4^{--} \longrightarrow Ag_2CrO_4$$
silver chromate
(red)

Procedure. The titration requires three stages :

(a) *Preparation of standard approx N/30 silver nitrate solution* (*known solution*) :

Weigh accurately about 1.42 gm silver nitrate and transfer it to a 250 ml measuring flask. Dissolve the salt in distilled water and make up the volume.

Let the weight of $AgNO_3$ be w gm. The normality of $AgNO_3$ will be $\dfrac{w \times 4}{169 \cdot 9}$

(b) *Standardisation of sodium chloride solution with the help of known $AgNO_3$ solution* :

Fill the burette with known $AgNO_3$ solution and note the initial reading. Now pipette out 25 ml. of given sodium chloride solution in a clean conical flask. Add 1 ml potassium chromate indicator and now add slowly $AgNO_3$ solution from the burette shaking the flask constantly. A white precipitate of AgCl is obtained. After the addition of a few ml $AgNO_3$, a red colour appears in the flask but disappears quickly upon shaking. As the end point is approached, the red colour disappears slowly. Continue the addition of $AgNO_3$ solution drop by drop till a permanent reddish brown colour is obtained. The colour persists even after shaking. This will be the end point. Note the reading and repeat the titration until two concordant readings are obtained.

Suppose the volume of $AgNO_3$ solution used in the titration is V_1 ml.

Since,

$$\text{Volume} \times \text{normality of NaCl} = \text{Volume} \times \text{normality of } AgNO_3$$

$$25 \times N_1 = \frac{V_1 \times w \times 4}{169 \cdot 9}$$

$$N_1 = \frac{V_1 \times w \times 4}{25 \times 169 \cdot 9}$$

$$\text{Thus normality of NaCl solution} = \frac{V_1 \times w \times 4}{25 \times 169 \cdot 9}$$

(c) *Determination of the strength of given silver nitrate solution* (*unknown solution*) :

Fill the burette with unknown silver nitrate solution. Pipette out 25 ml of same sodium chloride solution in the conical flask. Add one ml of potassium chromate indicator and complete the titration exactly in the same way as you have done with known $AgNO_3$ solution.

Suppose V_2 ml of $AgNO_3$ solution is used in the titration.

Calculation.

Volume × normality of unknown $AgNO_3$ solution (N_2)

$$= \text{Volume} \times \text{normality of NaCl}$$

$$V_2 \times N_2 = \frac{25 \times V_1 \times w \times 4}{25 \times 169 \cdot 9}$$

$$N_2 = \frac{25 \times V_1 \times w \times 4}{V_2 \times 25 \times 169 \cdot 9}$$

Since equivalent weight of $AgNO_3$ is $169 \cdot 9$.

hence, $$S_2 = \frac{25 \times V_1 \times w \times 4 \times 169 \cdot 9}{V_2 \times 25 \times 169 \cdot 9}$$

$$\text{Strength in gm/litre} = \frac{4 \times \omega \times V_1}{V_2}$$

Result. The strength of given $AgNO_3$ solution is .... gm lit^{-1}.

Note. If $AgNO_3$ solution is given as an intermediate, one may prepare standard NaCl solution and the procedure of titration remains the same.

EXERCISE 18

> *To determine strength of given $AgNO_3$ solution being provided with an approx. N/30 NaCl solution using* **adsorption indicator.**

Theory. When a solution of silver nitrate is added to a chloride solution containing fluorescein as an indicator, a white precipitate of AgCl is initially formed. A primary adsorbed layer is formed owing to adsorption of Cl^- ions by AgCl. This secondary adsorption holds oppositely charged ions like Ag^+ ion. Thus when Ag^+ ions become in excess, then by secondary adsorption try to hold negative ions like NO_3^-. If fluorescein is present in solution, the negative fluorescein ion is more strongly adsorbed than the nitrate ion. Thus a pink complex is formed between Ag^+ and fluoresceinate ion. At the end point, therefore, when even a trace of Ag^+ becomes in excess, a pink or red solution is obtained.

Procedure. This includes three stages :

(a) *Preparation of standard $AgNO_3$ solution.* Prepare as described earlier.

(b) *Standardisation of sodium chloride solution with the help of known $AgNO_3$ solution.* Fill the burette with $AgNO_3$ solution. Pipette out 25 ml of given sodium chloride solution into a conical flask. Add 10 drops of fluorescein indicator and now add $AgNO_3$ solution from the burette slowly rotating the flask constantly. With the progress of the titration white precipitate of AgCl settles to large extent and a faint pink colour begins to appear in the solution. As the end point approaches, the pink colour becomes more and more pronounced. Continue addition of drops of $AgNO_3$. When the end point is reached, the whole precipitate suddenly becomes pink or red.

The strength of NaCl is calculated in the same way as described in Mohr's method.

(c) *Determination of strength of unknown silver nitrate solution* : Fill the burette with AgNO solution. Pipette 25 ml. of the NaCl solution in conical flask, add fluorescein indicator and perform the titration as above.

Calculation. Calculate as in Exercise 17.

Note. In the titration using fluorescein as an indicator, the solution should not be concentrate. Good resu are obtained when solutions are approx. N/30 or low. Further, the solutions should not be acidic.

In place of fluorescein, dichlorofluorescein may be used as an indicator.

EXERCISE 19

> ***To titrate silver nitrate against a thiocyanate solution* (Volhard's method).**

In this method, a given thiocyanate solution is standardised with the help of prepared standard silver nitrate solution using *ferric alum* as an indicator. The standardised thiocyanate is then titrated against given silver nitrate and then the strength of the latter is calculated as usual.

Theory. When a solution of potassium or ammonium thiocyanate is added to a silver nitrate solution containing some ferric alum, a white precipitate of silver thiocyanate is produced initially. But when all the Ag^+ ions have been precipitated out, even a slight excess of thiocyanate added further, gives a reddish-brown colouration due to the formation of ferric thiocyanate.

$$AgNO_3 + NH_4CNS \longrightarrow AgCNS + NH_4NO_3$$
White ppt.

or

$$Ag^+ + CNS^- \longrightarrow AgCNS$$
$$Fe^{3+} + 3CNS^- \longrightarrow Fe(SCN)_3$$
Red colour

Procedure. The titration involves three steps :

(a) *Preparation of approx. N/20 standard silver nitrate solution (known solution)* :

Prepare as described earlier.

(b) *Standardisation of potassium thiocyanate solution* :

Fill the burette with given potassium thiocyanate solution. Pipette out 25 ml of the known silver nitrate solution in a conical flask. Add 5 ml of dil. nitric acid and 1 ml of ferric alum indicator. Now add thiocynate solution from the burette dropwise and upon each addition of thiocyanate swirl the flask gradually. A white precipitate of silver thiocyanate is produced and also a red colour appears temporarily. This colour disappears when the solution is shaken. At the end point, the precipitate settles at the bottom while a permanent reddish supernatant liquid is formed in the flask. Note the burette reading and repeat the titration to get two concordant values.

Suppose the normality of known $AgNO_3$ solution is $\dfrac{4 \times w}{169 \cdot 9}$

where w = Mass of $AgNO_3$ in 250 ml solution.

Let V_1 be the volume of thiocyanate required to titrate 25 ml of $AgNO_3$ solution.

∴ Volume × normality of thiocyanate

$$= \text{Volume} \times \text{Strength of } AgNO_3$$

∴ $\quad V_1 \times$ normality of thiocynate $= \dfrac{25 \times w \times 4}{169 \cdot 9}$

Normality of thiocyanate $= \dfrac{25 \times w \times 4}{V_1 \times 169 \cdot 9}$

(c) *Determination of the strength of the given $AgNO_3$ solution* :

Fill the burette with standardised thiocyanate solution. Take 25 ml of unknown silver nitrate solution in conical flask. Add 5 ml of dil. nitric acid and 1 ml of the indicator and proceed with the titration as described above.

Calculation. Suppose V_2 is the volume of thiocyanate required for unknown $AgNO_3$ solution.

Volume × normality of unknown $AgNO_3$

$$= \text{Volume} \times \text{normality of thiocyanate}$$

$$25 \times \text{normality of } AgNO_3 = \dfrac{V_2 \times 25 \times w \times 4}{V_1 \times 169 \cdot 9}$$

$$\text{Normality of unknown AgNO}_3 = \frac{V_2 \times 25 \times w \times 4 \times 169 \cdot 9}{25 \times V_1 \times 169 \cdot 9}$$

i.e., Strength of unknown $AgNO_3$ in gm/litre

$$= \frac{4 \times w \times V_2}{V_1}$$

Result. The strength of unknown $AgNO_3$ solution is ....,........ gm lit^{-1}.

Complexometric Titrations

Many metals form complexes with such reagents as contain appropriate ligands. EDTA (Ethylenediamine tetracetic acid) is one such reagent which forms complexes with metals. In form of its disodium salt, it is used to estimate Ca^{++} and Mg^{++} ions.

Dissociation contant values indicate that EDTA behaves like a dicarboxylic acid.

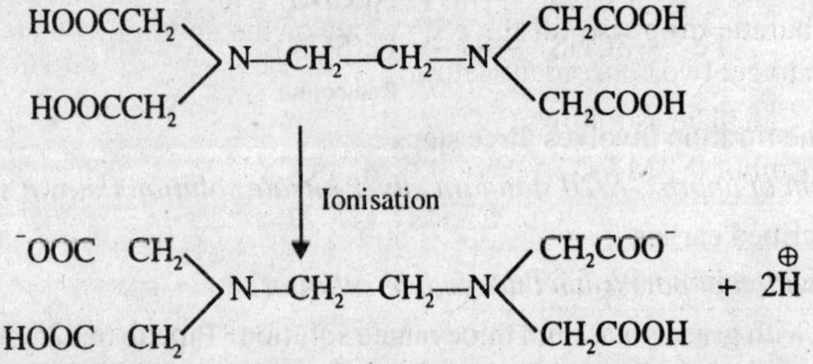

i.e., two of its carboxyl groups are stongly acidic and the other two hydrogens are released during formation of complex as shown :

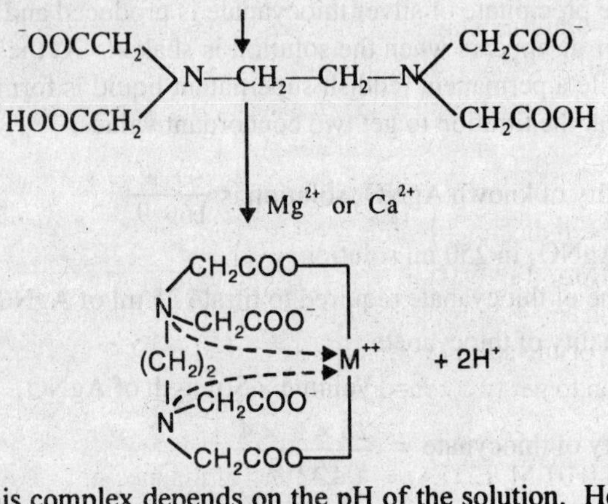

The ionisation of this complex depends on the pH of the solution. Hence in this titration, pH sensitive indicators are used. In the estimation of Ca^{++} and Mg^{++} with EDTA, an azo dye called Eriochrome Black-T is used as an indicator. This forms a metal indicator complex, the stability of which is lower than that of the metal EDTA complex. The solution is initially red. As the titration proceeds, the metal ions form more stable complex with EDTA, hence the indicator anion goes into solution. As the indicator anion accumulates, the colour changes from red to blue at the end point. Since the action of the indicator and formation of metal—EDTA complex is governed by pH hence pH of the solution is kept nearly constant by adding a suitable buffer.

EXERCISE 20

| To estimate calcium with EDTA. |

1. *0.1 M calcium solution* : Use A. R. $CaCO_3$. Mol. weight of $CaCO_3 = 100 \cdot 09$ hence for 0·
M solution 10 gm. $CaCO_3$ are needed per litre.

To prepare 250 ml solution, weigh 2.5 gm calcium carbonate and transfer it to a 250 ml measuring flask. Add dil. hydrochloric acid in drops till there is effervescence and the salt completely dissolves. Now add distilled water to make up the volume.

2. *0·1 M EDTA solution* : For this, use dry disodium salt of EDTA (mol. wt. 372·25). To prepare 0.1 M solution 37.2 gm (approx.) of the salt is needed in one litre.

To prepare 250 ml solution, weigh 9.3 gm of salt and dissolve it in distilled water and make up the volume.

3. *Ammonia Buffer Solution* : Dissolve 70 gm of ammonium chloride in 570 ml concentrated ammonia and dilute to one litre with distilled water.

4. *Indicator Solution* : Dissolve 0·4 gm Eriochrome—Black T in 100 ml methanol.

Procedure. In a 250 ml conical flask, pipette out 10 ml of prepared calcium carbonate solution. Add 20 ml distilled water, 19 ml buffer solution and 5-6 drops of the indicator solution. Add EDTA solution from the burette dropwise till the red colour of the solution changes to permanent blue. Repeat the titration to get two concordant values.

Calculation : Calculate as follows :

$$1 \text{ ml of } 0·1 \text{ M EDTA} = 4·008 \text{ mg of Ca}$$

EXERCISE 21

> **To estimate Magnesium with EDTA.**

Prepare following reagents :

1. *0·01 M Magnesium sulphate Solution.* Use A.R. $MgSO_4.7H_2O$ (Mo. wt. 246·5). For a 0·01 M solution 2·46 gm salt is needed per litre.

To prepare 250 ml solution, weigh 0·62 gm salt, dissolve it in distilled water and make up the volume in 250 ml measuring flask.

2. *0·01 M EDTA Solution* : Weigh 0·93 gm disodium salt of EDTA and prepare 250 ml solution in distilled water.

3. *Buffer Solution* : Prepare as descibed in case of calcium.

4. *Indicator* : Same indicator which is used in calcium.

Procedure of titration. Take 10 ml. of prepared magneiusm solution in a conical flask. To this add 10 ml. buffer solution and 4-5 drops of the indicator. Now run in EDTA solution from the burette gradually till the colour of the solution changes from red to blue and there remains not even a red tinge. Repeat the titration to get two concordant values.

Calculation

$$1 \text{ ml } 0·01 \text{ M EDTA} = 0·243 \text{ mg of magnesium}$$

The hardness of water

Water contaminated with Ca or Mg salts is called hard water as it gives hardly any lather with soap. When impurities are in form of bicarbonate salts, they are easily removed by boiling (boiling decomposes bicarbonates to insoluble carbonates), water is said to be *temporary hard*. When they are in form of chlorides or nitrates etc. it is called permanent hard.

Expression for hardness

Since these salts are in trace amounts, hardness is expressed in parts per million (PPM). Although both, Ca and Mg may be present, hardness is expressed in form of only calcium carbonate. This means number of grams of $CaCO_3$ present in 10^6 gms of water. This is equivalent to number of milligrams of this salt present in a litre of the sample water (in 1000 gm of water).

EXERCISE 22

To determine the total, permanent and temporary hardness of a given sample of water by
complexometric method using EDTA (disodium salt of ethylene-diamine tetraacetic acid).

Indicator. Eriochrome-Black-T indicator is used. It joins with metal ions forming wine red colour.
When titrated with EDTA solution, metal ions leave the indicator and form chelate with EDTA. At the
end point, blue colour is observed.

The entire process can be discussed stepwise.

(a) *Preparation of standard EDTA solution.* EDTA can not be used as primary standard as it
contains a trace of moisture. Its formula is $Na_2H_2C_{10}H_{12}O_8N_2 2H_2O$ (mol. wt. 372·25). Its 0·01 M to
0·05 M can be conveniently used.

To prepare 0·01 M, weigh 9.3 gm of EDTA, dissolve in distilled water and make up the volume
to 250 ml water. It can be standardised with zinc chloride or calcium chloride. To standardise with
$CaCl_2$, weigh accurately 0.2500 gm of anhydrous $CaCO_3$ (dried at 105°C in oven) and transfer it in
150 ml. of flask containing 100 ml of distilled water. Add nearly 6N HCl dropwise until all $CaCO_3$
dissolves and gives clear solution of $CaCl_2$. Shake it well and make up the volume to 250 ml by
distilled water. The strength of this solution is 0.02 N.

Now, pipette 25 ml of $CaCl_2$ solution in titration flask. Add ~10 ml of buffer solution, 5-6 drops
of the indicator (refer Ex. 20) and titrate it against EDTA solution taken in burette. Note the volume
where solution turns to blue and find the exact normality of EDTA solution using formula :

$$N_{EDTA} V_{EDTA} = 25 \times 0·02$$

This normality will be used in calculation.

(b) *Determination of total hardness of water.*

Pipette out 25 ml of sample water in a titration flask, add ~ 5 ml of buffer solution (pH = 10) and
4-5 drops of indicator. Titrate it against EDTA solution. At end point, colour changes from red to
blue. Repeat it and note the readings.

(c) *Titration for permanent hardness of water.*

Take another 25 ml of sample of same water and boil it for about 15-20 minutes. This will make
temporary hardness to precipitate. Filter it off and titrate the filtrate left as described above.

(a) *Calculation for total hardness*

No. of millimoles of EDTA = its molarity × volume used

= MV

No. of millimoles of calcium carbonate = MV in 25 ml of water

No. of gms of $CaCO_3$ in 1 litre = $100 \times 40 \times MV$

No. of miligrams in 1000 gms of water = $1000 \times 100 \times 40$ MV

This is equivalent to total hardness in P.P.M.

Calculate similarly for permanent hardness and then temporary hardness.

10

Gravimetric Analysis

General Operations. The gravimetric analysis is that branch of analytical chemistry in which the end product of the analysis is weighed and this weighed mass has direct relationship with the substance to be quantitatively analysed. In most of the operations of this analysis, the material to be analysed is precipitated by some suitable reagent which after drying is weighed. The different steps in the gravimetric analysis along with precipitation have been summarized below with the apparatus used therein.

1. Precipitation. The substance whose gravimetric estimation is to be done is taken in the soluble form. To this solution, we add some reagent which precipitates the desired component under certain experimental conditions. These conditions are often :

(*i*) The solution from which estimation is to be done is diluted by distilled water (if concentrated). This avoids the coprecipitation.

(*ii*) The solution is heated gently before adding the precipitating reagent. This saves the precipitate to go in colloidal state and also helps in giving good crystals. If the conditions permit, it is desirable to heat the solution and the precipitating reagent both.

(*iii*) It is suggested to add the precipitating reagent slowly and with constant stirring as it keeps the degree of saturation small and this provides large crystals.

(*iv*) The precipitating reagent is added so much as to precipitate the constituent completely. However, it should not be in too much quantity as the precipitate may partially be soluble in this stage.

(*v*) The crystalline precipitate so obtained is digested (heated gently on water bath) for nearly an hour. In some cases, where the possibility of post prec[,] itation does not exist it requires overnight digestion.

In brief, we take the solution in pyrex beaker subject it to slight heating. Now to it, we add slowly hot precipitating reagent while stirring with glass rod. When assessed the precipitation of the constituent to be complete, we leave the solution as such to allow the precipitate to settle. To assume that the desirable matter has been completely precipitated, we apply complete precipitation test.

Complete Precipitation Test. When all the precipitate has settle down, we add few drops of the precipitating reagent by the side of beaker without disturbing the solution at all. If in the upper part of solution which were clear, no further precipitate appears, it is supposed that complete precipitation has occurred. If precipitate comes, we add some more quantity of it and once again test after the precipitate is being settled.

2. Filtration. After precipitating the substance completely, the next step is filtration which simply involves the isolations of the precipitated form quantitatively from the solution. For this purpose, we use a medium which permits the solution to separate completely from the precipitate. For this, we can use porous media. These are commonly filter paper, sintered glass crucibles, asbestos mats and platinum mats etc. in which the first two are of common applications. Both these techniques have been discussed separately here.

(*a*) *Filtration by Sintered Crucible.* Sintered crucible is a unique type of crucible which has porous base (containing small pores which can permit only solution).

This crucible is attached to the mouth of a conical flask by the help of rubber padding. The flask has one outlet which is attached with the lower mouth of filter pump by rubber tube. The main mouth of filter pump is attached with tap. When fast current of water passes through pump, it takes some air of the conical vessel reducing its pressure which affects the quick filtration.

Though these crucibles provide the most useful filtration media but these have certain limitations. After filtration is being achieved, it requires cleaning of the pores of the crucible. This becomes tedious particularly where the porous medium becomes clogged and cannot be cleaned easily. If this process of cleaning is advanced this affects the porous surface.

(*b*) *Filtration by Filter Paper.* The filter paper is supposed to be the most widely accepted filtering medium. For this purpose, varieties of the filter papers are available with respect to the size of the pores in it. The most commonly used filter paper is Whatman No. 42.

The filter paper is wrapped properly and fitted in a funnel air-tightly. The solution containing precipitate is poured slowly which permits the solution leaving only precipitate behind.

This process of filtration is not only slower but also fails to give promising results in micro estimations. Silica of the filter paper is drawn away slightly while pouring the solution and that reacts with the carbon during ash treatment. By and large, it affects the weighing product.

3. **Washing.** In fact, washing and filtration are done simultaneously. The washing is done in order to remove the *undesirable* ions sticking on the precipitate which may give higher yield. It is usually done with the help of wash bottle which provides the fast current of the water (by blowing with mouth) that agitates the precipitate thoroughly and the ions sticking on it are also disturbed. These undesirable ions being lighter take time to settle and meanwhile, these can be filtered off.

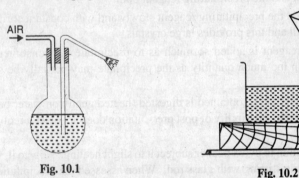

Fig. 10.1   Fig. 10.2

After the precipitate has been completely washed, we filter the supernatant portion (upper half) without washing and then precipitate is washed by fast current of hot water containing some percentage of precipitating reagent or other suitable reagent. After each washing the filtrate obtained is taken to test ions which are to be removed. So long as these ions remain coming in filtrate, the washing is required. When these stop, it shows that no ions are now in the precipitate. *The excess of washing is also harmful as it surely takes away slight part of the precipitate to the filtrate.* The washing and filtration techniques have been shown in the diagrams.

After the complete washing has been achieved the next step is to transfer the precipitate completely in the filter paper attached to the funnel. The technique is most simple and is itself evident from the figure given alongside.

4. **Drying.** The drying of the precipitate is done by placing the funnel containing filter paper with precipitate on the metallic chimney which is kept on tripod stand with

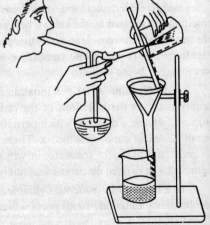

Fig. 10.3. Transferring of the precipitate.

wire gauze. Now the low flame of the burner is maintained to heat gently. The mouth of funnel is covered with a piece of paper in which small holes are made to permit the water vapours to go out.

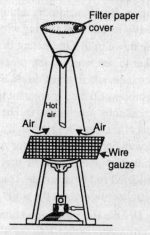

Fig. 10.4. Drying of the precipitate on an air-cone.

The drying may more successfully be done by putting it on steam or in electric oven. If the washing is done with water having nitric acid, the filter paper gets cracked on drying. Washing should be done only by water finally.

5. **Ignition and Weighing.** The dried precipitate, as much as possible, is collected on glazed paper and covered with inverted funnel. Some portion of the precipitate still remains sticking on the filter paper. Ignite this paper to remove carbon completely and then convert back the precipitate by suitable reagent as it is mostly reduced in ignition. The whole operation is called *ash-treatment* which is done as follows :

Fold the paper and hold it by means of pair of tongs and burn completely in non-sooty flame. The ash obtained so, is collected in a silica crucible which is then heated strongly until the ash becomes white. Finally, get this reduced precipitate to the original condition. For example, some of $BaSO_4$ sticking to filter paper is converted to BaS after ash treatment. It is then treated with one drop of conc. HCl and then with 1-2 drops of conc. H_2SO_4 to convert it to $BaSO_4$.

Now transfer the precipitate kept on glazed paper to the crucible and weigh the crucible finally.

Note. The crucible is weighed empty 2-3 times after heating and cooling to get the constant weight before we are going to use it in the analysis.

The burden of ash treatment can be removed simply by replacing silica crucible with sintered one but as discussed earlier it is not suitable in all cases.

EXERCISE 1 : ESTIMATION OF BARIUM

> *To estimate Barium as barium suphate in barium chloride solution.*

Discussion. The precipitation is done by adding hot dilute H_2SO_4 (approximately 0·5 M) to a hot solution of barium salt. The reaction in case of $BaCl_2$ solution is :

$$BaCl_2 + H_2SO_4 \longrightarrow BaSO_4 \downarrow + 2HCl$$

Barium sulphate has very low solubility in water (about 3 mg per litre at room temperature). The presence of mineral acid increases the solubility due to formation of acid sulphate ion HSO_4^-. Even then, the precipitation is done in weakly acid medium to prevent the precipitation of carbonate, chromate and phosphate. Acid decomposes carbonate and chromate. The phosphate remains soluble in acidic medium. To avoid coprecipitation, both, salt solution as well as H_2SO_4 are kept hot. $BaCl_2$ and $Ba(NO_3)_2$ may be coprecipitated and to avoid these, precipitant is added slowly with constant stirring. The precipitate is washed with cold wate.·.

$BaSO_4$ is not easily decomposed. It decomposes at 1400°C as follows:

$$BaSO_4 \xrightarrow{1400°C} BaO + SO_3$$

But here due to carbon of the filter paper, it is reduced just above 600°C

$$BaSO_4 + 4C \longrightarrow BaS + 4CO$$

During ignition of the filter paper, care is taken that the filter paper should be charred without inflaming and in ample supply of oxygen to convert most of the carbon to carbon dioxide. Small conversion of $BaSO_4$ to BaS is unavoidable. To convert such BaS to $BaSO_4$ we perform the ash treatement by adding conc. HCl and then conc. H_2SO_4.

Process. Dilute the *given solution* of $BaCl_2$ by adding equal volume of distilled water. Make the solution slightly acidic by adding 5 ml of dil. HCl. Heat it to boiling and add to it hot dil. H_2SO_4 slowly with constant stirring with a glass rod. A white precipitate is formed. Allow the precipitate to settle down until supernatant liquid becomes clear. Add a little more of dil. H_2SO_4 by the side of beaker to see whether the precipitation is complete or not. If the supernatant liquid gives white precipitate, add more of dil. H_2SO_4 and allow the solution to become clear and test it again. Repeat this process till the supernatant liquid gives no precipitate. *Avoid use of dil. H_2SO_4 in excess.* Cover the beaker by watch glass and heat (but not to boiling) for about half an hour to digest the precipitate. Decant the clear liquid through a filter paper (Whatman No. 42) and wash the precipitate by decantation with hot water and test for Ag^+ and SO_4^{2-} ions in the filtrate. After repeated washing, if the last filtrate does not give white precipitate with $AgNO_3$ or $BaCl_2$ solution, the Cl^- and SO_4^{2-} ions are considered to be removed. Transfer the precipitate on filter paper. Use policeman to remove any precipitate sticking on the walls of beaker.

Dry the filter paper on chimney or in electrical oven. When the precipitate gets dried, *keeping the precipitate inside, fold the filter paper in the form of a packet* and put it in the crucbile. Ignite the filter paper over low flame *but do not allow the filter paper to burn with flames.* When the filter paper is completely charred, raise the flame of burner to burn off carbon to clear ash. Cool the crucible, add one drop of conc. HCl and one drop of conc. H_2SO_4 and heat the crucible gently so that white fumes are removed. Now heat the crucible strongly and then cool it in air. Put the crucible in desiccator and weigh accurately.

Calculation. $BaSO_4 \equiv BaCl_2.2H_2O \equiv Ba$

                                      233·43   244·31   137.36

$\therefore$ Conversion factor $= \dfrac{137·36}{233·43} = 0·5884$

Weight of Barium in given solution $= 0·5884 \times$ Wt. of ppt.

To report $BaSO_4$, only write the mass of the precipitate.

Precautions

1. Precipitation is carried out in dilute solution and in hot condition.
2. Some dilute HCl must be added before precipitaiton.
3. Ash treatment is done when crucible is cool.

Questions and their Explanations

Q. 1. *Explain why $BaSO_4$ is precipitated in presence of dil. HCl though it is less soluble in dil. H_2SO_4 than in dil. HCl ?*

Expl. Dilute HCl is added before precipitation in order to :

(*a*) prevent the precipitation of barium salts of carbonate, phosphate and chromate etc. (if they are present as impurities) along with $BaSO_4$ because they are decomposed in dil. HCl.

(*b*) form bigger crystals of $BaSO_4$ which helps in filtration.

Q. 2. *Why is it necessary to add slolwy hot dil. H_2SO_4 to hot dil. $BaCl_2$ solution with constant stirring ?*

Expl. During precipitation of $BaSO_4$, all barium salts especially chloride, chromate and nitrate are carried down with $BaSO_4$ (*i.e.*, coprecipitation). This causes low result. The error due to co-precipitation can be reduced to a large extent :

(*a*) by increasing the volume of supplied $BaCl_2$ solution by adding distilled water.

(*b*) by adding slowly hot dil. H_2SO_4 to hot dil. $BaCl_2$ solution with constant stirring by a rod.

Q. 3. *What is the reason for low result in Barium estimation ?*

Expl. Due to coprecipitation of $BaCl_2$ with $BaSO_4$, some of $BaCl_2$ is not completely converted to $BaSO_4$. Since molecular weight of $BaCl_2$ is less than that of $BaSO_4$, the presence of $BaCl_2$ in precipitate will lower the yield.

Q. 4. *The filter paper ash is treated with a drop of conc. HCl and a drop of conc. H_2SO_4. Explain the necessity of this treatment.*

Expl. During ignition of filter paper (even at 600°C), some of the precipitate is reduced to BaS by carbon of filter paper which will cause low result. The reaction is :

$$BaSO_4 + 4C \longrightarrow BaS + 4CO$$

Hence in order to reconvert BaS to $BaSO_4$, the ash is treated with conc. HCl to give $BaCl_2$ which easily changes to $BaSO_4$ by adding conc. H_2SO_4. The reactions are :

$$BaS + 2HCl \longrightarrow BaCl_2 + H_2S \uparrow$$
$$BaCl_2 + H_2SO_4 \longrightarrow BaSO_4 \downarrow + 2HCl$$

Q. 5. *How can you minimise the reduction of precipitate ?*

Expl. The reduction can, however, be reduced by the following ways :

(*a*) The filter paper should burn without flames.

(*b*) The carbon of filter paper should be burnt off at low temeprature.

(*c*) BaS also changes back to $BaSO_4$ by heating the crucible tilted slightly with small flame from the back side of crucible so that there is ready approach for air into crucible.

Q. 6. *Explain why $BaCl_2$ solution is preferred over $Ba(NO_3)_2$ solution for estiamtion of Ba as $BaSO_4$?*

Expl. Since barium nitrate is more easily coprecipitated than barium chloride with $BaSO_4$ so if the original solution contains barium nitrate, the result of estimation will be very low. This is why $BaCl_2$ is more frequently used in barium estimation rather than $Ba(NO_3)_2$.

EXERCISE 2 : ESTIMATION OF SULPHATE

To estimate sulphate as $BaSO_4$.

Principle of process. Sulphate is estimated as $BaSO_4$. The precipitation is carried out by adding hot soln. of $BaCl_2$ to the boiling solution of sulphate acidified with HCl. The precipitate is washed, dried, ignited and weighed as $BaSO_4$.

Process. To the *supplied solution* of sulphate, add about 100 ml of distilled water and 2 ml conc. HCl. Now, add very slowly hot 5-10% solution of $BaCl_2$ with constant stirring. Allow the precipitate to settle until the supernatant liquid becomes clear. In order to ensure the complete precipitation, add a few drops of the same $BaCl_2$ solution to the clear liquid. If any precipitate appears, add more $BaCl_2$ soln. and test again for the complete precipitation. Repeat this process till the supernatant liquid gives no white precipitate. Cover the beaker by watch glass and heat the contents of beaker (do not boil) for 20-30 minutes. Allow the precipitate to settle and once again test the complete precipitation.

Filter by decantation. The precipitate is washed by decantation with hot water until free from barium ions (test with dil. H_2SO_4) and chloride ions (test with $AgNO_3$). Now transfer the ppt. to filter paper and dry it.

The ignition of the precipitate is done similarly as in the estimation of barium as barium sulphate.

Calculation.

$$BaSO_4 \equiv SO_4^{2-}$$
$$233 \cdot 43 \qquad 96 \cdot 06$$

$$\text{Conversion factor} = \frac{SO_4^{2-}}{BaSO_4} = \frac{96 \cdot 06}{233 \cdot 43} = 0 \cdot 4115$$

$$\therefore \quad \text{Wt. of } SO_4^{2-} \text{ ions} = 0 \cdot 4115 \times \text{wt. of ppt.}$$

EXERCISE 3 : ESTIMATION OF SILVER

> *To estimate Ag as AgCl in AgNO₃ solution.*

Principle of process. AgCl is precipitated in presence of dil. HNO_3 by adding slowly hot dil. HCl to hot $AgNO_3$ solution with constant stirring. As the precipitate is sensitive to light, precipitation is done in *subdued* light. Direct sunlight or bright light decomposes AgCl to metallic silver.

$$2AgCl \xrightarrow{\quad hv \quad} 2Ag + Cl_2$$

The precipitate is washed with cold water containing 0·01 per cent HNO_3 in order to prevent peptization of the precipitate. Dry and separate it from filter paper, treat the ash with one drop of conc. HNO_3 and conc. HCl in order to reconvert Ag to AgCl, followed by gently heating.

$$3Ag + HNO_3 + 3HCl \longrightarrow 3AgCl + 2H_2O + NO$$

The precipitate in crucible is not strongly heated because losses may occur due to decomposition and volaltilization. The precipitate may melt and stick on the walls of crucible.

Process. To the *supplied solution* of $AgNO_3$, add an equal volume of distilled water and 1 ml conc. HNO_3 (if the $AgNO_3$ solution already contains HNO_3, do not add conc. HNO_3). Heat to about 70°C (*i.e.*, just less than boiling) and add hot dil. HCl slowly with vigorous stirring so that precipitation may occur. *Avoid excessive use of dil. HCl. Do not expose contents of beaker to bright light.* It is better to *wrap* the beaker by black paper. Warm the contents until the precipitate settles and then test the supernatant liquid with a little dil. HCl to ensure the complete precipitation.

Filter and wash the precipitate thoroughly with water containing a little dil. HNO_3 (1 ml conc. HNO_3 in 100 ml water) until chloride ions are removed (test with $AgNO_3$) and then finally wash precipitate on filter paper twice with water (not containing dil HNO_3) to remove nitric acid. *Avoid excessive washings.* Dry the precipitate near 100°C. Transfer *as much of precipitate as possible on a glazed paper* and burn the filter paper to a clear ash and then cool. *Moisten* the ash with one drop of conc. HNO_3 and one drop of conc. HCl and heat quite gently over a flame and then cool. Transfer the ppt. from glazed paper to the crucible. Place the wire gauze on a tripod stand. Keep clay pipe triangle with its corners downward on the wire gauze. Place the crucible on clay pipe triangle and heat the crucible containing precipitate over very low flame, holding the burner *horizontally and moving it.* Do not allow the precipitate to melt. Cool and weigh. Repeat this process until a constant weight is obtained.

Calculations :

$$\underset{143 \cdot 34}{AgCl} \equiv \underset{169 \cdot 89}{AgNO_3} \equiv \underset{107 \cdot 88}{Ag}$$

$$\text{Conversion factor} = \frac{107 \cdot 88}{143 \cdot 34} = 0 \cdot 7526$$

∴   Mass of Ag in supplied volume of $AgNO_3$ solution

$$= 0 \cdot 7526 \times \text{wt. of ppt. (AgCl)}$$

Questions and their Explanations

Q. 1. *Explain why it is necessary to add nitric acid to silver nitrate solution before the precipitation of silver chloride ?*   (*Kanpur B.Sc. II ; Agra B.Sc.*)

Expl. HNO_3 is added before precipitation because :

(*a*) it will oxidise Cu(I) and Hg(I) if present. It will also destroy cyanides and thiosulphates.

(*b*) HNO_3 helps in coagulation of colloidal particles of precipitate.

(*c*) HNO_3 prevents the precipitation of other silver salts especially carbonate and phosphate (as these may be present in the form of impurities in solution).

Q. 2. *Explain why precipitate of AgCl is washed with water containing HNO₃ in the estimation of Ag as AgCl ?*   (*Kanpur B.Sc.II, ; Agra B.Sc. II*)

Expl. The washing of precipitate with pure water may change AgCl to colloidal form which may pass through the filter paper. Hence, the precipitate is washed with water containing HNO_3 for the following advantages :

(a) The formation of colloidal AgCl is prevented.

(b) HNO_3 has no action on precipitate.

(c) HNO_3 can be easily removed.

Q. 3. *Explain why it is necessary to use water which does not contain HNO_3 for final washing?*

(*Kanpur B.Sc. II*)

Expl. In final washing, water alone is used to remove HNO_3 completely from filter paper. If it is not removed, the filter paper on drying,becomes fragile (easily broken) and it will be very difficult to handle the filter paper and to remove the precipitate from it.

Q. 4. *Explain why the precipitate should not be exposed to direct sunlight or bright light ?*

Expl. The precipitate should be protected from direct sunlight or bright light because it is decomposed to metallic silver and Cl_2. This will cause low result.

$$2AgCl \longrightarrow 2Ag + Cl_2$$

Due to formation of Ag, the precipitate becomes somewhat blackish in colour.

Hence precipitation should be done in subdued light in order to prevent the decomposition of precipitate.

Q. 5. *Explain why it is necessary to avoid the strong heating of the precipitated AgCl ?*

(*Kanpur B.Sc. II, Agra B.Sc. II*)

Expl. On strong heating :

(a) AgCl will melt and percolate through the walls of crucible and will thus be difficult to remove it from crucible.

(b) AgCl is decomposed to metallic Ag and also it is volatilized. This will cause low result.

Hence the strong heating of precipitate is avoided.

Q. 6. *Explain why it is necessary to use both HNO_3 and HCl acids in treatment of ash ?*

(*Kanpur B.Sc. II*)

Expl. While igniting the filter paper, a little amount of precipitate present in filter paper is reduced to metallic Ag by carbon of filter paper. Hence in order to oxidise Ag to AgCl, it is necessary to treat the ash with one drop of conc. HNO_3 and one drop of conc. HCl followed by gentle heating. The reaction is

$$3Ag + 4HNO_3 \longrightarrow 3AgNO_3 + 2H_2O + NO$$
$$AgNO_3 + HCl \longrightarrow AgCl \downarrow + HNO_3$$

EXERCISE 4 : ESTIMATION OF CHLORIDE

To estimate chloride as silver chloride.

Principle of process. Chloride ions are generally estimated as AgCl. The precipitation is carried out by adding hot $AgNO_3$ solution to chloride solution acidified with dil. HNO_3. The precipitate is washed, dried, ignited and weighed as AgCl.

Process. To the *supplied solution* of chloride, add an equal volume of distilled water and 0·5-1·0 ml of conc. HNO_3. Now heat the contents to about 70°C. Add very slowly an excess of hot 2-5% $AgNO_3$ solution with vigorous stirring. Heat the solution to boiling with constant stirring so that the colloidal AgCl coagulates. Stop heating and allow the precipitate to settle. In order to ensure the complete precipitation, add a few drops of $AgNO_3$ solution to the clear supernatant liquid and see whether any precipitate appears or not. If any precipitate appears, add more of $AgNO_3$ solution and test again. This process is repeated till no further precipitate appears.

This precipitate is washed, dried, ignited and weighed in the very same way as in the estimation of silver as AgCl.

Calculations

$$AgCl \equiv Cl^-$$
$$143 \cdot 34 \qquad 35 \cdot 46$$

$$\text{Conversion factor} = \frac{Cl^-}{AgCl} = \frac{35 \cdot 46}{143 \cdot 34} = 0 \cdot 2474$$

∴   Wt. of Cl^- ions $= 0 \cdot 2474 \times$ Wt. of ppt.

EXERCISE 5 : ESTIMATION OF ZINC

To estimate Zinc as Zinc oxide from the Zinc chloride or Zinc sulphate solution.

Principle of process. Zinc is precipitated as basic carbonate having no definite composition and changes to zinc oxide on ignition. The condition of precipitation is to add hot dil. Na_2CO_3 solution drop by drop to boiling solution of zinc salt until there is slight turbidity then boil and add few drops of phenolphthalein followed by slow addition of hot Na_2CO_3 solution until slight pink colour is obtained.

Process. Add an equal volume of distilled water to the *supplied solution* ($ZnCl_2$ or $ZnSO_4$) and boil. Add hot dil. solution of Na_2CO_3 (10%) drop by drop with constant stirring until there is slight turbidity. Now boil the solution and after it add a few drops of phenolphthalein. Now add more quantity of hot Na_2CO_3 solution with constant stirring till a light pink colour is obtained. Boil the contents again and allow the precipitate to settle.

Filter off the clear liquid through filter paper by decantation and wash the precipitate with hot water by decantation until free from CO_3^{--} ions (test with red litmus paper) and Cl^- ions (if the original soln. was of $ZnCl_2$) or sulphate ions when original solution was of $ZnSO_4$. Transfer whole of the precipitate to filter paper and dry it.

After drying, transfer as much of precipitate as possible on a glazed paper and cover it by funnel or watch glass. Soak the filter paper with saturated solution of NH_4NO_3 and dry it. Ignite the filter paper in a weighed crucible at low temperature. Cool the crucible and then transfer the precipitate. Heat first gently and then strongly for 15-20 minutes. $ZnCO_3$ is converted to ZnO. Cool and weigh. Repeat the process of heating, cooling and weighing until a constant weight is obtained.

Calculation

$$ZnO \equiv Zn$$
$$81 \cdot 38 \qquad 65 \cdot 38$$

$$\text{Conversion factor} = \frac{Zn}{ZnO} = \frac{65 \cdot 38}{81 \cdot 38} = 0 \cdot 8034$$

∴   Wt. of zinc in the given volume of solution $= 0 \cdot 8034 \times$ Wt. of ZnO.

Questions and their Explanations

Q. 1. *Explain why it is preferred to precipitate zinc as carbonate to sulphate ?*

Expl. The carbonate, nitrate and oxalate of zinc are easily converted to zinc oxide by igniting them in air, whereas the sulphate of zinc changes to ZnO with difficulty. That is why zinc is preferably precipitated as carbonate to sulphate.

Q. 2. *Explain why in the estimation of zinc as zinc oxide, it is necessary to*

(a) *add hot solution of zinc salt ?*

(b) *avoid an excessive use of alkali ?*

(c) *use phenolphthalein ?*

(d) *remove ammonium salts if any present, from the original solution of zinc salt ?*

Expl. (a) The precipitate is less soluble in hot water than in cold water hence precipitation is done in hot conditions.

(b) The precipitate is slightly soluble in Na_2CO_3 solution and the alkalies are also coprecipitated with the precipitate if large excess of alkali is used.

(c) The phenolphthalein is used to check the excessive addition of Na_2CO_3 solution.

(d) Since in the presence of ammonium salts, zinc carbonate is not completely precipitated, the solution of zinc should be free from ammonium salts.

Q. 3. *Explain why in the estimation of zinc as zinc oxide, it is necessary to treat the filter paper with concentrated solution of NH_4NO_3 before incineration.*   *(Agra B.Sc. II, ; K.U.)*

Expl. A little amount of precipitate adhering to filter paper may get reduced to metallic zinc by carbon of filter paper. The zinc is volatile at the flame of burner. This will cause low result. Hence the paper is treated with conc. NH_4NO_3 solution in order to check the reduction of precipitate and the loss due to volatilization. Ammonium nitrate converts zinc to zinc oxide as follows :

$$ZnCO_3 + 2C \longrightarrow Zn + 3CO$$
$$ZnO + C \longrightarrow Zn + CO$$
$$Zn + NH_4NO_3 \longrightarrow ZnO + N_2 + 2H_2O$$

EXERCISE 6 : ESTIMATION OF COPPER

To estimate copper as cupric oxide in a solution of copper sulphate.

Principle of process. Copper is precipitated as $Cu(OH)_2$ by adding hot solution of NaOH or KOH to hot solution of copper salt containing a little Cl^- ions. The precipitation is done in hot condition and in presence of HCl in order to enhance the rapid formation of granular precipitate. Use of phenolphthalein checks excessive addition of alkali. The treatment of ash with a drop of conc. HNO_3 is necessary to check cupric oxide from reduction during ignition. The precipitate on heating changes to black or dark brown cupric oxide.

$$CuSO_4 + 2NaOH \longrightarrow Na_2SO_4 + Cu(OH)_2 \downarrow$$
$$\text{Blue}$$

$$Cu(OH)_2 \xrightarrow{\text{heat}} CuO + H_2O$$
$$\text{Brown}$$

Process. To the *supplied solution of $CuSO_4.5H_2O$*, add an equal volume of distilled water and 1 ml of conc. HCl. Cover the beaker by watch glass and boil the contents of beaker. Now add few drops of phenolphthalein. Add slowly a hot 10% solution of NaOH with constant stirring, until clear liquid becomes light pink in colour. A blue precipitate of $Cu(OH)_2$ is obtained. Cover the beaker and boil for 3-5 minutes. On boiling, $Cu(OH)_2$ partially changes to dark brown CuO. After the precipitate has settled, immediately pour off the upper liquid through filter paper and wash the precipitate by decantation with hot water until last filtrate becomes free from alkali (*i.e.*, no effect on red litmus paper). If any precipitate sticks on the sides of beaker and on the tip of glass rod, wash it down with hot water using policeman. Transfer the precipitate to filter paper and continue the washing till the filtrate becomes free from SO_4^{--} and Cl^- ions (test with $BaCl_2$ and $AgNO_3$ solution respectively).

Dry the precipitate and transfer as much of precipitate as possible on a glazed paper and cover it with the funnel. Burn the filter paper to a clear ash. Cool the crucible and add one drop of conc. HNO_3 and now heat first gently then on full flame. Cool and then add precipitate to the crucible. Heat strongly, cool and weigh. Repeat this process until a constant weight is obtained.

Calculations

$$CuO \equiv Cu \equiv CuSO_4.5H_2O$$
$$79{\cdot}54 \quad 63{\cdot}54 \quad\quad 249{\cdot}69$$

$$\text{Conversion factor} = \frac{Cu}{CuO} = \frac{63 \cdot 54}{79 \cdot 54} = 0 \cdot 7988$$

$\therefore$ Wt. of copper in the supplied volume of solution

$$= 0 \cdot 7988 \times \text{wt. of ppt.}$$

Questions and their Explanations

Q. 1. *Explain why the following are necessary during estimation of copper :*

(a) *precipitation is done in hot condition ?*

(b) *precipitation is carried out in presence of Cl$^-$ ions ?*

Expl. [(a) and (b)]. The precipitation is done in hot condition and in presence of Cl$^-$ ions in order to make the precipitate granular. Cl$^-$ ions also check the formation of basic salts.

Q. 2. *Explain how and why the large excess of alkali is avoided ?*

Expl. The large excess of alkali is avoided by using phenolphthalein which at once indicates any excess of alkali added because phenolphthalein is colourless in acidic medium and pinkish in alkaline medium.

Since NaOH or KOH has solvent action on $Cu(OH)_2$ and removal of alkali from precipitate is also very difficult, excessive use of alkali is avoided.

Q. 3. *Explain why filtration of Cu(OH)$_2$ should be done immediately after precipitation ?*

Expl. The ppt. sticks on the sides of beaker so firmly that it cannot be removed easily. Hence filtration should be done in hot conditions immediately after precipitation. If the precipitate becomes dry, it can not be removed even with policeman.

Q. 4. *Explain why it is necessary to treat the ash with one drop of conc. HNO$_3$?*

Expl. During ignition of filter paper, some of the precipitate sticking on the filter paper is reduced to cuprous oxide and metallic copper. Hence ash is treated with conc. HNO_3 in order to convert copper to CuO. The reactions are :

Reduction :

$$2CuO + C \longrightarrow Cu_2O + CO$$
$$CuO + C \longrightarrow Cu + CO$$

Oxidation :

$$Cu_2O + 6HNO_3 \longrightarrow 2Cu(NO_3)_2 + 2NO_2 + 3H_2O$$
$$Cu + 4HNO_3 \longrightarrow Cu(NO_3)_2 + 2H_2O + 2NO_2$$

$$2Cu(NO_3)_2 \xrightarrow{\text{heating}} 2CuO + 4NO_2 + O_2$$

EXERCISE 7 : ESTIMATION OF IRON

> *To estimate Iron as Ferric-oxide in a solution of ferrous sulphate or ferrous ammonium sulphate.*

Principle of process. The ferric ions are precipitated by NH_4OH as hydrous ferric oxide (also called as hydrated ferric oxide), $Fe_2O_3.xH_2O$, having no definite composition as it contains variable amount of water (x at the maximum is 40-50). Hence, in original solution, the ferrous ions are first converted to ferric state by oxidising agents like Br_2 water, H_2O_2 or conc. HNO_3. The original solution must be free from other ions like Al^{3+}, Cr^{3+} which are precipitated as hydroxides along with iron hydroxide, and also free from oxalate, arsenate and phosphate which form insoluble salts of iron in alkaline medium.

The ppt. on heating at about 1000°C changes to Fe_2O_3. The reactions in the process are :

$$6FeSO_4 + 3H_2SO_4 + 2HNO_3 \longrightarrow 3Fe_2(SO_4)_3 + 4H_2O + 2NO$$

$$Fe_2(SO_4)_3 + 6NH_4OH \longrightarrow 2Fe(OH)_3.xH_2O + 3(NH_4)_2SO_4$$

$$\underset{\text{(ppt.)}}{Fe_2O_3.xH_2O} \xrightarrow[\substack{1000°C}]{\text{at about}} Fe_2O_3 + xH_2O$$

Process. Add equal volume of distilled water to the *supplied solution* of ferrous salt and heat to boiling. Add 1 ml of conc. HNO_3 and boil gently. The solution becomes blackish-brown at first and then turns to bright yellow. Bright yellow colour indicates that all ferrous ions have been completely oxidised to ferric state. Add slowly with stirring, a freshly distilled or pure filtered NH_4OH solution (1 : 1) until there is distinct smell of ammonia. Boil off excess of ammonia and allow the precipitate to settle After the precipitate is settled, immediately pour off upper liquid through the filter paper. Wash the precipitate in a beaker by decantation, *i.e.*, add 10-20 ml of hot solution of NH_4NO_3 (1-2 %). Stir the mixture completely, allow it to settle and then decant through the filter paper. Repeat this process of washing until filtrate is almost free from SO_4^{2-} and Cl^- ions (if the original solution contains HCl). Transfer the precipitate to filter paper. *Do not fill the filter cone more than 3/4th with the precipitate.* Wash the precipitate again with the washing solution until becomes completely free from SO_4^{2-} ions or both SO_4^{2-} and Cl^- ions. Do not transfer the precipitate to filter paper until it is thoroughly washed, and should not remain for long time on filter paper because on filter paper, the precipitate dries and forms cracks through which the washing will run down without coming in contact with the precipitate.

Dry and ignite the filter paper keeping the precipitate inside, without inflaming it. Remove the carbon at very low temperature (keep burner horizontally) with free approach of air in the crucible and then heat on full flame for 15-20 minutes. Cool and weigh. Repeat this process until two successive weighings are constant.

Calculation

$$Fe_2O_3 \equiv 2FeSO_4 \equiv 2[FeSO_4.(NH_4)_2SO_4.6H_2O] \equiv 2Fe$$
$$159·68 \qquad\qquad 2 \times 392·16 \qquad\qquad\qquad 111·68$$

$$\text{Conversion factor} = \frac{2Fe}{Fe_2O_3} = \frac{111·68}{159·68} = 0·6994$$

∴   Wt. of iron in the supplied volume of solution

$$= 0·6994 \times \text{wt. of ppt.}$$

Questions and their Explanations

Q. 1. *Explain why dil. H_2SO_4 or dil. HCl is added to the aqueous solution of ferrous salts ?*

Ans. Dil. H_2SO_4 or dil. HCl is added to solution in order to check the hydrolyis of ferrous salts.

Q. 2. *Explain why ferrous ammonium sulphate solution is treated with conc. HNO_3 before its precipitation as hydroxide ?*   (*Agra B.Sc. II*)

Ans. HNO_3 oxidises ferrous into ferric state. This treatment is necessary as ferrous hydroxide is slightly soluble in water, whereas ferric hydroxide is least soluble and is completely precipitated.

$$FeSO_4(NH_4)_2SO_4.6H_2O \longrightarrow FeSO_4 + (NH_4)_2SO_4 + 6H_2O$$

$$6FeSO_4 + 3H_2SO_4 + 2HNO_3 \xrightarrow{\text{heating}} 3Fe_2(SO_4)_3 + 4H_2O + 2NO$$

Q. 3. *What is the composition of the precipitate obtained by adding NH_4OH solution to ferrous salt solution oxidised by HNO_3 ?*

Expl. The precipitate corresponds to hydrated ferric hydroxide, $Fe_2O_3.xH_2O$ and not to $Fe(OH)_3$ strictly. 'x' is as large as 40 or 50. The precipitate contains variable amount of water depending upon the conditions of precipitation.

Q. 4. *Explain the following in the estimation of iron and give equations wherever necessary :*

(a) *Why does a blackish brown colour appear on addition of conc. HNO_3 to ferrous ammonium* *sulphate solution.*

(b) *Why is precipitation carried at the boiling stage of solution ?*

(c) *How is complete oxidation indicated ?*

Expl. (a) During oxidation of ferrous salt NO is formed. Nitric oxide combines with ferrous sulphate forming unstable blackish brown complex, $FeSO_4.NO$.

$$6FeSO_4 + 3H_2SO_4 + 2HNO_3 \longrightarrow 3Fe_2(SO_4)_3 + 4H_2O + 2NO$$
$$FeSO_4 + NO \longrightarrow FeSO_4.NO$$

(b) Firstly, the precipitate is in colloidal form. Since the coagulation (*i.e.*, precipitation of colloidal particles) is helped by raising the temperature, precipitation is carried out at the boiling point and the solution is boiled for short time after precipitation.

(c) During oxidation, the solution becomes blackish-brown at first and then turns to bright yellow. Bright yellow colour is an indication of the complete oxidation of ferrous ions.

Q. 5. Explain the following instructions :

(a) *The precipitate is washed with hot 2% NH_4NO_3 solution.*   (*Agra B.Sc. II*)

(b) *Do not wash the precipitate with NH_4Cl solution.*

(c) *Do not leave the precipitate for a long time on filter paper and do not transfer the precipitate to filter paper until completely washed.*

Expl. (a) Ammonium nitrate helps to prevent the precipitate from passing into colloidal form. It also prevents the precipitate from becoming slimy. NH_4NO_3 is volatilized when precipitate is ignited.

(b) The washing of precipitate with NH_4Cl solution is not generally recommended as during ignition of precipitate volatile $FeCl_3$ is formed and it will cause low result.

(c) The precipitate should be completely washed before transferring it to filter paper and should not become dry otherwise cracks are developed in the precipitate through which the washing solution will run down and take too much time for washing the precipitate.

Q. 6. *Explain why the precipitate is not heated at temperature higher than 1000°C and why carbon of filter paper is consumed at a low temperature ?*

Expl. The precipitate on ignition at about 1000°C changes to Fe_2O_3. Heating of precipitate at higher temprature produces more stable Fe_3O_4. Fe_3O_2 and even Fe are also formed if carbon of filter paper is not consumed at a low temperature. Fe_3O_4 once formed hardly changes back to less stable Fe_2O_3 upon continued heating. That is why precipitate should not be heated above 1000°C.

Q. 7. *Explain why sometimes, the addition of ashless filter paper pulp after precipitation is recommended ?*

Expl. The addition of ashless filter paper pulp has the following advantages :

(a) It prevents formation of Fe_3O_4 during ignition of precipitate.

(b) It helps in filtration of the solution.

(c) It helps in making the precipitate porous on ignition so that steam and volatile substances are easily expelled out from precipitate.

EXERCISE 8 : ESTIMATION OF ALUMINIUM

To estimate Al or Al_2O_3 in potash alum or ammonium aluminium sulphate.

Principle of process. Aluminium is precipitated as hydrated oxide, $Al_2O_3.xH_2O$ by adding dil. NH_4OH solution in presence of NH_4Cl or HCl. The precipitate on ignition changes to Al_2O_3 and then weighed as such.

Process. To the *supplied 25 ml solution* of soluble aluminium salt, add 2-5 gm of solid NH_4Cl

and a few drops of 20% alcoholic solution of methyl red. Heat the contents of beaker just to boiling. Add NH_4OH solution (1 : 1) dropwise with constant stirring until the colour of solution changes to distinct yellow. Boil for 2 minutes and add some filter pulp or a pinch of tannic acid to beaker and filter at once by decantation. Wash the precipitate with hot 2% solution of NH_4NO_3 by decantation until free from Cl^- and SO_4^{2-} ions. But it is not necessary that SO_4^{2-} and Cl^- ions should be completely removed because the impurity in precipitate is due to ammonium salts, e.g., NH_4Ci or $(NH_4)_2SO_4$ which are volatile when precipitate is ignited. Transfer the ppt. to filter paper and dry it. Ignite the precipitate along with filter paper in a weighed silica crucible at 1200°C. Cool and weigh. Repeat the process of heating, cooling and weighing until a constant weight is obtained.

Calculation

$$Al_2O_3 \equiv 2Al$$
$$101.94 \quad 53.94$$

$$\text{Conversion factor} = \frac{2Al}{Al_2O_3} = \frac{53.94}{101.94} = 0.5291$$

∴ Wt. of Al $= 0.5291 \times$ wt. of ppt.

Questions and their Explanations

Q. 1. *Explain why in the estimation of Al as Al_2O_3 it is necessary to :*

(a) *use methyl red indicator ?*

(b) *add NH_4Cl before precipitation ?*

(c) *filter the solution as hot and in presence of filter pulp.*

(d) *wash the precipitate with hot 2% NH_4NO_3 or NH_4Cl solution and not with hot water alone?*

Expl. (a) The precipitation of aluminium hydroxide begins at pH = 4.14 and is complete at pH 6.5-7.5. The precipitate dissolves at pH ~10.8 (*i.e.*, in highly alkaline medium). That is why large excess of NH_4OH must be avoided. In order to adjust the pH range 6.5-7.5, methyl red indicator is used.

(b) Due to buffer effect of NH_4Cl, it fixes up the pH necessary for precipitation by NH_4OH. It also helps the coagulation of colloidal precipitate and minimises the coprecipitation of other ions such as Ca^{2+} and Mg^{2+} etc.

(c) The filtration becomes easier if filter paper pulp is added to the solution before filtration and solution is filtered hot.

(d) Since the precipitate is soluble in hot water, it is washed with hot 2% solution of NH_4NO_3 or NH_4Cl. NH_4NO_3 is also volatile at ignition temperature and NH_4Cl does not cause any volatilization of aluminium.

Q. 2. *Explain why the precipitate is ignited at very high temperature ?*

Expl. Al_2O_3 is hygroscopic. If $Al(OH)_3$ is heated below 1200°C, it will absorb moisture and result will be high. Hence the precipitate is ignited at or above 1200°C to get non-hygroscopic Al_2O_3 and weighing should be done immediately after cooling the crucible in a desiccator.

EXERCISE 9 : ESTIMATION OF CHROMIUM

To estimate chromium as chromic oxide, Cr_2O_3.

Principle of process. For the estimation of chromium as chromic oxide, the chromium in the original solution must be present in the form of chromic salt. Chromium is precipitated as chromic hydroxide by means of slight excess of dil. NH_4OH in presence of ammonium salt *e.g.*, NH_4Cl or NH_4NO_3 etc. The chromic hydroxide when ignited is converted into Cr_2O_3.

If chromium is present as chromate or dichromate, we first reduce chromate or dichromate to chromic salt by heating the original solution with conc. HCl (or dil. H_2SO_4) and a little alcohol and then chromium is precipitated as above.

Process. (*a*) **Precipitation :**

(*i*) *When chromium is present as chromic salt e.g.* $Cr_2(SO_4)_3$, take about 25 ml of original solution and dilute it by adding the same volume of distilled water, add a little solid NH_4Cl and then boil the contents. Now, add a freshly prepared boiling dil. NH_4OH solution little by little with stirring till NH_4OH is just in excess. Avoid a large excess of NH_4OH. Boil off the excess of ammonia. Allow the precipitate to settle.

$$Cr_2(SO_4)_3 + 6NH_4OH \longrightarrow 3(NH_4)_2SO_4 + 2Cr(OH)_3$$

(*ii*) *When chromium is present as dichromate e.g.* $K_2Cr_2O_7$. Take 25 ml of solution of potassium dichromate, add 5 ml of conc.HCl or dil H_2SO_4 and 15 ml of C_2H_5OH. Heat the contents until smell of acetaldehyde and alcohol vapours ceases and solution becomes green. Dilute the solution by adding 50 ml of distilled water and proceed for precipitation of chromium as before.

$$K_2Cr_2O_7 + 4H_2SO_4 + 3C_2H_5OH = K_2SO_4 + Cr_2(SO_4)_3 + 7H_2O + 3CH_3CHO$$

(yellow)   (green)

(*b*) **Washing and ignition.** Wash the precipitate by decantation with hot 2% NH_4NO_3 solution until free from SO_4^{2-} ions then transfer the precipitate to filter paper, wash 2-3 times and dry it.

Ignite the filter paper along with precipitate into a weighed crucible on a low flame until whole of carbon is converted to ash. Cool and add a drop of conc. HNO_3. Heat first gently and then strongly. Cool the crucible in air, transfer it to dessicator and then weigh. Repeat heating, cooling and weighing until weight becomes constant.

$$2Cr(OH)_3 \xrightarrow{\text{heating}} Cr_2O_3 + 3H_2O$$

Calculation

$$Cr_2O_3 \equiv 2Cr \equiv K_2Cr_2O_7 \equiv Cr_2(SO_4)_3$$
$$152 \cdot 02 \quad 104 \cdot 02 \quad 294 \cdot 21 \quad 392 \cdot 02$$

$$\text{Conversion factor} = \frac{2Cr}{Cr_2O_3} = \frac{104 \cdot 02}{152 \cdot 02} = 0 \cdot 6842$$

∴   Wt. of chromium = $0 \cdot 6842 \times$ weight of Cr_2O_3.

EXERCISE 10 : ESTIMATION OF LEAD

To estimate lead as lead sulphate.

Principle of process. Lead is precipitated by dil. H_2SO_4 and estimated as $PbSO_4$. The original solution is made free from HNO_3 and HCl, as precipitate is slightly soluble in these. Since $PbSO_4$ is almost insoluble in C_2H_5OH, precipitation is done in presence of alcohol. Also due to lesser solubility in cold than in hot water, precipitate is washed with cold water containing an equal volume of alcohol. During ignition, the precipitate adhered in filter paper is reduced to PbS, so the ash is treated with one drop of conc. HNO_3 and one drop of conc. H_2SO_4 to convert PbS into $PbSO_4$.

Process. (1) **Precipitation**

(*a*) *When Pb is originally present as chloride or nitrate :*

To the supplied solution, add an excess of dil. H_2SO_4 and heat the mixture on water bath as far as possible then over a free flame until white dense fumes of H_2SO_4 are evolved. This process removes completely HCl or HNO_3 present in solution. Cool and then dilute with a mixture of water and ethyl alcohol and allow the precipitate to settle.

(*b*) *When original solution is lead acetate :*

To the supplied solution, add twice its volume of ethyl alcohol and an excess of dil. H_2SO_4. Allow it to stand for at least an hour.

(2) Washing and ignition

The precipitate is washed by decantation with cold water containing an equal volume of ethyl alcohol and finally with pure alcohol until free from SO_4^{2-} ions (test with $BaCl_2$ solution).

Dry the precipitate and transfer as much of it as possible to a glazed paper and cover the precipitate with a funnel. Ignite the filter paper to a clear ash in a weighed crucible. Moisten the ash with a drop of conc. HNO_3 and one drop of conc. H_2SO_4 and heat gently until white fumes are evolved. Cool and add the precipitate to the crucible and heat again at about 500-600°C otherwise loss will occur due to decomposition and volatilization. Cool and weigh. Repeat the process of heating, cooling and weighing until a constant weight is obtained.

Calculation

$$Pb \equiv PbSO_4 \equiv Pb(NO_3)_2 \equiv Pb(CH_2COO)_2$$
$$207\cdot21 \quad 303\cdot27 \quad\quad 331\cdot23 \quad\quad\quad 325\cdot29$$

$$\text{Conversion factor} = \frac{Pb}{PbSO_4} = \frac{207\cdot21}{303\cdot27} = 0\cdot6833$$

∴ Wt. of Pb in the supplied volume of solution

$$= 0\cdot6833 \times \text{wt. of ppt.}$$

Questions and their Explanations

Q. 1. *Explain why in the estimation of Pb as $PbSO_4$?*

(a) the precipitation is done in presence of ethyl alcohol.

(b) precipitate is washed with a mixture of cold water and ethyl alcohol.

(c) the ash is treated with conc. HNO_3 and conc. H_2SO_4.

(d) the precipitate in crucible is heated at 500—600°C.

(e) the original solution of $Pb(NO_3)_2$ or $PbCl_2$ is evaporated off with excess of dil. H_2SO_4 whereas this treatment is not employed during precipitation of $Pb(CH_3COO)_2$?

Expl. (a) $PbSO_4$ is almost insoluble in ethyl alcohol hence precipitation is carried out in presence of alcohol.

(b) Due to slight solubility of $PbSO_4$ (0·046 gm/litre at 20°C) in water, the result may be low. Hence the precipitate is washed with a mixture of cold water and alcohol.

(c) During ignition of filter paper, some of the precipitate present in filter paper may reduce to PbS by carbon of filter paper. Hence ash is treated with one drop of conc. HNO_3 and one drop of conc. H_2SO_4 to oxidise PbS. The changes that take place are :

$$PbSO_4 + 4C \longrightarrow PbS + 4CO$$
$$PbS + 2HNO_3 \longrightarrow Pb(NO_3)_2 + H_2S$$
$$Pb(NO_3)_2 + H_2SO_4 \longrightarrow PbSO_4 + 2HNO_3$$

(d) The precipitate is not heated above 500—600°C, otherwise losses will be there due to decomposition and volatilization.

(e) $PbSO_4$ is more soluble in HCl and HNO_3 than in dil. H_2SO_4. Hence from original solution HNO_3 and HCl must be removed. That is why the original solution of $Pb(NO_3)_2$ or $PbCl_2$ is evaporated off with excess of dil. H_2SO_4 in order to remove HNO_3 or HCl completely. This treatment is not necessary with $Pb(CH_3COO)_2$ solution as it is not soluble in acetic acid.

11

Few Inorganic Preparations

1. Sodium Ferrioxalate, Na$_3$ [Fe(C$_2$O$_4$)$_3$]. 9H$_2$O

IUPAC Name- Sodium trisoxalato ferrate (III) nona-hydrate

It is light green in appearance, in solution the ions give deep apple green colour. Its crystals are octahedral and the oxalate ligand behaves bidentate.

Theory- The compound is obtained by the interaction of sodium oxalate and ferric hydroxide. The sodium oxalate for this purpose is prepared in situ by allowing sodium hydroxide to react with oxalic acid. The ferric hydroxide is prepared by allowing sodium hydroxide to react with ferric chloride. The chemical equations are shown as below.

(i) $2NaOH + \underset{\overset{|}{COOH}}{COOH} \longrightarrow \underset{\overset{|}{COONa}}{COONa} + 2H_2O$

(ii) $3NaOH + FeCl_3 \longrightarrow Fe(OH)_3\downarrow + 3NaCl$

Finally:

$3\underset{\overset{|}{COO\ Na}}{COONa} + Fe(OH)_3 \longrightarrow Na_3[Fe(C_2O_4)_3] + 3NaOH$

Reagents Required

Ferric Chloride = 5 gms

Oxalic Acid = 6.5 gms

Sodium hydroxide = 2 gms and 7 gms separately

Procedure: Weigh all the three chemicals in amounts given above. Take about 25 ml of water in a beaker and heat it but not to boiling. Now add the oxalic acid with constant stirring with a glass rod. This gives clear aqueous solution of oxalic acid. Now to it, add 2 grams of sodium hydroxide pallets and stir slowly to make clear solution. This is solution number 1 (sodium oxalate).

In a separate beaker dissolve the ferric Chloride in 10 ml of water and add to it 7 gram of sodium hydroxide by stirring with a glass rod. This is solution no.2.

Add solution no.2 to the hot solution no. 1. Stir for 5 minutes. It should give a clear solution. If any residue settles at the bottom, remove it by filtration. Allow the solution to cool, where crystals of sodium ferrioxalate are obtained. If no crystals are obtained, concentrate the solution by heating and then cool to obtain the crystals. Remove the supernetant liquid and wash the crystals with ethyl alcohol and weigh when dry.

Ideal yield = 6 to 8 gms.

Determination of Composition By Permanganometry

Theory: Sodium ferrioxalate reacts with sulphuric acid to produce corresponding amount of oxalic acid which can be titrated against KMnO$_4$ in acidic medium, the strength of oxalic acid is a measure the strength of sodium ferrioxalate. KMnO$_4$ acts as a self indicator and the reaction that takes place is shown below.

$2Na_3[Fe(C_2O_4)_3] + 6H_2SO_4 \longrightarrow 6(COOH)_2 + Fe_2(SO_4)_3 + 3Na_2SO_4$

$2KMnO_4 + 3H_2SO_4 + 5(COOH)_2 \longrightarrow K_2SO_4 + 2MnSO_4 + 8H_2O + 10CO_2$

Reagents and Their Preparation

(i)   Prepare N/25 oxalic acid solution by weighing 0.630 gm and dissolve it in distilled water and make upto 250 ml.

(ii)   Prepare approximately N/25 $KMnO_4$ solution by dissolving it about 1.26 gm in water and make up to 1 litre.

(iii) Dissolve about 3.89 gm of the complex obtained in distilled water and make up its volume to 250 ml.

(iv)   Dilute H_2SO_4.

Experimental Procedure

Pipette 25ml of oxalic acid solution in a titration flask and 10 ml of dilute H_2SO_4. Run in from the burette the $KMnO_4$ solution slowly until the solution acquires just slight pink colour. Note the reading , say V_1 ml.

Now take 25 ml of prepared solution of the complex and also to it add 10 ml of dilute H_2SO_4. Titrate this against $KMnO_4$ similarly and note the end point, say V_2 ml.

Calculation

The strength of oxalic acid, produced from sodium ferrioxalate will be

$$= \frac{4 \times 0.63 \times V_2}{V_1} \text{gms litre}$$

Calculate the corresponding amount of sodium ferrioxalate in 1 litre.

$$Na_3[Fe(C_2O_4)_3] \equiv (COOH)_2 . 2H_2O$$
$$388.88 = 126$$

The strength of sodium ferrioxalate will be

$$= \frac{4 \times 0.63 \times V_2}{V_1} \times \frac{388.88}{126} \text{gms/litre}$$

2. Nickel Dimethyl Glyoxime Complex, $[Ni(DMG)_2]$

IUPAC name of DMG: 2, 3 butanedione dioxime.

Other name of DMG: Chugaev's Reagent

Theory: When alcoholic solution of DMG is allowed to react with a nickel salt in a slightly ammoniacal medium, a scarlet red precipitate of Ni—DMG complex is obtained. Nickel ammonium sulphate is taken which in aqueous solution dissociates into nickel, ammonium and sulphate ions.

$$NiSO_4 + 2(CH_3C = NOH)_2 + 2NH_4OH \longrightarrow$$

Reagents and their Preparation

(i)   Nickel ammonium sulphate = 4 gms

(ii)   1% alcoholic solution of DMG = 50 ml

(*iii*)   Conc. ammonia (1 : 1) = 50 ml

(*iv*)   Ethyl alcohol = 10 ml

   (*a*)   Weigh 4 gms of nickel ammonium sulphate and dissolve it in 100 ml of distilled water. If it is a clear solution, it is okay. If slight turbidity appears, add few drops of dilute HCl while shaking. It will give a clear solution.

   (*b*)   Dimethyl glyoxime is very sparingly soluble in water and therefore, its alcoholic solution is preferred. Dissolve 0.5 gram of DMG in 50 ml of ethyl alcohol. This is nearly 1% solution.

   (*c*)   Concentrated ammonia solution- Take nearly 25 ml of liquor ammonia in a stoppered bottle and add 25 ml of distilled water in it. Stopper it tightly to prevent the escape of ammonia gas.

Experimental Procedure

In 500 ml of Pyrex beaker, (resistant to heat) transfer the nickel ammonium sulphate solution in it. Now add 30 ml of DMG solution slowly with slight stirring. Now add slowly and slowly ammonia solution with a constant stirring. A scarlet precipitate slowly and slowly appears. Continue adding ammonia with stirring until a distinct smell of ammonia from the solution is obtained. The scarlet red precipitate covers the entire volume of solution.

To attain the crystalline precipitate, digest on water-bath by slow heating for about one hour. The precipitate will settle at the bottom leaving supernetant liquid clear and colourless.

Filler the precipitate while hot in a Gooch or sintered crucible. Wash the precipitate with warm water containing few drops of DMG and finally with ethyl alcohol. Allow it to dry and weigh.

3. Copper Tetrammine Complex, $[Cu(NH_3)_4] SO_4$

IUPAC Name: Tetrammine Copper (II) Sulphate

Other name- Tetrammine cupric sulphate

Theory: When copper sulphate reacts with liquor ammonia, we get in initital stage a pale blue precipitate of copper basic sulphate which dissolves in more of liquor ammonia to yield a dark blue solution of copper tetrammine ions:

$$CuSO_4 + 4NH_4OH \longrightarrow [Cu(NH_3)_4]SO_4 + 4H_2O$$

Reagents Required

(*i*)   Copper sulphate = 5 gms

(*ii*)   Concentrate ammonia solution = 10 ml

(*iii*) Ethyl Alcohol = 50 ml

Experimental Procedure

Weigh 5 gm of crystalline copper sulphate ($CuSO_4 .5H_2O$), powder it and dissolve it in about 25 ml of water. A slight turbid solution due to hydrolysis of copper sulphate is obtained. Add few drops of dilute sulphuric acid and shake it well to get its clear solution. Now take 5 ml of liquor ammonia in a test tube and add to it 4-5 ml of water. Add to this ammonia solution slowly in copper sulphate solution with a constant stirring by a glass rod. This gives to pale blue precipitate in initial stage which subsequently dissolves on the addition of more ammonia to give finally a clear deep blue solution. The solution must give smell of ammonia. Now add nearly 40 ml of ethyl alcohol with a constant stirring until the blue colour is nearly discharged. Allow, the solution to stand for 30 minutes. The needle shaped blue crystals of copper tetrammine sulphate are obtained. Drain the supernetant liquid and wash the precipitate with a small amount of ethyl alcohol. Dry and weigh crystals finally.

If these crystals are heated gently to 150°C, they decompose to give apple green powder of composition $CuSO_4 \cdot 2NH_3$

4. Cuprous Chloride, Cu_2Cl_2

Theory- Cuprous chloride is obtained by heating the mixture copper sulphate, copper turnings and concentrated HCl:

$$CuSO_4 + 2HCl + Cu \longrightarrow Cu_2Cl_2 + H_2SO_4$$

Reagents Required

(i)   Crystalline copper sulphate = 10 gms

(ii)   Sodium Chloride = 5 gms

(iii) Copper Turnings = 10 gms

(iv)   Sodium bisulplute = ~ 4 gms

(v)   Conc. HCl = 30 ml

Experimental Procedure:

Weigh all the first four (i to iv) chemicals separately and transfer these into a round bottomed flask. Now add near by 20 ml of conc. hydrochloric acid. Heat this mixture and keep an inverted funnel on the mouth of the flask. If the contents become viscous during heating, add some more acid. Continue heating until the blue colour of the copper sulphate disappears and a straw colour is produced. Take a beaker of 500 ml and fill nearly 400 ml of water. Add 4 gms of sodium bisulphite it. Now without any lapse of time, pour the contents of the flask into it. A white precipitate of chloride cuprous settles at the bottom. Wash the precipitate 2-3 times with water by decantation.

As cuprous chloride darkens on exposure to air to cupric chloride, keep the precipitate tightly corked.

Ideal yield = ~ 4 gms

Alternative Method

Dissolve 6 gms of copper sulphate and 1.8 gm sodium chloride in 5 ml of hot water. Take a separate beaker and in it dissolve 1.4 gm of sodium bisulphite and 0.9 gm of solid sodium hydroxide in 5 ml of cold water. Add this bisulphite and NaOH solution. Slowly and slowly to copper sulphate solution with a constant stirring. Cool the contents where a white precipitate of cuprous chloride settles at the bottom. Decant the supernetant liquid and wash the precipitate with water. Dry and store it in a stoppered bottle.

5. Chrome Alum, $K_2SO_4 \, Cr_2(SO_4)_3 \cdot 24\,H_2O$

It is a double salt of chromium and potassium and exists as a dark violet crystalline solid.

Theory: Potassuim dichromate with sulphuric acid oxidises ethyl alcohol to acetaldehyde. Both, potassuim sulphate and chromium sulphate are also produced. Sulphates upon crystallisation produce chrome alum.

$$K_2Cr_2O_7 + 4H_2SO_4 + 3C_2H_5OH \longrightarrow K_2SO_4 + Cr_2(SO_4)_3 + 3CH_3CHO + 7H_2O$$

Reagents Required

(i)   Potassuim dichromate (solid) = 10gm

(ii)   Conc. sulphuric acid = 10 ml

(iii) ethyl alcohol = 8 ml

Experimental Procedure

Take about 80 ml of water in a beaker and add very slowly say drpopwise, concentrated sulphuric acid. Now to it, add 10 gms of powdered solid potassuim dichromate. Keep this beaker in an icy cooled water to avoid any rise in temperature. Finally, add 8 ml of ethyl alcohol slowly. Wait for reaction to complete.

Concentrate this solution at about 50–60°C on a water bath for 3-4 hours. Allow the contents to cool where dark-violet crystals of chrome alum are produced. Removed the upper liquid by decantation and dry the precipitate.

Ideal yield = 10 gm.

6. Sodium Thiosulphate, $Na_2S_2O_3 . 5H_2O$

Common Name:

'Hypo' because of its use in photography

Theory: It is obtained by the interaction of free sulphur and sodium sulphite.

$$Na_2SO_3 + S \longrightarrow Na_2S_2O_3$$

Reagents Required

(i)   Sodium sulphite crystalline = 10 gm

(ii)   Role or powdered sulphur = 2 gm

Experimental Procedure

Dissolve 10 gms of sodium sulphite in 50 ml of water taken in a round bottomed flask and add to it 2 gms of powdered sulphur. Heat the contents nearly to boiling for about an hour. During this period, most of the sulphur combines with sodium sulphite leaving a clear solution. The unreacted sulphur being insoluble in water remains as a residue. Remove this residue by filtration. Concentrate this solution for about half an hour and allow it to cool where crystals of sodium thiosulphate are produced.

Remove the supernetant liquid by decantation and allow the crystals to become dry. The crystals are transparent and monoclinic

Ideal yield = 3.5 gm.

7. Ferrous sulphate, $FeSO_4 . 7H_2O$

Common Name: Green Vitriol

Theory: It is usually obtained from Kipp's Waste. The Kipp's apparatus is used for obtaining H_2S in the laboratory.

Experimental Procedure:

During the production of H_2S gas in the laboratory, the solution from the lowest globe of Kipp's apparatus is taken out from time to time to make it nearly 100 ml. Transfer this solution in a large porcelain dish and add some iron fillings. Heat the contents in fume-cupboard for about half an hour. The hot solution is filtered and filtrate is evaporated until crystallisation begins. Allow the contents to cool where crystals of ferrous sulphate grow. Filter it these and dry.

8. Carbonato tetra ammine cobaltic nitrite (Tetra ammine carbonato Cobalt (III) nitrate), $[Co(NH_3)_4 . CO_3]NO_3$

Add a solution of ammonium carbonate (i.e., 20 gm in 100 ml of water and 50 ml of conc. ammonia to the cobalt nitrate (20 gm) solution made in water (20 ml)). Draw a stream of air through the contents for about three hours till the solution changes from deep blue to red. Concentrate this solution to 50 ml and add one gram of amm. carbonate after every 15 minutes in all five or six times.

solution to 50 ml and add one gram of amm. carbonate after every 15 minutes in all five or six times.

Filter and concentrate the filtrate to near about 40 ml and once again add nearly 2 gm ammonium carbonate. Keep this solution for sometime for crystallisation. Carmine red crystals are formed which are collected in Buckner funnel. Wash these crystals first by small amount of cold water and finally by alcohol. Dry in air.

9. Ferrous Ammonium Sulphate (Mohr's Salt), $[FeSO_4.(NH_4)_2SO_4.6H_2O]$

Weigh almost 9 gm of ferrous sulphate and 4-5 gm of ammonium sulphate. Put these together in a beaker and add hot distilled water (containing few drops of H_2SO_4) slowly by constant stirring just sufficient to dissolve them completely. Cover the beaker, boil for 2-3 minutes and then cool for sometime. Light blue crystals of ferrous ammonium sulphate separate out. A small crystal of ferrous amm. sulphate when added during cooling, assists the crystallisation. Yield about 7 gm.

10. Cuprous oxide (Cu_2O)

It is usually prepared by reducing copper sulphate in alkaline medium by glucose. The reactions in the preparation taking place are :

$$CuSO_4 + 2NaOH \longrightarrow Cu(OH)_2 + Na_2SO_4$$
$$C_6H_{12}O_6 + 2Cu(OH)_2 \longrightarrow Cu_2O + C_5H_{11}O_5.COOH + 2H_2O$$
gluconic acid

Dissolve nearly 25 gm of powdered copper sulphate in a requisite quantity of water (100—125 ml) Heat the contents to boiling. Add to it now 15 gm of glucose followed by 10 ml of conc. NaOH by constant stirring. Continue the heating till the blue colour of copper sulphate disappears. Leave the solution to cool. This provides bright reddish precipitate of cuprous oxide after sometime.

11. Magnesium Sulphate, $(MgSO_4.7H_2O)$

It is conveniently prepared by the action of dil. H_2SO_4 upon magnesium carbonate. For this, we take mineral magnesite which has chiefly magnesium carbonate. The reaction taking place is :

$$MgCO_3 + H_2SO_4 \longrightarrow MgSO_4 + H_2O + CO_2$$

Take 10 gm of mineral and powder it finely. Make its paste by distilled water. Now add dil. H_2SO_4 gradually by stirring till the further addition of acid does not give effervescence. The acid should not be in excess, if it becomes, add a little mineral again. Heat the contents to boiling and filter. The filtrate is evaporated till the point of crystallisation. Cooling separates the crystals of magnesium sulphate (Epsom salt) completely.

12. Sodium cobaltinitrite, $Na_3[Co(NO_2)_6]$ (Sodium hexanitrito cobaltate III)

Dissolve 10 gm of $NaNO_2$ in 15–20 ml of warm water and after cooling, dissolve 3 gms. of cobalt nitrate by stirring. To it, add 2 ml of glacial acetic acid. Transfer this solution to a conical flask and draw the current of air through the contents for more than half an hour. Filter off the precipitate and to the filtrate, add 10–15 ml of alcohol and leave the solution to stand for about four hours ; the crystals of sodium cobaltinitrite separate out. These are further washed by alcohol and finally dried. The yield is 5 gm.

Reactions :

$$Co(NO_3)_2 + 2NaNO_2 \longrightarrow Co(NO_2)_2 + 2NaNO_3$$
$$CH_3COOH + NaNO_2 \longrightarrow CH_3COONa + HNO_2$$
$$2HNO_2 + 3NaNO_2 + Co(NO_2)_2 \longrightarrow NO\uparrow + Na_3[Co(NO_2)_6] + H_2O$$
Sod. Cobaltinitrite
yellow

Note. To draw the current of air, we close the conical flask with a cork having two delivery tubes. The mouth of one is left in the open air and other end is dipped in the contents. The outer delivery tube is kept quite up from the flask solution and other end of it is joined with the filter pump.

13. Sodium chloride from common salt

Prepare 500 ml saturated solution of common salt in distilled water. Place the clear solution in a wide beaker. Pass hydrochloric acid gas from the mouth of an inverted funnel dipping just below the surface of liquid.

Filter, wash with pure conc. hydrochloric acid, partially dry with filter paper then in air and finally by heating in a porcelain dish, to remove last traces of hydrochloric acid.

One impurity in common salt is magnesium chloride which by its deliquescence, causes the salt to become damp when exposed to the air. The hydrochloric acid gas increases chloride ion concentration and the solubility product is soon reached, i.e., solution becomes supersaturated hence pure sodium chloride is precipitated. Due to small concentration Mg^{2+} ions remain in solution.

14. Potash alum, $K_2SO_4.Al_2(SO_4)_3.24H_2O$

Dissolve 2·5 gm K_2SO_4 and 10 gm $Al_2(SO_4)_3$ in 40 ml of water in a beaker. Add 2-3 drops of H_2SO_4 to make the solution clear. If still the solution has some milkiness, filter. Take the filtrate into a porcelain dish and evaporate to reduce the bulk to half. Cool and allow it to remain undisturbed for 5-8 hours. Octahedral crystals of potash alum will be obtained. Dry them between filter papers.

$$K_2SO_4 + Al_2(SO_4)_3 + 24H_2O \longrightarrow K_2SO_4.Al_2(SO_4)_3.24H_2O$$

15. $K_2Cr_2O_7$ from chromium acetate

Chromium acetate is not very soluble in water. Prepare the saturated solution of chromium acetate in dil. HCl. Add to it, excess of NaOH (solid) and Br_2 water and heat. The green solution turns to yellow indicating the formation of sodium chromate. Take a little of this solution into a test tube, acidify with acetic acid and add lead acetate. If yellow ppt. of lead chromate is obtained, it shows complete conversion of chromium acetate to sodium chromate. Now acidify the yellow solution with conc. H_2SO_4.

$$2Na_2CrO_4 + H_2SO_4 \longrightarrow Na_2Cr_2O_7 + Na_2SO_4 + H_2O$$

On cooling, less soluble $Na_2SO_4.10H_2O$ crystallises out and is removed by filtration. To the filtrate containing sodium dichromate, add a sufficient amount to KCl and heat.

$$Na_2Cr_2O_7 + 2KCl \longrightarrow K_2Cr_2O_7 + 2NaCl$$

NaCl precipitates out from hot solution and is removed by filtration. On cooling the filtrate, the crystals of $K_2Cr_2O_7$ (orange-red) separate out. Melting point of $K_2Cr_2O_7$ crystal is 398°C.

16. Potassium trisoxalato aluminate.

A solution of 3-4 gm aluminium sulphate (or of its any salt of equivalent amount) is treated with a solution of 1·2 gm of sodium hydroxide. The precipitated aluminium hydroxide is filtered, washed and boiled with a solution of 3·84 gm of potassium hydrogen oxalate (or a mixture of 2.76 gm. of potassium oxalate monohydrate and 1.89 gm of oxalic acid dihydrate) in about 40 ml of water. Any aluminium hydroxide which does not dissolve is filtered out and filtrate is evaporated to crystalliza-tion. The yield is nearly quantitative.

17. Potassium trisoxalato chromate.

To a solution of 2·3 gm of potassium oxalate monohydrate add 5·5 gm of oxalic acid dihydrate in 80 ml of water 1·9 gm of powdered potassium dichromate is added in small portions with vigorous stirring. When the reaction is complete, the soluton is evaporated to crystallize. Potassium trisoxalato chromate form deep green crystals. The yield is 4-5 gm (about 90%).

18. Sodium Ammonium Hydrogen Phosphate (Micro-cosmic salt), $Na(NH_4)HPO_4.4H_2O$

Calculate molecular proportions of disodium hydrogen phosphate and amm. chloride to make 8·5 gm of the mixture (*i.e.*, 6·5 gm Na_2HPO_4 and 2 gm NH_4Cl). Dissolve Na_2HPO_4 by adding it in portions to 10 ml of boiling water in a boiling tube and shake during solution. Add ammonium chloride, dissolve, filter if necessary and allow to crystallise. By fractional crystallisation free the substance from NaCl.

$$Na_2HPO_4 + NH_4Cl \longrightarrow Na(NH_4)HPO_4 + NaCl$$

The yield is 4·5 gm.

Microcosmic salt is efflorescent. It decomposes at 316°C leaving sodium metaphosphate which unites with metallic oxides forming orthophosphates, some of which are coloured. It is, therefore, used as a dry test for metals.

$$Na(NH_4)HPO_4 \longrightarrow NaPO_3 + NH_3 + H_2O$$
$$NaPO_3 + CuO \longrightarrow CuNaPO_4$$

(blue bead)

Silica is insoluble in metaphosphate bead,

$$CaSiO_3 + NaPO_3 \longrightarrow CaNaPO_4 + SiO_2$$

Microcosmic salt is found present in human urine (microcosmohuman), hence its name.

19. Potassium chlorochromate, $CrO_2Cl(OK)$.

Take 5 gm of finely powdered potassium dichromate in a big silica dish, add 2 ml of conc. HCl and heat on water bath to 60—70°C. Hydrochloric acid may be added, a little at a time, until all the salt has dissolved. Allow it to remain undistrubed for overnight. The red crystals of potassium chlorochromate will appear. Separate the crystals and dry them on a porous plate. The yield is 5 gm.

$$K_2Cr_2O_7 + 2HCl \longrightarrow 2CrO_2Cl(OK) + H_2O + 2KCl$$

Red

20. Chrome red, $PbCrO_4.PbO$

Dissolve 20 gm of lead nitrate in water. Dissolve 12 gm of potassium chromate in water in a separate test tube. Mix these solutions in a beaker with stirring. The precipitate of lead chromate (chrome yellow) so obtained is washed well by decantation and transferred to a beaker containing 5% NaOH solution. Boil until colour of the precipitate changes to bright red. Filter the precipitate of chrome red and dry it between the folds of filter paper.

$$Pb(NO_3)_2 + K_2CrO_4 \longrightarrow PbCrO_4 + 2KNO_3$$
$$2PbCrO_4 + 2NaOH \longrightarrow PbCrO_4.PbO + Na_2CrO_4 + H_2O$$

Red

21. Barium thiocarbonate, $BaCS_3$

Take about 50 ml of distilled water in a conical flask and heat it. To it, dissolve 16 gms of barium hydroxide and divide the solution into two portions. In first portion pass H_2S for about 10 minutes. The pH of this solution should become around 8. Add second portion to first one and cool the contents. To it, add about 3 ml of *carbon disulphide carefully. Close the flask with air tight stopper and shake it 3–4 minutes everytime for in all 30 minutes. This gives yellow crystalline precipitate of barium thiocarbonate. Filter the precipitate first with cold water, then with 50% ethyl alcohol and finally with alcohol alone. Dry at 110°C and weigh. Reactions taking place are:

$$Ba(OH)_2 + H_2S \longrightarrow BaS + 2H_2O$$
$$BaS + CS_2 \longrightarrow BaCS_3$$

---

* CS_2 is highly inflammbale.

12

Qualitative Organic Analysis

Introduction

The identification of organic compounds by qualitative tests involves a study of various chemical characteristics and then a careful correlation of observed facts. The general procedure adopted for systematic analysis consists of following steps :

1. Preliminary examination
   - (a) Physical state
   - (b) Colour
   - (c) Odour
2. Ignition test
3. Solubility behaviour
4. Test for unsaturation
5. Detection of elements
6. Class Determination — Test for functional group
7. Determination of melting or boiling point
8. Specific tests
9. Preparation of derivatives.

The data collected under above heads if carefully interpreted, give a definite clue regarding the given compound.

1. Preliminary Examination

Physical State, Colour and Odour. A careful observation of the physical state, colour and characteristic smell of the compound gives useful information regarding the nature of the compound. For example, the colour of a compound is related to its chemical constitution or else to some impurities in it. Similarly, there are certain groups of compounds which have similar smell. The following table is helpful to some extent in correlating the physical characteristics and the chemical nature.

Colourless solids	:	Carbohydrates, simple acids, some phenols, amides and anilides.
Coloured solids	:	Nitro compounds, amines, phenols, quinones.
Colourless liquids	:	Alcohols, aldehydes, ketones, simple hydrocarbons and simple acids.
Coloured liquids	:	Nitro compounds, phenols and amines.
Carbolic smell	:	Phenols.
Smell of bitter almonds	:	Benzaldehyde and nitrobenzene.
Fruity-pleasant smell	:	Esters.
Spirituous smell	:	Alcohols.
Pungent smell	:	Formic acid or formalin.

2. Ignition Test

Take a small portion of the given compound on a metallic spatula and ignite it over nonluminous flame and note the changes :

Observation	Inference
(a) Burns with a sooty flame	: May be an aromatic compound.
(b) Yellow and non-sooty flame	: May be an aliphatic compound.
(c) Chars	: May be a carbohydrate, hydroxy acid or sulphonic acid.
(d) Smell of ammonia	: Nitrogenous compound like urea.

Note. Some of the aliphatic compounds like $CHCl_3$, CCl_4 and C_2Cl_6, owing to high percentage of carbon, give sooty flame.

3. Solubility Behaviour

The behaviour of the compound towards various solvents like water, dilute caustic soda, dil. HCl and conc. H_2SO_4 also reveals its nature. Take a small portion of the substance and note its solubility in above solvents :

Observation	Inference
(a) Soluble in hot water and solution is acidic to litmus	: Salts of aromatic bases. Lower aliphatic acids, hydroxy acids or polyhydroxy phenols.
(b) Soluble in hot water and solution is acidic	: Higher acids or some phenols.
(c) Soluble in cold water and solution is neutral	: Carbohydrates or alcohols.
(d) Soluble in hot water and solution is neutral	: Starch, quinones.
(e) Soluble in cold dil. $NaHCO_3$ with effervescence	: Carboxylic acids.
(f) Soluble in cold dil NaOH	: Carboxylic acids or phenols.
(g) Soluble in hot NaOH and give off ammonia	: Amides, ammonium salts.
(h) Soluble in cold conc. H_2SO_4	: Aromatic hydrocarbons or phenols.
(i) Soluble in hot conc. H_2SO_4 and charring occurs	: Carbohydrates, aldehydes, ketones or hydroxy acids.
(j) Soluble in dil. HCl	: Amines.

4. Test for Unsaturation

(a) **Test with Bromine.** Take a solution or suspension of the compounds in water and add to it few drops of bromine water. You may also take a solution/suspension of the compound in carbon tetrachloride and add to it a solution of bromine, shake well and note the changes :

Observation	Inference
(i) Decolourisation :	Unsaturated compounds containing $\diagup C = C \diagdown$ bond
(ii) A precipitate is obtained :	Amines and phenols.

(b) **Test with Alkaline KMnO₄ Solution.** Dissolve a small amount of the substance in water or acetone. Add to it. 1 ml of dil. sodium carbonate solution and a few drops of potassium permanganate solution and shake well. *If the colour of the permanganate disappears, the given compound may be unsaturated. Heating is forbidden in this test.*

Detection of Elements

All orgainc compounds contain carbon. In addition, they may contain hydrogen, oxygen, nitrogen, sulphur and halogens. No attempt is made to detect the presence of carbon, hydrogen and oxygen; their presence is usually inferred from chemical reactions. The detection of nitrogen, sulphur and halogens is carried out by converting them into water soluble ionic compounds. This has to be done very carefully because further analysis of the organic compound is largely based on the type of the element present in it.

Tests for Nitrogen, Sulphur and Halogens

Nitrogen, sulphur and halogens are usually detected by *Lassaigne's test.*.

Lassaigne's test. (Sodium Extract). In this test, these elements are converted to water soluble sodium compounds upon fusion with metallic sodium.

In actual procedure, a freshly cut small piece of sodium is flame first dried by pressing between the folds of filter paper and then put into an ignition tube. The ignition tube is heated very gently until sodium melts. The ignition tube is removed from the flame and a small quantity of the size of a rice-grain (or 2-3 drops if liquid) of the given compound is put inside it so that it may come in contact with the molten sodium. Heating is again continued first very gently and then strongly until the bottom of the tube is red hot. It is then plunged into about 10 ml of distilled water contained in a beaker or porcelain dish. The ignition tube if not broken completely, is crushed with the help of a glass rod. The contents in the dish are stirred with glass rod, boiled for about five minutes and filtered. The filtrate is called sodium extract with which the qualitative tests for nitrogen, sulphur and halogens are performed as ahead. The filtrate should be water clean and alkaline. If it is dark in colour, it indicates incomplete fusion and the process is repeated with fresh ignition tube.

1. *Test for nitrogen.* Take about 2–3 ml of sodium extract (fusion solution) in a test tube and add to it 0.1–0.2gram of powdered ferrous sulphate. This gives a dirty green precipitate of ferrous hydroxide. If no precipitate is formed, add a few drops of dilute sodium hydroxide solution to get the precipitate. Now heat this mixture gently with shaking to boiling and without cooling, add dilute sulphuric acid to dissolve the precipitate and add some more to make the solution acidic. *A prussian blue colour or precipitate confirms the presence of nitrogen.* If the colour or precipitate does not appear atonce allow it to stand for ten minutes to obtain so. In absence of nitrogen, the solution will remain pale yellow due to the presence of iron compound.

2. *Test for sulphur,* (*i*) To about 1 ml of sodium extract, add 1 to 1.5 ml of freshly prepared sodium nitroprusside solution. If a purple colour is obtained, sulphur is present. This colour fades away slowly.

(*ii*) Acidify 1 ml of sodium extract with excess of acetic acid and then add a few drops of lead acetate solution. A black precipitate indicates the presence of sulphur.

3. *Test for Halogens.* After ascertaining the presence or absence of nitrogen and sulphur, test for halogens should be performed. For it two separate procedures may be adopted :

(*a*) *When nitrogen and sulphur are absent.* Acidify 1-2 ml of sodium extract with dilute nitric acid and to it, add 2-3 ml of silver nitrate solution and observe :

(*i*) a white precipitate soluble in NH_4OH indicates the presence of chlorine.

(*ii*) a pale yellow precipitate, sparingly soluble or soluble in excess of NH_4OH indicates the presence of bromine.

(*iii*) a yellow precipitate insoluble in NH_4OH indicates the presence of iodine.

Bromine and Iodine. Acidify 1-2 ml of sodium extract with dilute HNO_3 and 1 ml of chloroform and excess of chlorine water and shake well. If the chloroform layer becomes yellow or brown bromine is present. If the layer turns violet, presence of iodine is indicated.

(b) When *nitrogen* and/or *sulphur present.* Take about 2-3 ml of sodium extract and acidify with some dilute nitric acid. Now evaporate the solution to almost 1 ml. This will expel hydrogen cyanide and hydrogen sulphide. Dilute the solution with equal volume of water and test for halogens as given in part 'a'.

Chemistry involved in Lassaigne's test. When an organic compound is fused with sodium, the nitrogen of the compound is converted into cyanide, sulphur into sodium sulphide and halogens into sodium halides. These sodium salts being soluble in water come into aqueous solution. These salts are ionisable and hence the presence of cyanides, sulphide or halide ions is easily detected through the usual tests.

$$Na + C + N \longrightarrow NaCN$$
$$\text{Sodium cyanide}$$

$$2Na + S \longrightarrow Na_2S$$
$$\text{Sodium sulphide}$$

$$Na + X \longrightarrow NaX$$
$$\text{Sodium halide}$$

$$(X = Cl, Br, I)$$

When nitrogen and sulphur both are present, sodium thiocyanate is formed.

$$Na + C + N + S \longrightarrow NaCNS$$

In the test for nitrogen, when ferrous sulphate is added to sodium extract, it reacts with the alkali (produced by reaction between Na and H_2O) and dirty green precipitate of ferrous hydroxide is produced. This reacts with sodium cyanide to give sodium ferrous cyanide which with ferric ion produces bluish ferric ferrocyanide.

$$FeSO_4 + 2NaOH \longrightarrow Fe(OH)_2 \quad + Na_2SO_4$$
$$\text{Ferrous hydroxide}$$
$$\text{(dirty green)}$$

$$Fe(OH)_2 + 6NaCN \longrightarrow Na_4Fe(CN)_6 \quad + 2NaOH$$
$$\text{Sodium ferrocyanide}$$

$$3Na_4Fe(CN)_6 + 4FeCl_3 \longrightarrow Fe_4[Fe(CN)_6]_3 \quad + 12NaCl$$
$$\text{Ferri-ferrocyanide}$$
$$\text{(Prussian blue)}$$

In the test for sulphur, a violet coloured complex is produced upon addition of sodium nitroprusside to the sodium extract.

$$Na_2S + Na_2[Fe(NO)(CN)_5] \longrightarrow Na_3[Fe(O = N - S - Na)(CN)_5]$$
$$\text{Sodium nitroprusside} \qquad\qquad \text{Purple coloured complex}$$

When the sodium extract is treated with lead acetate, a black precipitate of lead sulphide insoluble in acetic acid is produced.

$$Na_2S + (CH_3COO)_2Pb \longrightarrow PbS + 2CH_3 COONa$$
$$\text{Black precipitate}$$

In the test for halogens, when sodium extract is boiled with conc. HNO_3, sodium cyanide, if present, is decomposed and is eliminated as HCN gas. If NaCN is not decomposed, it gives white precipitate of $AgNO_3$ even in absence of halogens hence it is confusing.

After decomposing NaCN, $AgNO_3$ is added thereby silver halide is precipitated.

$$NaCN + HNO_3 \longrightarrow NaNO_3 + HCN$$
$$NaX + AgNO_3 \longrightarrow AgX + NaNO_3$$
$$\text{precipitate}$$

The chloroform test for bromine and iodine is based on the principle that chlorine liberates bromine and iodine from sodium bromide and iodide. Free bromine and iodine are soluble in chloroform, therefore, chloroform layer becomes yellow if bromine is present and it becomes violet if there is iodine.

$$2NaBr + Cl_2 \longrightarrow 2NaCl + Br_2$$
$$2NaI + Cl_2 \longrightarrow 2NaCl + I_2$$

TEST FOR FUNCTIONAL GROUPS

The determination of functional group depends largely on the correct determination of elements. Once the functional group present in the organic compound is known, one is able to find out the name of the probable compound with the help of M.P. or B.P. Next procedure is to study some specific reactions of the compound and then to confirm its name by preparing suitable derivatives. It may be noted that element carbon, hydrogen and oxygen are not tested for. This clearly means that if no other element is present, functional group containing carbon, hydrogen and oxygen are determined.

The test for functional groups should be performed according to the following way :

[A] When only Carbon, Hydrogen and/or Oxygen are present

S.No.	Class	Functional group	Tests
1.	Carboxylic acid	$-C\diagfrac{O}{OH}$	(i) **Litmus Test.** Shake a small amount of the given substance in water and dip a blue litmus paper in it. If the litmus paper turns red, it shows presence of carboxylic group (Phenols being acidic in nature also give this test).
			(ii) **Sodium bicarbonate test.** Add a pinch of the given substance if solid (or a few drops if liquid) to 5 ml of cold 50% solution (aqueous) of $NaHCO_3$ and shake gently. Effervescence (or slow up and down movement of particles) indicates presence of carboxylic group.
2.	Phenols	Phenolic —- OH	(i) Blue litmus changes to red but no effervescence with $NaHCO_3$.
			(ii) **$FeCl_3$ Test.** To 1 ml of aqueous or alcoholic solution of original substance (O.S.) add 2-3 drops of neutral or very dilute solution (aqueous) of ferric chloric. A blue, green, red or violet colour indicates presence of phenolic group (— OH).
			(iii) **Libermann's Test.** All phenols in which para position is free respond to this test.
			Take 0.2 gm of O.S. and a few crystals of $NaNO_2$ in a dry test tube and heat gently for a minute. Cool the mixture, add 1 ml of conc. H_2SO_4 and shake. A deep green or blue colour develops, upon dilution with water the solution turns red. To it, add excess of dil. NaOH solutions, the red solution again becomes deep green or blue. It indicates the presence of phenolic group.

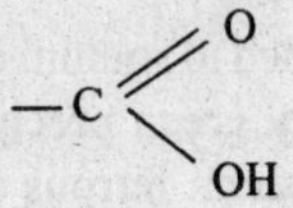

			(iv) **Phthalein Test.** Take 0.2 gm each of O.S. and phthalic anhydride in a dry test tube, add 0.5 ml of conc. H_2SO_4 and heat for a minute. Cool. Make it alkaline with dil. NaOH solution Pour a few drops of this solution in 20 ml water taken in a beaker. Characteristic colour shows the presence of phenolic group, e.g. Pink colour — Phenol, o-Cresol Blue colour — Catechol, m-Cresol Fluorescent green — Resorcinol No colouration — p-Cresol.
3.	Carbohydrate	....	(i) **Blackening with H_2SO_4.** Heat 0.2 gm of O.S. with 1 ml conc. H_2SO_4. An immediate charring and blackening of the solution shows presence of carbohydrate.
			(ii) **Molisch's Test.** To 2 ml aq. solution of O.S. add 1 ml of Molisch's reagent and shake well. Now add carefully 1-2 ml of conc. H_2SO_4 (taken in a separate test tube) from the side of the test tube. Allow to stand. A reddish violet ring at the junction of two liquids shows the presence of carbohydrate.
			(iii) To 1 ml aqueous solution of O.S. add 1-2 drops of anthrone solution, shake and warm the mixture gently. A green or bluish green colour shows the presence of carbohydrate.
4.	Aldehydes	$-C\!\!\diagup^{\displaystyle O}_{\diagdown H}$	(i) **Schiff Test.** To 1 ml aqueous or alcoholic solution of O.S. add 2 ml of Schiff reagent and shake. A deep red or violet colour shows the presence of aldehydic group (Never heat during this test).
			(ii) **Tollen's Test.** To about 2 ml of Tollen's reagent add about 0.5 gm of O.S. and heat the mixture on water bath. Formation of silver mirror or blackish precipitate indicates the presence of aldehydic group. (This test is also given by reducing sugars).
			(iii) **Fehling Solution Test.** Add Fehling solution B to A until a blue precipitate first formed is redissolved to give a deep blue solution. To it add a little of O.S. and heat for 2 minutes. Reddish brown precipitate shows the presence of aldehydic group. (This test is also given by reducing sugars.)
5.	Ketones	$\diagdown_{\diagup}C=O$	(i) No colour with Schiff reagent.
			(ii) **2, 4 Dinitrophenylhydrazine Test.** Add 1-2 drops (0.5 gm) of O.S. to 2 ml of 2, 4 dinitrophenylhydrazine solution (Brady's regent). Shake vigorously. Heat and cool. A red, yellow or orange, coloured precipitate indicates the presence of ketonic group.

6.	Esters		(i) **Hydroxamic acid Test (Feigl Test).** Take a rice grain (R. G.) or two drops of the compound in a test tube, add 1 ml of 5% methanolic solution of hydroxylamine hydrochloride. Now, add a few drops of 2N methanolic KOH untill the solution is just alkaline. Boil, cool and acidify with dil. HCl solution. If Deep red or violet colour is produced, it indicates the presence of ester group.
			(ii) **Hydrolysis Test.** Dissolve 0.5 gm of O.S. in 2 ml alcohol. To it, add 4-5 drops of dil. NaOH solution and 2-3 drops of phenolphthalein. A pink colour is produced. Now put the test tube in boiling water bath for 5 min. If the pink colour either fades or disappears, this shows presence of ester group.
7.	Alcohols	— OH	(i) In a dry test tube, take 2 ml of O.S., add 1 gm of anhydrous sodium sulphate (to absorb water if already present) and filter. To the filtrate, add a small piece of sodium. Effervescence due to evolution of H_2 shows the presence of an alcohol.
			(ii) To 1 ml of O.S. or its aqueous solution, add a few drops of cerric ammonium nitrate. A pink or red colour shows the presence of an alcohol.
8.	Hydrocarbons	Nil	No specific group test.

Note : **R.G.** means rice grain.

O.S. means original substance.

[B] When nitrogen is present

S.No.	Class	Functional group	Tests
1.	Amides		To 0.2 gm or 0.5 ml of O.S., add 1 ml of aq. NaOH and heat, Smell of ammonia shows the presence of an amide.
2.	Primary amines	— NH_2	(i) **Carbylamine Test.** Take a small amount of substance, 4 ml of alcoholic solution of KOH and 2 drops of chloroform in a test tube shake and heat gently. An intolerable offensive odour of carbylamine shows the presence of primary amine.
			(ii) **Nitrous Acid Test.** Take 0.2 gm of O.S. in 2 ml of dil. HCl and cool. Now add 10% aq. $NaNO_2$ solution. A brisk effervescence shows the presence of an aliphatic primary amine.
			(iii) **Diazotisation Test.** Dissolve 0.2 gm or 0.5 ml of O.S. in 2-3 ml of dil. HCl. Cool under tap water. Now add 2 ml of 2.5%

3.	Secondary amines	>NH
4.	Tertiary amines	$>$N:
5.	Anilides	ArNHCOR

NaNO$_2$ solution (aqueous), cool again and add 0.5 ml of alkaline β-naphthol solution. An orange-red or a red dye shows the presence of **aromatic** primary amine.

(*i*) Do not give carbylamine test.

(*ii*) **Nitrous Acid Test.** Take 0.2 gm or 0.5 ml of O.S., add 2 ml of dil. HCl and cool under tap water. Add 2 ml of 2.5% NaNO$_2$ solution. Yellow oily drops are obtained. This shows the presence of secondary amine. (This test is given by both aliphatic and aromatic secondary amines.)

(*iii*) **Libermann's Nitroso Reaction.** Take 0.2 gm or 0.5 ml of O.S., add 2 ml of dil. HCl and cool under tap water. Add 2.5% aqueous solution of NaNO$_2$ gradually with shaking until yellow oil separates at the bottom. Decant off the aqueous layer. Now, take 2 drops of this oily nitroso compound in a dry test tube, add 0.5 gm of phenol and warm gently for a few seconds. Cool and add 1 ml of conc. H$_2$SO$_4$. A greenish blue colour is obtained which changes red upon dilution with water. On adding excess of NaOH solution, the greenish blue colour is restored. This shows the presence of secondary amine.

(*i*) Do not give carbylamine test.

(*ii*) Dissolve 0.2 gm or 0.5 ml of O.S. in 2-5 ml of dil. HCl and cool under tap water. Add 2-5 ml of 2.5% aqueous solution of NaNO$_2$ gradually with shaking and observe :

(*a*) No reaction indicates the presence of aliphatic tertiary amine.

(*b*) Production of green or brown coloured salt indicates the presence of aromatic tertiary amine.

(*i*) **Carbylamine Test.** Heat a mixture of 0.5 gm of O.S., 4 ml of alcoholic KOH solution and 1 ml chloroform. Bad smell of carbylamine comes out.

(*ii*) **Tafel's Test.** Take 0.2 gm of O.S., add 5 ml of conc. H$_2$SO$_4$ and shake. Add 0.2 gm of powdered potassium dichromate. A red or violet colour which changes to green on standing, shows the presence of an anilide.

(*iii*) **Hydrolysis Test.** Take 0.2 gm of O.S., add 5 ml of dil. HCl, boil and cool. Add 2ml of 2.5% NaNO$_2$ solution, cool again and add 0.5 ml of alkaline β-naphthol solution. An orange-red dye shows the presence of an anilide. (Anilides are hydrolysed on boil-

			ing with HCl and thus liberate primary amino group which gives the *dye test*.)
6.	Nitro compounds	$—NO_2$	(i) **Mulliken-Barker Test.** Take 0.2 gm or 2-3 drops of O.S. in a test tube, add 4-5 ml of alcohol, a pinch of zinc dust and nearly 10 drops of 10% $CaCl_2$ or (NH_4Cl solution). Boil and cool. Filter it directly into 2 ml. Tollen's reagent, taken in another test tube. On warming, a grey or black precipitate or silver mirror is produced which confirms presence of nitro group.
			(ii) **Azo Dye Formation.** In a boiling tube, take 0.2 gm or 3 drops of O.S., 3 ml conc. HCl, 2 ml water, and 1 gm of solid stannous chloride (or tin granules). Heat the contents on water bath for few minutes. Filter and cool. To the filtrate, add 2.5% aq. $NaNO_2$ solution drop by drop to complete the diazotization. Cool again and add 1-2 ml of alkaline β-naphthol solution. An orange-red or red dye shows the presence of nitro group. (The dye is formed due to liberation of primary amino group by reduction of nitro group.)

[C] When Nitrogen and Sulphur are present

1.	Thioureas	$—HN—C—NH—$ with $\overset{\|}{\underset{S}{}}$	(i) Simple thiourea upon boiling with NaOH solution, gives off ammonia.
			(ii) Boil a little amount of the substance with dil. NaOH solution, cool and add lead acetate. A brown or black colour of precipitate indicates the compound to be a thiourea.
2.	Amino Sulphonic acid	$—NH_2$ and $—SO_3$ groups	Dissolve a little of the O.S. in dil. HCl. Cool and add $NaNO_2$ solution so as to complete the diazotisation. Now, add alkaline solution of β-naphthol. Red or orange dye is formed.

[D] When Halogen is present

1.	Halogenated (a) Aliphatic halogen compounds, RX	....	Boil 0.5 ml of O.S. with 5 ml of alcoholic NaOH solution for about 10 minutes. Cool, dilute the solution with water, add excess of dil. HNO_3 and then $AgNO_3$ solution. A precipitate of silver halide is obtained.
	(b) Aromatic halogen compounds	....	Test depends on the type of the additional functional group present.

Determination of Melting or Boiling Point

This is the most important step in the identification of an organic compound. Pure organic compounds have fixed melting or boiling points and hence this determination helps to identify the individual compound once the nature of the functional group present in it is ascertained.

Determination of melting point

The compound whose melting point is to be determined, is powdered thoroughly on a porous plate with the help of a spatula. A capillary tube of approximately 2″ length is sealed at one end by heating in a Bunsen flame. It is then filled up to about 1 cm length with the powdered substance. The capillary is then attached to the lower end of the thermometer as shown in Fig. 12.1. The thermometer is now placed in a Thiele's tube filled with paraffin oil or concentrated sulphuric acid such that the liquid convers atleast the filled length of the capillary. Cork used is split one to allow for expension of air.

The flask is gently heated and rise in temperature is observed carefully. The tempetature at which the substance begins to liqueify is noted. The temperature at which the solid has completely changed into liquid is also noted. This range of temperature is recorded as M.P. range of the substance.

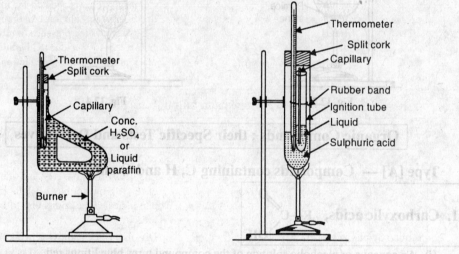

Determination of boiling point

The boiling point of a liquid may be recorded in either of the two ways depending on the availability of the appliances.

(*i*) **First Method.** A few drops of the liquid whose boiling point is to be determined is taken in an ignition tube. A capillary tube sealed at the upper end is put inside the ignition tube and the latter is attached to the lower part of the thermometer with the help of a rubber thread. The thermometer along with the ignition tube is placed inside a pyrex test tube in such a way that the liquid inside the ignition tube is covered by concentrated H_2SO_4 taken in the pyrex tube as shown in Fig. 12.2.

The test tube is heated slowly and the rise of bubbles inside the capillary is carefully observed. The temperature at which a regular and speedy stream of bubbles begins to escape is taken to be the boiliing point of the liquid. This is recorded.

(*ii*) **Second Method.** As shown in Fig. 12.3, the liquid whose boiling point is to be determined is taken in a distillation flask. Some pumice stones or porcelain pieces are added to avoid bumping. A thermometer is fitted in the flask through a cork. The delivery tube of the flask is connected to a water condenser. If the condenser is not provided, vapours can be also collected in a test tube.

The flask is heated and rise in temperature is carefully observed. The liquid begins to distil over after some time. The bulk of the liquid distils over within a certain temperature range which remains nearly constant throughout the distillation. This temperature range is taken to be the boiling point of the liquid.

This method is useful because in addition to determination of the boiling point, one can also purify the given liquid by distillation. Now-a-days, the boiling point is more conveniently determined by test-tube like flask with an outlet at the top from where the vapours are collected in a tube (Fig. 12.4).

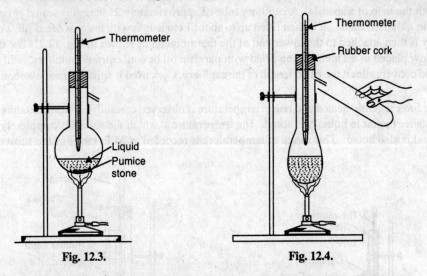

Fig. 12.3.   **Fig. 12.4.**

Organic Compounds ; their Specific Tests and Derivatives

Type [A] — Compounds containing C, H and O only

1. Carboxylic acids, R—C$\overset{\displaystyle O}{\underset{\displaystyle OH}{}}$

(*i*) An aqueous or alcoholic solution of the compound turns blue litmus red.

(*ii*) Add about 1/2gram (a few drops if liquid) of the substance to a saturated solution of sodium bicarbonate. There will be effervescence due to evolution of CO_2 gas and the substance will dissolve in it.

$$RCOOH + NaHCO_3 \longrightarrow RCOONa + H_2O + CO_2\uparrow$$

Characteristic derivatives (General method of preparation) :

(*i*) *Amide derivative.* In a dry flask fitted with a reflux condenser, take a mixture of 1 gm of the acid and 4-5 ml of thionyl chloride. Reflux the mixture for about 30 minutes, cool and pour the acid chloride thus formed into 10-15 ml of conc. ammonia. Amide is precipitated. Filter, wash with water and recrystallise from rectified spirit, dry and determine the M.P.

$$RCOOH + SOCl_2 \longrightarrow RCOCl + SO_2\uparrow + HCl\uparrow$$
$$RCOCl + NH_3 \longrightarrow \underset{\text{ppt.}}{RCONH_2\downarrow} + HCl$$

(*ii*) *Anilide derivative.* To about 1 ml acid chloride, add 5-6 ml of acetone, 1 ml aniline and 10 ml dil. NaOH solution. Shake the mixture well and cool. Filter the precipitated anilide, wash well with water and allow it to dry.

$$RCOCl + H_2NC_6H_5 \longrightarrow RCONHC_6H_5 \downarrow + HCl$$
anilide

List of some acids
Aliphatic

Liquids		Solids	
B.P $^{\circ}$C	Name	M.P. $^{\circ}$C	Name
100–101°	Formic acid	97	Glutaric acid
118	Acetic acid	101	Oxalic acid (crystalline)
140	Propionic acid	133	Malonic acid
163	n-Butyric acid	150-51	Adipic acid
		153	Citric acid
			(Crystalline acid
			M.P. 100°)
		169	Tartaric acid
		185	Succinic acid

Aromatic

Solids

M.P. $^{\circ}$C	Name	M.P. $^{\circ}$C	Name
121	Benzoic acid	159	Salicylic acid
133	Cinnamic acid	195	Phthalic acid

Specific tests for some acids
(a) Alphatic acids

LIQUIDS

1. **Formic acid**   HC$\diagdown$ (B.P. 100-101°C)

   Pungent smell

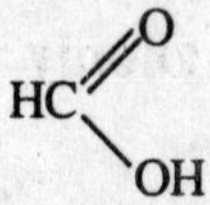

   (i) Reduces mercuric chloride solution : Add 1 ml mercuric chloride to 2 ml of formic acid and warm. A white or grey precipitate is obtained.

   (ii) Reduces Tollen's reagent : Add a few drops of neutral solution of the acid to 1 ml of Tollen's reagent and warm. A silver mirror or grey precipitate is obtained.

   (iii) Decolourises alkaline $KMnO_4$ solution.

   (iv) Warm a little of the acid with conc. H_2SO_4. Carbon monoxide is evolved which burns with a blue flame.

   (v) Reduces Fehlings solution

   Amide derivative — M.P. 195°C

   Anilide derivative — M.P. 47°C.

2. **Acetic acid,**   CH$_3$C$\diagdown$ (B.P. 118°C)

   (Vinegar smell)

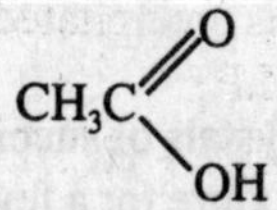

   (i) To a neutral solution of the acid, add a few drops of dil. $FeCl_3$ solution. A red colour of ferric acetate is obtained.

   (ii) On warming a mixture of the acid, absolute alcohol and a few drops of conc. H_2SO_4 a fruity smell of ethyl acetate is produced.

(*iii*) Does not have reducing properties.

Amide derivative : M.P. 82°C

| SOLIDS |

3. **Oxalic acid,**   COOH   **(M.P. 101°C)**
   | .2H$_2$O
   COOH

Colourless crystals soluble in cold water

(*i*) Decolourises KMnO$_4$ solution. To an aqueous solution of the acid, add dil. H$_2$SO$_4$ and a few drops of KMnO$_4$ and then warm. Decolourisation occurs.

(*ii*) To a neutral solution of the acid, made by adding NaOH, add calcium chloride solution. A white precipitate of calcium oxalate in cold is obtained. Anhydrous oxalic acid melts at 187°C temperature.

Anilide derivative : M.P. 245°C

4. **Citric acid,**   CH$_2$COOH   **(M.P. 153°C)**
   (anhydrous) | ⁄OH
                            C
                            | ＼COOH
                            CH$_2$COOH

Crystalline acid *M.P. 100°C;* colourless solid

(*i*) A neutral solution of the acid when added to calcium chloride solution, gives white precipitate of calcium citrate after boiling the solution.

(*ii*) Heat 0.2 gm of the acid with ½ ml of conc. H$_2$SO$_4$. Carbon monoxide and CO$_2$ are evolved and the solution turns yellow. Cool and dilute the solution with water and make it alkaline with NaOH solution. Add a few ml of freshly prepared solution of sodium nitroprusside. An intense red colour is produced.

Amide derivative :   M.P. 210°C.

5. **Tartaric acid,**   CHOHCOOH   **(M.P. 169°C)**
                              |
                              CHOHCOOH

(*i*) Warm a little of the acid with conc. H$_2$SO$_4$. Charring takes place and CO, CO$_2$ and SO$_2$ are evolved.

(*ii*) A neutral solution of the acid, gives a white precipitate of calcium tartarate with CaCl$_2$ solution.

(*iii*) A neutral solution of the acid when added to Tollen's reagent and the mixture is then warmed. A silver mirror is produced.

(*iv*) *Fenton's test.* To an aqueous solution of the acid, add a few drops of freshly prepared ferrous sulphate solution, 2 drops of hydrogen peroxide and excess of NaOH solution. An intense violet colour is produced.

6. **Succinic acid,**   CH$_2$COOH   **(M.P. 185°C)**
   White crystalline solid |
                                    CH$_2$COOH

(*i*) *Fluorescein test.* In a dry test tube, take a few crystals of succinic acid, 1–2 crystals of resorcinol and ½ ml of conc. H$_2$SO$_4$. Heat the mixture for a few minutes, cool and then pour the contents in a beaker containing about 100 ml of water and to it add dil. NaOH in excess. A red colour is produced which exhibits green fluorescence.

(*ii*) If CaCl$_2$ solution is added to the neutral solution of the acid, a white precipitate is obtained after boiling the solution and scratching the sides of the test tube with a glass rod.

Amide derivative : M.P. 242°C

(b) Aromatic acids

1. **Benzoic acid,** (M.P. 121°C)

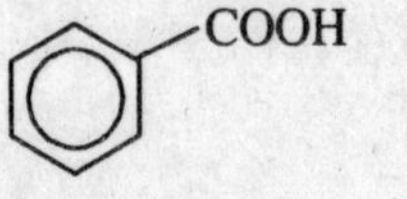

White crystalline solid soluble in hot water

A neutral solution of the acid made by adding NaOH, gives a buff coloured precipitate with ferric chloride solution.

Amide derivative : M.P. 128°C

2. **Cinnamic acid** —CH = CHCOOH (M.P. 133°C)

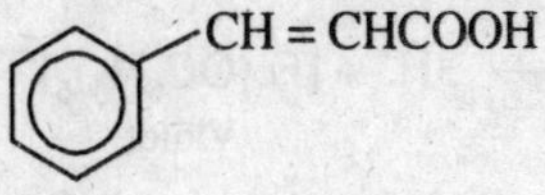

Pale yellow crystals, soluble in hot water

(i) On heating with acidified $KMnO_4$ solution, it is oxidised to benzaldehyde which can be detected by its smell of bitter almond.

(ii) Decolourises alkaline $KMnO_4$ solution.

Amide derivative : M.P. 142°C

3. **Salicylic acid** —OH —COOH (M.P. 159°C)

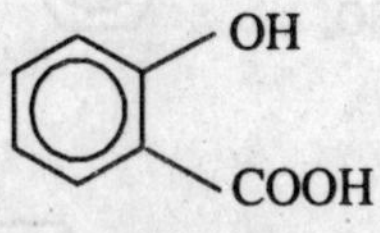

Colourless needle-shaped crystals soluble in hot water

(i) To an aqueous solution of the acid, add a few drops of $FeCl_3$— a violet or purple colour develops.

(ii) Gives a white or yellow precipitate with bromine water.

(iii) Warm a mixture of the substance, methyl alcohol and conc. H_2SO_4 in a test tube. The smell of oil of winter green (methyl salicylate iodex type) is obtained.

Amide derivative : M.P. 139°C

Bromo derivative : M.P. 225°C

4. **Phthalic acid,** —COOH —COOH (M.P. 195°C)

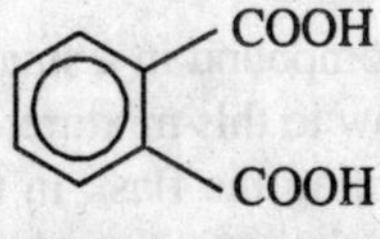

Colourless thin plates soluble in hot water

(i) Gives fluorescein test. Take phthalic acid in place of succinic acid and proceed as in the test for succinic acid.

(ii) *Phenolphthalein test.* Take a few crystals of the acid, add an equal amount of phenol and ½ ml of conc. H_2SO_4 in a test tube. Heat gently, cool, dilute with water and then add dil. NaOH solution in excess. A pink colour is obtained.

Derivative

Phthalimide. Heat 1 gm of the acid in a dry test tube untill it melts. Cool and add to it 1 gm of urea and heat it again for 15-20 minutes. Cool and then wash well with water. M.P. is 205°C.

Note. A neutral solution of the acid is prepared as follows :

To 0.5 gm of the acid, add NH_4OH drop by drop untill the solution has a distinct smell of ammonia. Now boil the solution to drive out excess of ammonia. The remaining solution will be the neutral solution of the acid.

2. Phenols

Group Tests :

(i) A solution or suspension of phenol in water turns blue litmus red but does not give effervescence with sodium bicarbonate solution.

(ii) **Colour with $FeCl_3$:** To about 1 ml of aqueous or alcoholic solution of phenol, add a few drops of neutral or very dilute solution of ferric chloride. A change in colour indicates the presence of a phenolic — OH group.

$$6C_6H_5OH + FeCl_3 \longrightarrow 3H^+ + [Fe(OC_6H_5)_6]^{3-} + 3HCl$$
$$\text{Violet}$$

(iii) **Liebermann's test :** All phenols in which para position is free, respond to this test.

Take a small amount of phenol and a few crystals of sodium nitrite in a dry test tube and heat it gently for a minute. Allow it to cool, add ½ ml of conc. H_2SO_4 and shake the test tube. A deep green or blue colour develops. Upon dilution with water, the solution turns red. Now add excess of dil. NaOH solution. The solution again becomes green or blue.

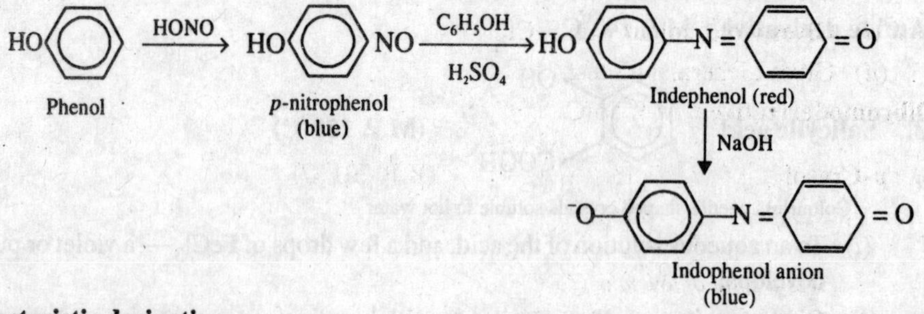

Phenol   *p*-nitrophenol   Indephenol (red)
                   (blue)

Indophenol anion
(blue)

Characteristic derivatives

(i) **Bromo derivative.** Dissolve about 0.5 gm of the compound in 2-3 ml of glacial acetic acid or acetone. Add to it, a solution of bromine (prepared in acetic acid) drop by drop with shaking till the solution becomes yellow. Warm it a little, cool and then pour the contents into 10 ml water. Filter the solid bromo-compound, wash it with water and recrystallise it from rectified spirit or methanol.

(ii) **Benzoate derivative.** Take 0.5 gm of the compound in a small conical flask. Add 5 ml acetone and 15 ml of dil. NaOH solution. Now to this mixture, add 2 ml benzoyl chloride drop by drop with constant shaking and cooling the flask in tap water or in ice. After allowing it a rest for 15 minutes, filter the solid derivative, wash and dry it on a porous plate.

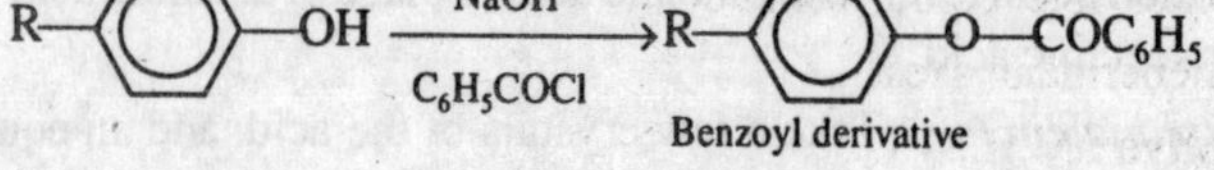

Benzoyl derivative

List of Some Phenols

Liquids B.P.°C.	Name	*Solids* M.P. °C	Name
182	Phenol	95	α-naphthol
191	o-Cresol	105	Catechol
201	p-Cresol	110	Resorcinol
202	m-Cresol	122	β-naphthol
		132	Pyrogallol
		169	Quinol

Specific tests for some phenols

Liquids with B. Pt.

1. **Phenol**   (B.P. 182°C)

It is a solid of low m.p. (42°C), gradually changes into pinkish liquid and causes caustic action on the skin.

   (*i*) Gives violet colour with $FeCl_3$
   (*ii*) Gives Liebermann's test
   (*iii*) Gives phenolphthalein test [Refer to the test (*ii*) of phthalic acid].

Derivatives :

Tribromophenol — M.P. 93°C.

Benzoate derivative — M.P. 68°C.

2. ***o*-Cresol**   (B.P. 191°C)

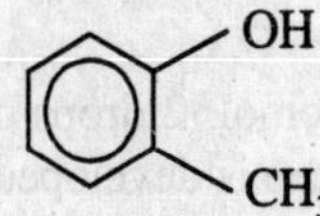

It is a solid of low m.p. (31°C)

   (*i*) Gives violet colour with $FeCl_3$
   (*ii*) Gives Liebermann's test.

Dibromoderivative : M.P. 56°C

3. ***p*-Cresol**   (B.P. 201°C)

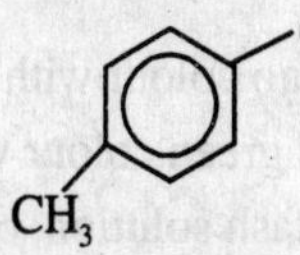

It is a solid of low m.p. (35°C)

   (*i*) Gives blue colour with $FeCl_3$.
   (*ii*) Does not give Liebermann's test.
   (*iii*) Does not give phthalein test.

Benzoate derivative : M.P. 71°C.

4. ***m*-Cresol**   (B.P. 202°C)

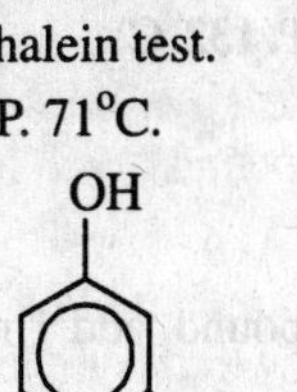

It is a liquid

   (*i*) Gives blue-violet colour with $FeCl_3$.
   (*ii*) Gives Liebermann's test.

Tribromoderivative : M.P. 84°C.

Solids with M.P.

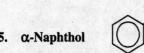

5. **α-Naphthol**   (M.P. 95°C)

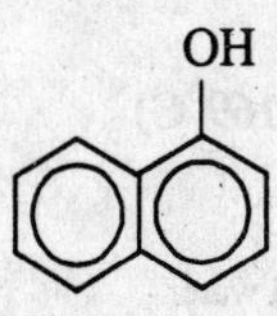

Dark violet coloured crystals

(i) Alcoholic solution of the compound gives blue violet colour with $FeCl_3$ but aqueous solution gives a white precipitate.

(ii) Add a little of the compound in titanic acid solution and then add a few drops of conc. H_2SO_4. A green colour is developed.

Derivatives

2, 4 Dibromo-derivative : M.P. $105^{\circ}C$.

Benzoate derivative : M.P. $56^{\circ}C$:

6. **Resorcinol,**   OH ... OH   **(M.P. $110^{\circ}C$)**

Pink coloured plates, soluble in water

(i) Gives violet-blue colour with $FeCl_3$.

(ii) Warm a mixture of a grain amount of resorcinol, 2 drops of chloroform and 1 ml of caustic soda solution. A red-green fluoroscene is developed.

Dibenzoate derivative : M.P. $117^{\circ}C$.

7. **β-Naphthol,**   OH   **(M.P. $122^{\circ}C$)**

Orange coloured plates

(i) An aqueous solution of the compound gives no colour with $FeCl_3$.

(ii) An alcoholic solution of the compound gives green colour with $FeCl_3$.

(iii) On warming with chloroform and caustic potash solution, blue colour is obtained.

Benzoate derivative : M.P. $107^{\circ}C$

8. **Pyrogallol,**   OH — OH — OH   **(M.P. $132^{\circ}C$)**

Dirty white crystals, soluble in water

(i) Gives yellow colour with $FeCl_3$.

(ii) To 1 ml of aqueous solution of the compound, add 1 ml of ferrous sulphate solution. Blue colour or precipitate is obtained.

(iii) Add 1ml aqueous solution of the compound to 2 ml Tollen's reagent and warm. Grey coloured precipitate or silver mirror is obtained.

Dibromo derivative : M.P. $158^{\circ}C$.

Tribenzoate derivative : $90^{\circ}C$.

9. **Hydroquinone or Quinol,**   OH ... OH   **(M.P. $169^{\circ}C$)**

White needle like crystals, soluble in cold water.

(i) Aqueous solution of the compound gives blue colour with $FeCl_3$.

(*ii*) Take a mixture of 0.2 gm of the substance, 0.2 gm of $K_2Cr_2O_7$ and 4 ml of dil. H_2SO_4 and warm. On cooling, greenish needles of quinhydrone are obtained.

(*iii*) Add a small amount of the substance to Tollen's reagent and warm. Grey coloured precipitate or silver mirror is obtained.

Dibenzoate derivative : M.P. $200^{\circ}C$.

3. Carbohydrates

Group tests :

(*i*) *Molisch's test* : Dissolve 0.5 gm of the compound in 2 ml of water, add 2-3 drops of Molisch reagent and shake well. Now add 2 ml of conc. H_2SO_4 very carefully along the side of the test tube and allow it to stand for sometime. A violet ring is produced at the junction of the two layers. On shaking the solution becomes violet or purple or a blue violet precipitate is obtained. This shows that the given compound is a carbohydrate [Molisch reagent is 20% solution of α-naphthol in alcohol or 10% solution in chloroform].

Note : Very little is known about the nature of violet coloured compound formed in this test. However, it is understood to be due to the formation of an unstable condensation product of α-naphthol with furfural (an aldehyde produced by the dehydration of the carbohydrate). This test is a characteristic of all carbohydrates and is very sensitive as well. Even fibre of the filter paper responds to this test.

(*ii*) To 1 ml aqueous solution of the compound, add 1-2 drops of anthrone solution, shake and warm the mixture gently. A green or bluish green colour is obtained.

(*iii*) On warming with conc. H_2SO_4, the carbohydrate chars; the solution blackens and there is effervescence.

Characteristic derivatives

(*i*) **Osazone.** Almost all carbohydrates which have a functional carbonyl group, give osazone with phenyl hydrazine. This serves as a suitable derivative for characterisation of carbohydrates.

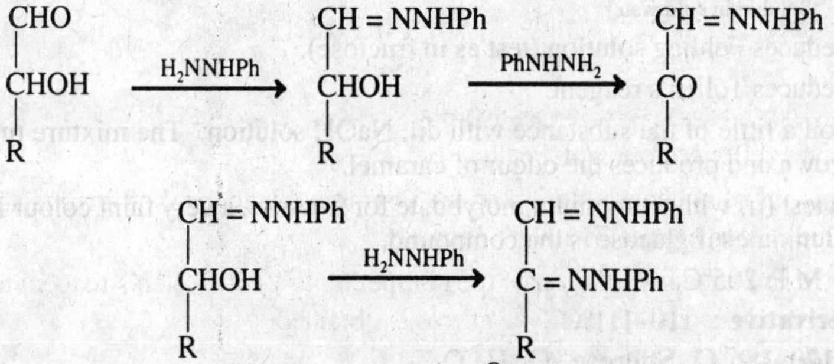

Dissolve 1 gm of carbohydrate in 5 ml of water. Add a solution of phenyl hydrazine in glacial acetic acid (2 ml of phenylhydrazine dissolved in about 3 ml of acetic acid). Shake the mixture well and heat it on a water bath for about 30 minutes. Cool the reaction mixture where osazone gradually begins to precipitate out. Filter and allow it to dry on a porous plate.

(*ii*) **Acetate derivative.** Reflux a mixture of 1 gm of the compound, 1–1.5 gm of anhydrous sodium acetate and 10 ml of acetic anhydride for about 30 minutes. Pour the contents in 50 ml of cold water. Filter the precipitated acetate derivative, wash with water and allow it to dry.

List of Some Carbohydrates

M.P. °C	Name	M.P. °C	Name
95–105	Fructose	decomposes	Maltose
146	(+) Glucose	180-186	Sucrose
166	Galactose	203 (dec.)	Lactose
		decomposes	Starch

Specific tests for some carbohydrates

1. M.P. 95–105°C Fructose, $CH_2OHCO(CHOH)_3CH_2OH$

Colourless solid soluble in cold water

(i) Reduces Fehling solution. Add 2 ml of aqueous fructose solution to 2 ml of Fehling solution (mix 1 ml each of Fehling solutions A and B) and boil. A red-brown precipitate of cuprous oxide is obtained.

The red precipitate is obtained due to formation of Cu_2O :

$$2Cu(OH)_2 + C_6H_{12}O_6 \longrightarrow C_5H_{11}O_5.COOH + Cu_2O \downarrow + H_2O$$

gluconic acid   Red ppt

(ii) To about 5 ml of 1% fructose solution in a boiling tube, add an equal volume of a freshly prepared 4% aqueous solution of ammonium molybdate and then 4-5 drops of glacial acetic acid. Heat the boiling tube in boiling water bath. A greenish-blue colour within 3-4 minutes is obtained.

(iii) *Rapid furfural test.* If in Molisch test, conc. HCl is used in place of conc. H_2SO_4, the violet colouration is shown at different rates by different carbohydrates.

Mix 1 ml of dilute solution of carbohydrate with 1 ml of 1% alcoholic solution of α-naphthol and 6-7 ml of HCl and boil. If immediate violet colouration is produced, the carbohydrate may be fructose or sucrose. In case of glucose, violet colour is produced after 1-2 minutes of boiling.

Osazone. M.P. 205°C

2. M.P. 146°C Glucose, $CHO(CHOH)_4CH_2OH$

Soluble in cold water

(i) Reduces Fehling solution (test as in fructose).

(ii) Reduces Tollen's reagent.

(iii) Boil a little of the substance with dil. NaOH solution. The mixture turns yellow then brown and produces the odour of caramel.

(iv) In test (ii) with ammonium molybdate for fructose, a very faint colour is obtained after 10 minutes if glucose is the compound.

Osazone : M.P. 205°C.

Acetate derivative : 110–111°C

3. M.P. 170–186°C Sucrose, $C_{12}H_{22}O_{11}$

White crystalline compound, soluble is cold water

(i) Does not reduce Fehling solution or Tollen's reagent.

(ii) Gives immediate violet colouration in rapid furfural test. [test (iii) for fructose].

(iii) Dissolve 0.5 gm of the compound in 5 ml water, add 2 ml dil. H_2SO_4 and boil for 5-10 minutes. Neutralise the excess of acid with Na_2CO_3 solution. Now this solution reduces Fehling solution because sucrose is hydrolysed into glucose and fructose by boiling with acid.

Derivative : Sucrose does not form osazone.

Octacetate derivative : M.P. 67°C.

4. **Does not melt**   **Starch** $(C_6H_{10}O_5)_n$.
but decomposes

White amorphous powder, soluble in hot water

   (*i*)   To 1 ml solution of the compound in water, add a few drops of iodine solution. A deep blue colour is obtained which disappears on boiling but reappears on cooling.

   (*ii*)   Reduces Fehling solution or Tollen's reagent.

---

4. Aldehydes, $R-C{<}^{O}_{H}$

Both aldehydes and ketones have carbonyl $>C = O$ functional group hence some of their properties are very common; they are characterised by similar derivatives. Since aldehydes contain a hydrogen attached to $>C = O$ group, they have reducing properties which ketones do not possess. Therefore, the two classes of compounds are described here under separate categories.

Tests for aldehydic group

   (*i*)   To 1 ml aqueous or alcoholic solution of the compound, add 2 ml of *Schiff's reagent* and shake. A wine-red or purple colour develops in cold.

   (*ii*)   *Reduction of Tollen's reagent.* To about 2 ml Tollen's reagent, add about 0.5 gm of the compound and heat the mixture on water bath. Formation of a silver mirror or blackish precipitate indicates the presence of an aldehydic group. (Tollen's reagent is prepared by adding some dil. NaOH to 1 ml of $AgNO_3$ solution. The white precipitate thus produced is dissolved by adding a few drops of ammonium hydroxide. This is the reagent and should be prepared fresh when needed.)

$$AgNO_3 + NaOH \longrightarrow AgOH\downarrow \ + NaNO_3$$
soluble in ammonia
$$AgOH + 2NH_3 \longrightarrow Ag(NH_3)_2OH$$
$$2Ag(NH_3)_2OH + RCHO \longrightarrow RCOONH_4 + 2Ag\downarrow + H_2O + 3NH_3$$
Silver mirror or grey precipitate

   NaOH is used to hasten up the reduction.

   (*iii*)   *Formation of phenylhydrazone.* To a mixture of 0.5 ml of phenylhydrazine, 0.5 ml of glacial acetic acid, 5 ml of water, add 2-3 drops of the compound and shake well. Yellow or cream coloured precipitate shows the presence of aldehydic group. As this test is of $>CO$ group, both aldehyde and ketone show.

Characteristic derivatives

   (*i*)   *Semicarbazone derivative.* Dissolve 0.5 gm of semicarbazide hydrochloride in 1 ml of water. Mix it with a solution of 1 gm of sodium acetate dissolved in minimum amount of water. To this mixed solution, add 0.5 gm of the compound and heat the mixture on boiling water bath for about 20 minutes. Upon cooling the solution, semicarbazone separates as a solid mass.

   (*ii*)   *2, 4-Dinitrophenyl hydrazone derivative.* Mix 0.5 gm of the compound with 2-4 ml of 2,4 dinitrophenyl hydrazine solution and heat it in boiling water bath for 15-20 minutes. Cool and filter the precipitated derivative.

List of Aldehydes

Liquids	
B.P. °C	Name
98	Formaldehyde (formalin)
97	Acetaldehyde (aqueous solution)
179	Benzaldehyde
196	Salicylaldehyde

Specific tests for Some Aldehydes

1. **Formaldehyde** $HC{\overset{O}{\underset{H}{\diagdown}}}$ **(B.P. 98°C)**

Although formaldehyde is a gas (B.P.–21°C) but it is given in the form of formalin (40% solution of HCHO in water). The solution has a pungent smell and it boils at about 98°C.

  (*i*) Reduces Fehling solution (test as in glucose).

Cu-tartarate complex + RCHO $\longrightarrow$ RCOOH + $Cu_2O\downarrow$ Red

  (*ii*) Reduces Tollen's reagent.

  (*iii*) Evaporate the aqueous solution. White residue of fishy odour.

  2, 4-Dinitrophenyl hydrazone derivative : M.P. 155°C.

2. **Acetaldehyde,** $CH_3C{\overset{O}{\underset{H}{\diagdown}}}$ **(B.P. 97°C)**

Usually a solution of acetaldehyde in water is supplied which boils at about 97-98°C. Pure acetaldehyde has b.pt. 21°C.

  (*i*) Reduces Fehling solution and Tollen's reagent.

  (*ii*) To ½ ml of aldehyde solution, add ½ ml of freshly prepared sodium nitroprusside solution and dil. NaOH in excess. A red colour is produced.

  (*iii*) On warming with dil. NaOH solution and I_2, gives yellow ppt of CHI_3

  Semicarbazone derivative : M.P. 89°C.

  2, 4-Dinitrophenyl hydrazone : M.P. 162°C.

3. **Benzaldehyde** [benzene ring with —CHO]   **(B.P. 179°C)**

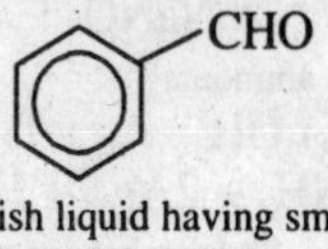

Colourless or yellowish liquid having smell of bitter almonds, immiscible with water

  (*i*) Restores the colour of Schiff's reagent very slowly.

  (*ii*) Does not reduce Fehling solution.

  (*iii*) *Cannizzaro's reaction.* When treated with dilute alkali, benzaldehyde gives benzyl alcohol and benzoic acid.

In a test tube, take 1ml of aldehyde and 2 ml of dil. NaOH solution (about 30%) and heat the mixture gently for a few minutes. Add sufficient water and decant off the oily layer if any. Now to the aqueous solution, add dil. HCl. A white precipitate of benzoic acid is formed on cooling.

  2, 4-Dinitrophenyl hydrazone derivative : M.P. 235°C

  Semicarbazone derivative : M.P. 214°C

5. **Ketones,** $\overset{R}{\underset{R}{\diagup}}C=O$

Group test :

  (*i*) To 0.5 gm of substance, add 2-3 ml of 2, 4-dinitrophenylhydrazine solution (Brady's re-agent) and heat the mixture on water bath for 5 minutes. A red, yellow or cream coloured precipitate indicates the presence of a carbonyl group.

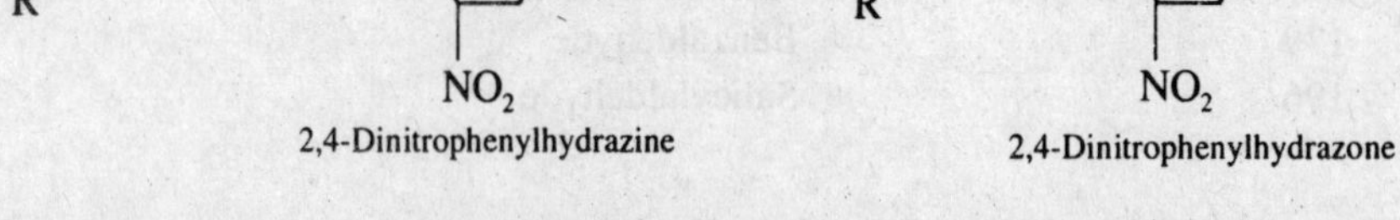

$$\overset{R}{\underset{R}{\diagup}}C=O + H_2N.NH{-}\bigcirc{-}NO_2 \longrightarrow \overset{R}{\underset{R}{\diagup}}C=N.NH\bigcirc{-}NO_2$$

NO_2   NO_2

2,4-Dinitrophenylhydrazine   2,4-Dinitrophenylhydrazone

(ii) Does not restore the colour of Schiff's reagent nor gives silver mirror with Tollen's reagent. Ketones containing CH_3CO group may restore the colour of Schiff's reagent very slowly. Those ketones which contain a CH_3CO group, give colour reaction with sodium nitroprusside and metadinitrobenzene.

(iii) To 1 ml aqueous solution of the compound, add 1 ml of sodium nitroprusside solution and a few drops of dil. NaOH. A wine-red colour is produced. (Benzophenone does not give this test).

Acetone or any other ketone having CH_3CO— group is converted by alkali to anion $CH_3COCH_2^-$ or $C_6H_5COCH_2^-$ etc. This anion reacts with nitroprusside ion to form a highly coloured anion.

$$[Fe(CN)_5NO]^- + RCO\!-\!CH_2^- \longrightarrow [Fe(CN)_5NO.CH_2OC.R]^{2-}$$

(iv) Take about 1 ml of compound and about 0.1 gm of finely powdered m-dinitrobenzene in a test tube. Add to it excess of dil. NaOH. A violet colour is produced.

Characteristic derivatives

Semicarbazone and 2, 4-dinitrophenyl hydrazone derivatives are prepared in the same way as in the case of aldehydes.

List of Some Ketones

Liquids		Solids	
B.P. $^\circ$C	Name	M.P. $^\circ$C	Name
56	Acetone	48	Benzophenone
80	Ethyl methyl ketone	95	Benzil
102	Diethyl ketone		
155	Cyclohexanone		
202	Acetophenone		

Specific Tests for some Ketones

1. **Acetone,** $\begin{array}{c} CH_3 \\ \\ CH_3 \end{array}\!\!\!>\!\!C=O$ **(B. P. 56°C)**

Colourless pleasant smelling liquid miscible with water

(i) Gives red colour with sodium nitroprusside in alkaline (dil. NaOH) medium.

(ii) *Iodoform test.* To 1 ml of acetone taken in a small beaker, add 10 ml of dil. NaOH or Na_2CO_3 solution and then add iodine solution drop by drop with shaking till the solution becomes yellow. Now heat the beaker at a very low flame for few minutes and cool. Yellow crystals of iodoform separate out. Filter and keep it as a derivative.

(iii) To 1 ml of acetone, add 0.1 gm powdered m-dinitrobenzene and then add an excess of NaOH solution and shake. A violet colour is obtained.

Derivatives

(i) 2, 4-Dinitrophenyl hydrazone : M.P. 128°C.

(ii) Iodoform—M.P. 119°C.

(iii) Semicarbazone—M.P. 188°C.

2.   Ethyl methyl ketone, $\begin{matrix} CH_3 \\ \\ C_2H_5 \end{matrix} \Big\rangle C = O$ **(B. P. 80°C)**

Colourless liquid, miscible with water

(*i*)   Gives wine-red colour with sodium nitroprusside solution and alkali.

(*ii*) Gives violet colour with *m*-dinitrobenzene and alkali.

(*iii*) Gives iodoform test.

Derivatives

Semi-carbazone — M.P. 135°C.

2, 4-Dinitrophenylhydrazone — M.P. 115°C

3.   Acetophenone,   **(B. P. 202°C)**

Colourless pleasent smelling liquid

(*i*)   Gives colour reaction with sodium nitroprusside and alkali.

(*ii*) Gives iodoform test.

(*iii*) Heat 1 ml of the substance with 1 ml of KOH and 1 ml of $KMnO_4$ solution for 5-10 minutes. Cool and acidify with dil. HCl. A white precipitate of benzoic acid is obtained.

Semi-carbozone derivative : M.P. 162°C

2, 4- Dinitrophenyl hydrazone : M.P. 240°C

Solid

4.   Benzophenone,   **(M. P. 48°C)**

Colourless crystalline solid pleasant smell. Insoluble in water.

(*i*)   Does not give colour with sodium nitroprusside.

(*ii*) Does not give iodoform test.

(*iii*) Dissolves in conc. H_2SO_4 giving a yellow solution.

2, 4- Dinitrophenylhydrazone — M.P. 238°C.

6.   Esters $R{-}C\begin{smallmatrix} \nearrow O \\ \searrow OR \end{smallmatrix}$

Group tests.

(*i*) *Hydroxamic acid test.* Take an amount equal to a rice grain or two drops of the compound in a test tube, add 1 ml of 5% methanolic solution of hydroxylamine hydrochloride. Now, add a few drops of 2N methanolic KOH till the solution is just alkaline and boil. Cool and acidify with dil. HCl and add 2-3 drops of $FeCl_3$ solution. A deep red or violet colour is produced. The reactions involved in the test are as follows :

$$RCOOR' + NH_2OH \longrightarrow \underset{\text{Hydroxamic acid}}{RCONHOH} + R'OH$$

$$3\underset{\overset{\|}{O}}{RC-NHOH} \rightleftharpoons 3\underset{\overset{|}{OH}}{R-C} = NOH + FeCl_3$$

$$\left[\begin{array}{c} R-C=N-OH \\ \overset{|}{O^-} \\ \text{Red colour} \end{array} \right]_3 Fe$$

(ii) *Hydrolysis test.* Dissolve 0.5 gm of ester in 1 ml of alcohol, add 4-5 drops of dil. NaOH solution and 2-3 drops of phenolphthalein. A pink colour is obtained. Now put the test tube in boiling water bath for about 5 minutes, the pink colour either fades or disappears.

Phenolphthalein is pink coloured so long as the medium is alkaline. When the alkali is consumed due to hydrolysis of ester, the pink colour disappears.

$$RCOOR' + NaOH \longrightarrow RCOONa + R'OH$$

Characteristic derivatives

(i) *Amide derivatives.* Dissolve 0.5 gm of ester in 10-15 ml of water and add 4-5 ml of conc. ammonia to it and shake. The amide is precipitated. Filter, wash and dry it on a porous plate. Methyl and ethyl esters give amides conveniently by this method.

$$RCOOR' + NH_3 \longrightarrow R'OH + RCONH_2$$

(ii) *Isolation of components.* The ester on hydrolysis gives an acid and alcohol or acid and a phenol. If these components can be isolated carefully, they serve as derivatives.

In a conical flask fitted with a reflux water condenser, place 2 ml of ester and about 20 ml of dil. NaOH and reflux the mixture for about 30–45 minutes until all oily drops in the flask disappear. Stop refluxing and distil off the alcohol. Now cool the residual liquid in the flask and acidify it with dil. HCl or dil. H_2SO_4. If a precipitate is obtained, it may be an aromatic acid. Filter it and wash with water, dry it on a porous plate and identify the acid.

[Most of the aliphatic acids are soluble in water hence they are not precipitaed].

List of some Esters

Liquids		Solids	
B.P. °C	Name	M.P. °C	Name
55	Methyl acetate	42	Phenyl salicylate
77	Ethyl acetate	54	Methyl oxalate
186	Diethyl oxalate		
196	Phenyl acetate		
198	Methyl benzoate		
213	Ethyl benzoate		
223	Methyl salicylate		

Specific Tests for some Esters

LIQUIDS

1. **Ethyl acetate, $CH_3COOC_2H_5$**   **(B. P. 77°C)**
   Pleasant smelling colourless liquid

   (i) Hydrolyse 1 ml of the ester by boiling with dil. alkali. Acetic acid and ethyl alcohol are produced. The presence of ethyl alcohol may be confirmed by iodoform test.

To the above hydrolysed solution, add 5 ml of 10% potassium iodide and 5 ml of freshly prepared sodium hypochlorite solution. Warm for a few minutes and cool. Yellow crystals of iodoform are produced. This has a characteristic smell. Filter and collect it as derivative also.

Derivative : Iodoform.

2. **Diethyl oxalate,** $\begin{array}{c} COOC_2H_5 \\ | \\ COOC_2H_5 \end{array}$ **(B. P. 186°C)**

Pleasant smelling colourless liquid

(*ii*) Readily gives oxamide with NH_3.

Make a suspension of the ester in water and to this, add conc. NH_4OH and shake. White precipitate of oxamide is rapidly obtained.

$$\begin{array}{c} COOC_2H_5 \\ | \\ COOC_2H_5 \end{array} + 2NH_3 \longrightarrow \begin{array}{c} CONH_2 \\ | \\ CONH_2 \end{array} + 2C_2H_5OH$$

(*iii*) Heat the mixture of 1 ml of ester and 1 ml of aniline for a few minutes and cool. White precipitate of oxanilide is obtained. Isolate it and keep it as a derivative.

Derivative : Oxanilide : M.P. 247°C.

3. **Ethyl benzoate of** [benzene ring]—$COOC_2H_5$ **(B. P. 213°C)**

Colourless pleasant smelling liquid, immiscible with water

Reflux 2 ml of ester with 4-5 ml of alcoholic KOH solution for about 20 minutes. Cool and acidify with dil. HCl. A white precipitate of benzoic acid is obtained. Filter and keep it as a derivative.

Derivative : Benzoic acid M.P. 121°C.

4. **Methyl Salicylate** [benzene ring with]—OH / —$COOCH_3$ **(B. P. 223°C)**

Characteristic smell of oil of wintergreen

(*i*) Gives violet colour with $FeCl_3$.

(*ii*) On hydrolysis by the method described in case of ethyl benzoate, it gives salicylic acid.

Derivative : Salicylic acid, M.P. 158°C.

Solids with M.P.

5. **Phenyl salicylate,** [benzene ring with]—OH / —COO—[benzene ring] **(M. P. 42°C)**

Colourless solid

(*i*) Does not give any colour with $FeCl_3$.

(*ii*) On hydrolysis with alcoholic KOH, gives salicylic acid.

Acetate derivative : M.P. 97°C

Benzoate derivative : M.P. 80°C

(Prepare as described in phenols)

6. **dimethyl oxalate,** $\begin{matrix} COOCH_3 \\ | \\ COOCH_3 \end{matrix}$ **(M. P. 54°C)**

(i) Reflux ½ gm of the compound with 2 ml of dil. NaOH solution for 20-25 minutes. Cool and divide the solution in two parts :

    (a) In one part, introduce a hot copper gauze. A pungent smell of HCHO is obtained due to oxidation of CH_3OH.

    (b) In the second part, add acidified $KMnO_4$ solution and warm — colour disappears showing the presence of oxalic acid.

(ii) It gives precipitate of oxamide when conc. NH_4OH is added to the ester. Filter and keep it as a derivative.

7. Alcohols

Group test

(i) *Effervescence with sodium.* In a dry test tube, take 2 ml of the given compound, add 1 gm of anhydrous sodium sulphate (to absorb water if already present) and filter. To the filtrate, add a small piece of sodium. There is effervescence with evolution of hydrogen if the substance is an alcohol.

$$2ROH + 2Na \longrightarrow 2RONa + H_2\uparrow$$

(ii) To ½ ml compound, add a few drops of ceric-ammonium nitrate solution. A red colour is obtained.

$$(NH_4)_2[Ce(NO_3)_6] + 2ROH \longrightarrow [Ce(NO_3)_4(ROH)_2] + 2NH_4NO_3$$
red colouration

(iii) Test for solid alcohols : Take the compound in a dry test tube and add acetyl or benzoyl chloride. The effervescence with evolution of HCl gas is noted.

$$ROH + ClCOCH_3 \longrightarrow ROCOCH_3 + HCl\uparrow$$
$$ROH + ClCOC_6H_5 \longrightarrow ROCOC_6H_5 + HCl\uparrow$$

Characteristic derivatives

(i) *p-Nitrobenzoate derivative.* In a boiling tube, take 1 ml of alcohol and 1 ml of *p*-nitrobenzoyl chloride and heat the mixture gently for a few minutes. Cool and pour the contents in 10 ml water. Filter the precipitate, wash with water and determine the m.p. in dry state.

$$O_2N-\!\!\!\bigcirc\!\!\!-COCl + HOR \longrightarrow O_2N-\!\!\!\bigcirc\!\!\!-COOR + HCl$$

(ii) *3, 5-Dinitrobenzoate.* Use 3, 5-dinitrobenzoyl chloride in place of *p*-nitrobenzoyl chloride and proceed as above.

$$\underset{NO_2}{\overset{NO_2}{\bigcirc}}-COCl + ROH \longrightarrow \underset{NO_2}{\overset{NO_2}{\bigcirc}}-COOR + HCl$$

List of some Alcohols

Liquids B.P. °C	Name	B.P. °C	Name
65	Methyl alcohol	118	n-Butyl alcohol
78	Ethyl alcohol	138	n-Amyl alcohol
82	Isopropyl alcohol	161	Cyclohexanol
97	n-Propyl alcohol	197	Ethylene glycol
83	tert-Butyl alcohol	205	Benzyl alcohol
108	iso-Butyl alcohol	290	Glycerol

Specific Tests for some Alcohols

B.P.

1. **65°C   Methyl alcohol, CH_3OH**

   Colourless liquid, miscible with water

   (i) In a dry test tube, take 1 ml compound and introduce in it a hot copper turnings ; pungent vapours of HCHO.

   (ii) Take ½ ml of compound, a few crystals of salicylic acid and a few drops of conc. H_2SO_4 in a test tube and heat gently. A characteristic smell of oil of wintergreen (iodex type) due to formation of methyl salicylate.

   3, 5-Dinitrobenzoate derivate : M.P. 107°C.

2. **78°C   Ethyl alcohol, C_2H_5OH**

   Colourless liquid ; spirituous smell, miscible with water.

   (i) *Iodoform test.* Take 1 ml compound and 10 ml of sodium carbonate or dil. NaOH solution in a boiling tube. Add iodine solution drop by drop with shaking till solution turns yellow. Warm in a water bath and then cool. Yellow crystals of iodoform begin to separate.

   (ii) *Oxidation.* Take ½ ml of alcohol, 2 ml of $K_2Cr_2O_7$ solution and 1 ml of conc. H_2SO_4 in a test tube and warm gently. The solution turns green and smell of acetaldehyde is obtained.

   (iii) 1 ml compd + $NaCH_3COO$ + conc. H_2SO_4 and heat → Fruity smell

   3, 5-Dinitrobenzoate derivative : M.P. 92°C.

3. **82°C   Isopropyl alcohol,** $\begin{matrix} CH_3 \\ \\ CH_3 \end{matrix}\Big\rangle CHOH$

   Colourless liquid, miscible with water

   (i) Gives iodoform test (proceed as described in case of ethyl alcohol).

   (ii) Oxidation (proceed as in ethyl alcohol).

   On oxidation, it gives acetone which can be detected by the colour reaction with sodium nitroprusside solution.

   3, 5-Dinitrobenzoate derivative : M.P. 122°C

4. **97°C   n-Propyl alcohol, $CH_3CH_2CH_2OH$**

   Colourless liquid, miscible with water.

   (i) Does not give iodoform reaction.

   (ii) Oxidation gives propionaldehyde and not acetone.

3, 5-Dinitrobenzoate derivative : M.P. 74°C

5.   197°C   Ethylene glycol, $\begin{array}{c} CH_2OH \\ | \\ CH_2OH \end{array}$

Colourless syrupy liquid, miscible with water

Add ½ ml of conc. H_2SO_4 to ½ ml substance and heat the mixture till a brown colour is obtained. Cool and add conc. NaOH solution to make the medium alkaline. A characteristic bad smell of aldehyde resin is obtained.

3, 5-Dinitrobenzoate derivative : M.P. 169°C

6.   205°C Benzyl alcohol,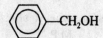

Colourless liquid, sparingly soluble in water

(i)   Reflux the mixture of 1 ml compound, ½ gm of $KMnO_4$ and 5 ml of dil. Na_2CO_3 solution for about 20 minutes, cool and add dil. HCl to acidify the solution. A white precipitate of benzoic acid is obtained. Filter, wash with water and collect it as a derivative.

(ii)   On heating with acidic $KMnO_4$ solution, smell of benzaldehyde is obtained.

3, 5-Dinitrobenzoate derivative : M.P. 113°C

7.   290°C   Glycerol, $\begin{array}{c} CH_2OH \\ | \\ CHOH \\ | \\ CH_2OH \end{array}$

Colourless syrupy liquid, miscible with water

(i)   Heat a few drops of the substance and a crystal of $KHSO_4$. An irritating smell of acrolein is obtained.

(ii)   To 2 ml borax solution, add 1-2 drops of phenolphthalein. A pink colour is developed. Now, add 2-4 drops of the compound and shake. Pink colour disappears but reappears on warming.

Tribenzoate derivative : 72°C.

8.   Hydrocarbons

Test

Hydrocarbons do not contain any functional group hence they are detected more by elimination than by direct functional group test. If no test for functional groups described above is positive, the compound may be a hydrocarbon. Attempt is made to identify it with the help of M.P. or B.P. and a suitable derivative. Generally a hydrocarbon is soluble in fuming sulphuric acid.

Characteristic derivatives

(i)   *Picrates.* A large number of hydrocarbons form addition compounds with picric acid which is known as picrate. It serves as a characteristic derivative.

Prepare a saturated solution of the compound in rectified spirit and mix it with equal volume of a solution of picric acid in alcohol. Warm this mixture in boiling water bath for a few minutes and cool. Yellow or orange coloured picrate begins to crystallise out. Filter and wash it thoroughly with water.

(ii)   *Nitro-derivative.* Take about 1 gm of the compound in a boiling tube and to this add drop by drop with shaking 5 ml of a mixture of conc. H_2SO_4 and conc. HNO_3 mixed in equal amounts. Cool it if necessary. Now heat the test tube in boiling water bath for about half an hour. Cool the contents and pour it into 20 ml of cold water or preferably in crushed ice. The nitro-compound separates as a yellow solid or oil. Separate the product, wash and determine its M.P. or B.P.

(The nitration method for different compounds may vary in minor way. However, the above method may be applied in most of the cases).

(iii) *Oxidation products.* The side chain of the aromatic hydrocarbons such as of toluene may be oxidised conveniently to give corresponding acid.

List of Some Hydrocarbons

Liquids			Solids	
B.P.°C	Name	M.P. °C	Name	
80	Benzene	70	Diphenyl	
110	Toluene	80	Naphthalene	
136	Ethyl benzene	95	Acenaphthene	
137	p-xylene	216	Anthracene	
139	m-xylene			
144	o-xylene			

Specific Tests for Some Hydrocarbons

Liquids

B.P.
1. 80°C Benzene,

Colourless liquid, petrol like smell, immiscible with water

Nitration : In above way gives m-dinitrobenzene, M.P. 90°C

2. 110°C Toluene,

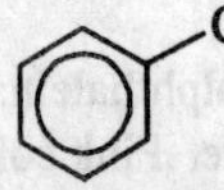

Colourless liquid, smells just like benzene. Lighter than water

(i) In a boiling tube, take ½ ml of substance, one crystal of copper sulphate, a pinch of manganese dioxide and ½ ml of conc. H_2SO_4. Heat the tube in boiling water bath for a few minutes. A characteristic smell of benzaldehyde (bitter almonds) is obtained.

(ii) Oxidation with alk. $KMnO_4$ gives benzoic acid which can be tested.

2, 4-Dinitroderivative : M.P. 70°C.

Solids

3. **M.P. 80°C Naphthalene**

Colourless crystalline solid smells like naphthalene balls. Insoluble in water.

Derivative : Picrate M.P. 149°C

O.S. + solid $KMnO_4$ + conc. H_2SO_4 and heat. Add resorcinol.

Pour the mix. in NaOH → Green fluorescence

4. **70°C Diphenyl (C_6H_5)$_2$**

Aromatic colourless crystalline solid. Faint pleasant smell.

Insoluble in water. Soluble in alcohol, ether, etc.

Derivatives : (i) Benzoic acid M.P. 121°C

(ii) p : p' — Dinitrodiphenyl M.P. 233°C

5. 95°C   Acenaphthene, $C_{10}H_6\Big\langle \begin{smallmatrix} CH_2 \\ | \\ CH_2 \end{smallmatrix}$

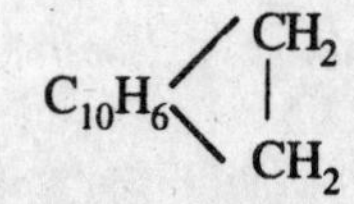

Colourless needle shaped crystals, insoluble in water, sparingly soluble in alcohol and soluble in toluene.

(*i*) Dissolve 2, 4 dinitrotoluene and acenaphthene separately in benzene and mix the two solutions. A solid of M.P. 60°C separates out.

Derivatives : (*i*) Acenaphthenequinone (on oxidation with potassium dichromate and conc. H_2SO_4) M.P. 261°C.

(*ii*) Picrate (orange) M.P. 161°C.

6. 100°C Phenanthrene,

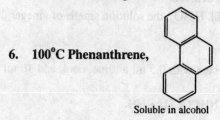

Soluble in alcohol

Derivative : Picrate M.P. 143°C

7. 216°C Anthracene,

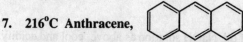

Colourless crystalline solid and insoluble in water.

Sparingly soluble in alcohol but readily soluble in benzene.

Add 0.5 gm of original substance to 2 ml of a saturated solution of picric acid in benzene and shake. A deep red solution is obtained due to formation of picrate.

Derivatives :   (*i*) Dibromoderivative (with Br_2 in CCl_4) M.P. 221°C

(*ii*) Picrate M.P. 138°C.

Type [B].   COMPOUNDS CONTAINING NITROGEN

[In addition to C, H and O]

1. Amides, $RCONH_2$

Group test. Heat 0.5 gm of compound with 2 ml of dilute caustic soda solution. Ammonia is evolved. This can be detected by its smell and by its action on moist red litmus paper which turns blue when brought near the mouth of the test tube.

$$RCONH_2 + NaOH \longrightarrow RCOONa + NH_3\uparrow$$

Characteristic derivatives

(*i*) *Isolation of acid.* About 1 gm of the compound is boiled with 5-10 ml of dil. NaOH solution as above. The solution is cooled and acidified with dil. H_2SO_4. The precipitated acid is filtered, washed and dried.

(*ii*) *Picrate derivative.* Many amides may be characterised as picrates which may be prepared in the same manner as described in case of hydrocarbons.

List of Some Amides and Imides

Solids M.P. °C	Name
82	Acetamide
125	Succinimide

128	Benzamide
132	Urea
139	Salicylamide
233-238	Phthalimide

Specific Tests for Some Amides

1.   M.P. 82°C Acetamide, CH_3CONH_2

White crystalline solid soluble in water.

(i)   When boiled with NaOH solution, ammonia is evolved.

(ii) When a little of the compound is boiled with dil. H_2SO_4, the solution smells of vinegar due to formation of acetic acid.

Derivatives. (i) Picrate, M.P. 107°C

(ii) Acetanilide. Heat 0.5 gm of the substance mixed with 0.5 ml aniline, cool, add 10 ml of water and filter the precipitated compound.

$$C_6H_5NH_2 + H_2NCOCH_3 \longrightarrow C_6H_5NHCOCH_3 + NH_3$$

2.   128°C Benzamide,

White crystalline compound soluble in hot water.

Boil ½ gm of compound with about 10 ml of dil. NaOH solution as above, cool and acidify the solution with dil. H_2SO_4. White precipitate of benzoic acid is obtained. Filter and collect it as a derivative.

Derivative : Benzoic acid M.P. 121°C

3.   132°C   Urea,   C = O

White crystalline solid, soluble in water.

(i)   When heated alone or with dil. NaOH, it gives out ammonia.

(ii) *Biuret test.* Take 0.2 gm of substance in a dry test tube and heat gently. The compound melts, ammonia is evolved and gradually the liquid solidifies due to the formation of biuret.

$$NH_2CONH \boxed{H + H_2N} CONH_2 \xrightarrow[-NH_3]{\text{heat}} NH_2CONHCONH_2 \text{ Biuret}$$

Dissolve this residue in a mixture of 2 ml of water and 2 ml of dil. NaOH solution and add 1 ml of $CuSO_4$ solution. A purple colour is obtained.

Derivatives : (i) *Urea nitrate.* M.P. 163°C.

Add ½ ml of conc. HNO_3 to 1 of ml conc. solution of urea in water, warm and cool. Urea nitrate is precipitated.

(ii) *Urea Oxalate.* M.P. 171°C.

Mix a concentrated solution of oxalic acid with conc. aqueous solution of urea and isolate the crystalline derivative.

4.   233-238°C Phthalimide, C_6H_4

Aromatic. Colourless crystalline solid, insoluble in water and benzene but soluble in alcohol.

(*i*) Heat a small quantity of original substance in a dry test tube. A smell of ammonia is obtained (the same test is also given by phthalamide).

(*ii*) Take 0.5 gm of original substance, add 0.5 ml of conc. H_2SO_4 and heat. Add 0.5 gm of resorcinol and heat the mixture to red brown. Cool and pour the mixture into 50 ml of water containing about 10 ml of aq. NaOH. A green fluorescence is obtained (similar to phthalamide) is obtained.

(*iii*) Take 0.5 gm of original substance and 10 ml of absolute alcohol in a test tube. Immerse the base of test tube in boiling water for 1 minute with constant shaking. The compound gets dissolved (unlike phthalamide which remains practically insoluble).

Derivatives : (*i*) *Phthalic acid,* M.P. 195°C.

(*ii*) *Benzyl phthalimide,* M.P. 115°C.

On adding alcoholic solution of KOH to alcoholic solution of original substance, crystals of potassium derivative are obtained which on heating with benzyl chloride forms benzylpthalimide.

2. Primary Amines

Group tests

(*i*) *Carbylamine test.* Take a small amount of the substance, 1 ml alcoholic caustic potash solution and a few drops of chloroform in a test tube, shake and heat gently. A very bad smell of carbylamine (isocyanide) is obtained. [After performing the test, destroy the isocyanide by adding excess of HCl to the reaction mixture and throw it out].

$$RNH_2 + CHCl_3 + 3KOH \longrightarrow RNC + 3KCl + 3H_2O$$
$$\text{alkyl Isocyanide}$$

(*ii*) *Distinction between aliphatic and aromatic primary amines.* Aromatic primary amines give diazonium salts with nitrous acid which can be detected by formation of red dye with alkaline phenol solution.

Dissolve 0.5 gm of the compound in 2 ml of conc. HCl in a boiling tube and dilute it with 4 ml water. Cool the solution in ice bath and diazotise it with a cold solution of sodium nitrite. Add an alkaline solution of β-naphthol. A red dye is obtained.

β-naphthol   Phenyl azo-β-naphthol
                                                  (Scarlet red dye)

Characteristic derivatives

(*i*) *Benzoyl derivative.* Mix 1 gm of the substance with 10 ml of dil. NaOH solution in a boiling tube. Cool and add 1-2 ml of benzoyl chloride drop by drop with shaking. Semisolid benzoyl derivative begins to separate out. Continue shaking for a few minutes, cool, filter the solid, wash well with water and recrystallise from spirit.

Benzoyl derivative

(ii) *Acetyl derivative.* Reflux gently a mixture of 1 gm of the substance with 4 ml acetic anhydride and glacial acetic acid mixture (2 ml acetic anhydride + 2 ml acetic acid) for about 20 minutes. Pour the reaction mixture into 20 ml of cold water and filter the precipitated solid. Recrystallise if necessary from water or aqueous alcohol.

$$RNH_2 + O \Big\langle \begin{array}{c} COCH_3 \\ COCH_3 \end{array} \longrightarrow \underset{\text{Acetyl derivative}}{RNHCOCH_3} + CH_3COOH$$

(iii) *Other derivatives.* In some cases, other derivatives like bromo-amines and picrates may be conveniently prepared.

Bromo-derivatives. Proceed as described in phenols.

Picrates. Described in hydrocarbons.

List of Some Primary amines

Liquids		Solids	
B.P. °C	Name	M.P. °C	Name
49	n-Propylamine	44-45	p-Toluidine
77	n-Butylamine	50	α-Naphthylamine
183	Aniline	57	p-Anisidine
185	Benzyl amine	102	o-phenylene-diamine
204	o-Toluidine	110-11°	β-Naphthylamine
203	m-Toluidine	127	Benzidine

Specific Tests for some Primary Amines

Liquids

B.P.

1. **183°C   Aniline,**

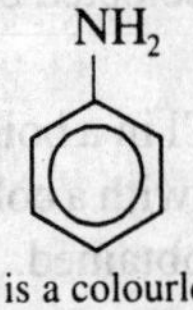

When distilled fresh, it is a colourless liquid but usually it turns reddish when exposed to atmosphere.

(i) Gives carbylamine test and forms red colourd dye when its diazotised solution ($NaNO_2$ + HCl) is treated with an alkaline solution of β-naphthol.

(ii) When treated with a solution of $K_2Cr_2O_7$ and conc. H_2SO_4, it gives a blue or black colour.

Derivatives. Acetanilide M.P. 114°C

2, 4, 6-Tribromoaniline M.P. 118°C.

Benzoate derivative M.P. 163°C.

2. **204°C   o-Toludine**

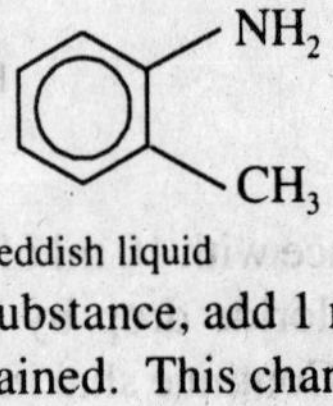

Reddish liquid

(i) To 0.5 ml of the substance, add 1 ml of 50% H_2SO_4 and a few crystals of $K_2Cr_2O_7$. A blue colour is obtained. This changes to purple upon dilution.

(ii) Gives carbylamine reaction and dye test.

Derivatives. Benzoate : M.P. 143°C.

Dibromoderivative : M.P. 50°C.

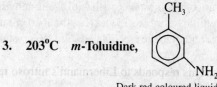

3. **203°C** **m-Toluidine,**

Dark red coloured liquid

(i) Gives carbylamine test.

(ii) Forms red dye.

Dissolve a little of the substance in 50% H_2SO_4 and add a few crystals of $K_2Cr_2O_7$. Yellowish brown colour is obtained. This upon addition of dil. HNO_3 changes to red.

Derivatives. Benzoyl derivative : M.P. 125°C.

Acetate derivative M.P. 65°C.

Solids

4. **M.P.**
   44 – 45°C **p-Toluidine,**

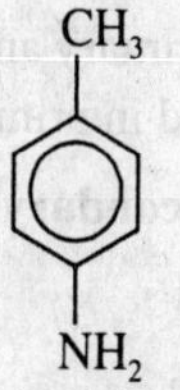

Dirty white crystals with a bad smell.

(i) Give carbylamine reaction and dye test.

Dissolve a little of substance in 5 ml water and add a few drops of bleaching powder solution. A yellow colour is obtained (aniline gives purple colour).

Benzoyl derivative. M.P. 158°C.

Picrate : M.P. 200°C.

5. **50°C** **α-Naphthylamine,**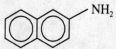

Dark red coloured solid with a pungent smell

(i) Dissolve a little of the substance in dil. HCl and add a few drops of $FeCl_3$ solution. A blue precipitate is obtained.

(ii) Give carbylamine reaction and dye test positive.

Derivatives. Picrate M.P. 161°C.

Acetate M.P. 159°C.

6. **110-111°C** **β-Naphthylamine**

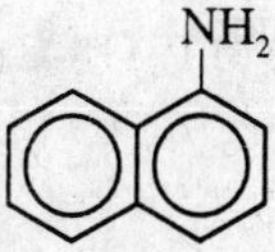

Pink coloured solid

(i) Give carbylamine test.

(ii) No colour with $FeCl_3$ solution.

Derivatives : Picrate : M.P. 195°C.

Acetate : M.P. 132°C.

3. Secondary Amines

Group tests

(i) Do not give carbylamine test reaction.

(ii) With nitrous acid, a nitrosoamine is obtained. This responds to Libermann's nitroso reaction.

Liebermann's nitroso reaction. Dissolve 0.5 gm of the given compound in 2 ml of dil. HCl, cool and add sodium nitrite solution gradually with shaking until yellow oil settles at the bottom. Decant off the aqueous layer.

Place 2 drops of the oily nitroso compound in a dry test tube, add 0.5 gm of phenol and warm gently for a few seconds. Cool and add 1 ml conc. H_2SO_4. A greenish blue colour is obtained which changes to red upon dilution with water. On of adding excess NaOH solution, the greenish blue colour is restored.

Derivatives

(i) *Acetyl derivatives.* Prepare as described in primary amines.

(ii) *Picrates* are prepared by the method described in primary amines.

Specific tests for some Secondary Amines

Liquids

1. **B.P. 193°C Methylaniline,**

Colourless liquid, characteristic smell

(i) Dissolve 2 drops of the compound in dil. HCl and add a few drops of $FeCl_3$. Greenish-blue colour is obtained.

(ii) Gives carbylamine reaction negative.

Derivatives

(i) Benzoyl derivative, M.P. 63°C.

(ii) Acetate, M.P. 102°C.

Solids

2. **M.P. 54°C Diphenylamine,**

White lustrous solid

(i) A solution of the compound in HCl gives a blue colour with HNO_3.

(ii) Does not give carbylamine reaction.

Derivatives : Acetate, M.P. 101°C.

4. Tertiary Amines $\begin{matrix} R \\ R-N: \\ R \end{matrix}$

Group tests

(i) Does not give carbylamine reaction.

(ii) Does not give acetyl and benzoyl derivatives.

(iii) Aromatic tertiary amines give *p*-nitroso derivative.

Dissolve a few drops of the substance in 2 ml of dil. HCl, cool in ice and add 1 ml of sodium nitrite solution drop by drop. A reddish solution is obtained. After allowing it to rest for a few minutes, add dil. NaOH solution. A green precipitate of *p*-nitroso derivative is obtained. Now, add 2 ml of ether dil. NaOH solution. A green precipitate of *p*-nitroso derivative is obtained. Now, add 2 ml of ether, cork the test tube and shake well. A deep green etherial layer is seen floating on the upper surface.

Derivatives

(i) *Picrates.* Use the method described earlier.

(ii) *Other derivatives.* Various other suitable derivatives may be prepared depending on the nature of compound. For example, in case of aromatic tertiary amines, nuclear substitution products such as bromo and nitro may be prepared.

Specific Tests for some Tertiary Amines

B.P.
193°C **Dimethyl aniline,**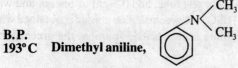

Colourless liquid

(i) Gives nitroso reaction described above.

Picrate derivatves : M.P. 163°C.

p-**Bromoderivative :** M.P. 55°C.

5. Anilides

C_6H_5—NHCOR

Group test

(i) Upon boiling with alkali, an anilide is hydrolysed to free amine and the acid. The presence of amine may be detected by any of the general tests applicable for primary amines *e.g.* carbylamine reaction may be applied to anilides.

Carbylamine test. Heat a mixture of 0.5 gm of the compound, 4 ml of alcoholic KOH solution and 1 ml of chloroform. A bad smell of carbylamine comes out.

Characteristic derivatives

(i) Bromoderivative : Prepare as described in phenols.

(ii) Nitroderivative : See in specific compound.

Specific Tests of Some Anilides

M.P.
1. **114°C** **Acetanilide,**   NHCOCH$_3$

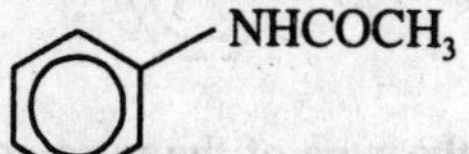

White thin plate like crystals. No smell.

(i) Gives carbylamine reaction test.

(ii) Heat a little of the substance with a pinch of $K_2Cr_2O_7$ crystals and a few drops of conc. H_2SO_4. A red colour changing to green is obtained.

Derivatives

(i) *p*-Bromoacetanilide, M.P. 167°C.

(ii) *p*-Nitroacetanilide, M.P. 210°C.

Dissolve 2 gms of acetanilide in 2 ml of glacial acetic acid by warming the mixture. Add 4 ml of conc. H_2SO_4 and cool the beaker. Now add 1 ml of conc. HNO_3 drop by drop with shaking. Allow the mixture to stand for about 20 minutes and then pour it into 20 ml of ice-water Filter the precipitated product, wash with water and dry.

2. 161°C Benzanilide,

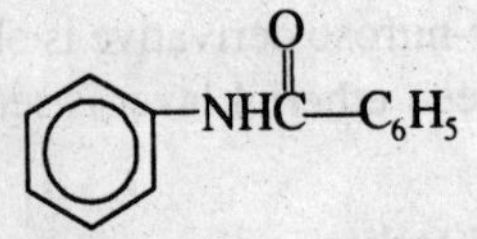

$$\text{C}_6\text{H}_5-\text{NHC}-\text{C}_6\text{H}_5 \quad (\text{with } \overset{\text{O}}{\|})$$

White crystalline solid

(i) Boil a little amount of the substance with alkali and chloroform. Bad smell of isocyanide is obtained.

(ii) Boil 1 gm of the substance with 2 ml of conc. alkali, cool and acidify with dil. HCl. A white precipitate of benzoic acid is obtained.

Derivatives

p-Bromoderivative, M.P. 204°C.

Benzoic acid, M.P. 121°C.

6. Nitro Compounds

Tests for nitro (—NO_2) group

(i) *Mulliken's test.* In a test tube, take about 0.2 gm of the compound, 4-5 ml of alcohol, 1 ml of calcium chloride or NH_4Cl solution add a pinch of zinc dust, boil, cool and filter. Add to the filtrate, 2 ml of Tollen's reagent. A grey or black precipitate or silver mirror is produced. The nitro group is reduced in the above process to —NHOH group which reduces Tollen's reagent.

$$RNO_2 + 4H \longrightarrow RNHOH + H_2O$$

(ii) *Reduction to amino compound and dye test.* In a boiling tube, take 0.2 gm of the compound, 2 ml of conc. HCl and 2 ml of water. Add to it, 1 gm of solid stannous chloride or tin granules. Heat the tube in boiling water bath for about 15 minutes. Cool, add sodium nitrite solution drop by drop to complete the diazotisation. Add 2 ml of alkaline and β-naphthol solution. A red dye is produced.

$$ArNO_2 + 6H \longrightarrow ArNH_2 + 2H_2O$$

$$ArNH_2 \xrightarrow[\text{HCl}]{NaNO_2} ArN_2Cl$$

$$ArN_2Cl \xrightarrow[\text{alkali}]{\beta\text{-naphthol}} dye$$

The presence of other functional groups may also be confirmed by the usual tests.

Characteristic derivatives

Preparation of derivatives depends on the type of the nitro compound detected. Usually, two types of nitro compounds are included in the syllabus of degree classes :

(a) Nitrohydrocarbons.

(b) Nitrophenols.

Derivatives of Nitrohydrocarbons

(i) The nitro compounds may be reduced to amino compounds which may be isolated and characterised.

(ii) Some of the nitrohydrocarbons may be further nitrated. For example, nitrobenzene may be nitrated to give m-dinitrobenzene.

Derivatives of Nitrophenols

The characteristic benzoyl and acetyl derivatives of phenols may be prepared by the same general methods as applied to simple phenols.

List of some Nitro Compounds

(a) **Nitro-Hydrocarbons**

Liquids		Solids	
B.P. °C	*Name*	M.P. °C	*Name*
210	Nitrotoluene	54	*p*-Nitrotoluene
219	*o*-Nitrotoluene	61	α-Nitronaphthalene
		90	*m*-dinitrobenzene

(b) **Nitrophenols**

Solids

M.P. °C	*Name*
44	*o*-Nitrophenol
97	*m*-Nitrophenol
144	*p*-Nitrophenol

Specific tests for some Nitro Compound

(a) **Nitro-hydrocarbons**

Liquids

B.P.
1. **210.9°C   Nitrobenzene,**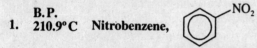

Pale yellow liquid which smells like bitter almonds and is immiscible with water.

Reduce nitrobenzene to aniline with tin and hydrochloric acid as described earlier. With this solution, perform carbylamine test and also dye test after diazotisation.

Derivative. *m*-Dinitrobenzene : M.P. 90°C.

To prepare this compound, mix 2 ml of conc. HNO_3 and 2 ml of conc. H_2SO_4. Add to it, 1 ml of nitrobenzene with shaking. Now heat the mixture for a few minutes and then pour it into cold water. At first thick oily layer settles which gradually solidifies. Filter, wash and dry.

2. **210°C   o-Nitrotoluene,**

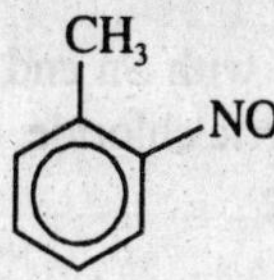

Yellowish liquid, smell resembling to that of nitrobenzene. Immiscible with water.

(i) When reduced with Sn + HCl, gives *o*-toluidine and this gives carbylamine reaction and dye test.

(ii) Reduce a small amount of the compound with Sn and HCl, dilute it with water and add 2-3 drops of $FeCl_3$ solution. Green colour appears.

Derivative. *o*-Nitrobenzoic acid : M.P. 147°C.

Heat a mixture of 1 ml of the compound, 2 ml of conc. $KMnO_4$ solution and 1 ml of KOH solution for 20 minutes. Cool, filter and acidify with dil. HCl. The acid is precipitated.

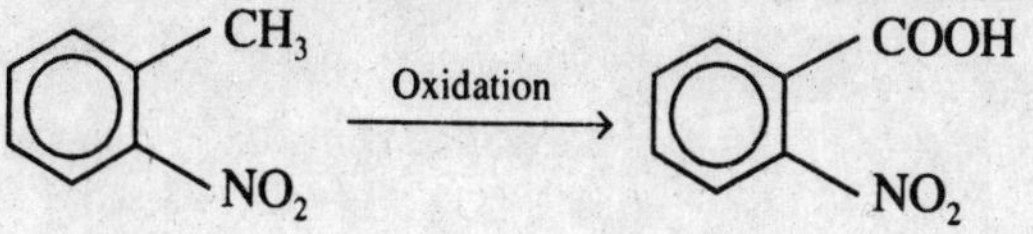

Solids

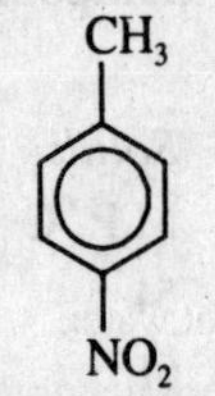

3. **M.P. 54°C** **p-Nitrotoluene**

Light yellow crystals

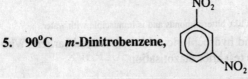

Upon reduction, it gives *p*-toluidine which responds positively to carbylamine test and dye test.

Derivative. *p*-Toluidine : M.P. 45°C.

Reduce *p*-nitrotoluene with Sn + HCl. Neutralise the acid left with dil. Na_2CO_3 solution and cool. The solid *p*-toluidine settles down. Filter and dry.

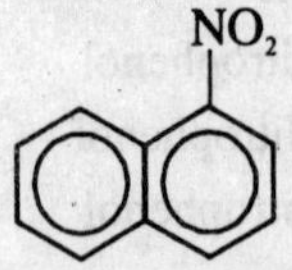

4. **61°C α-Nitronaphthalene,**

Light yellow crystals, smell resembles naphthalene.

(*i*) It dissolves in conc. H_2SO_4 giving red colour.

(*ii*) Upon reduction with Sn and HCl, it gives α-naphthylamine which can be detected by carbylamine test and dye test.

Derivative. α-Naphthylamine. M.P. 50°C.

5. **90°C m-Dinitrobenzene,**

Yellow crystalline solid.

(*i*) Dissolve 2 crystals of the compound in 1-2 ml of acetone and a few drops of dil. NaOH solution. A deep violet colour is obtained. Add acetic acid, the colour changes to red and disappears on addition of dil. HCl.

(*ii*) Reduce a small amount of the substance with Sn and HCl. Filter the solution, dilute the filtrate* and to this add a few drops of ferric chloride. A red colour is obtained.

Derivative. *m*-phenylene diamine, M.P. 63°C.

(*b*) **Nitro-phenols**

Solids

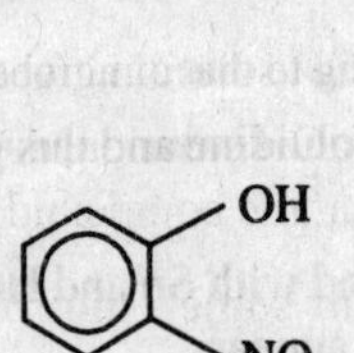

1. **M.P. 44°C** **o-Nitrophenol,**

Yellow crystalline solid, soluble in hot water and dissolves in alkali giving orange solution.

Derivatives. Benzoate derivative, M.P. 142°C.

Dibromoderivative, M.P. 117°C.

---

* In above reduction *m*-phenylene diamine is produced. The filtrate is neutralised with dil. Na_2CO_3 solution when the compound separates.

2. 97°C *m*-Nitrophenol,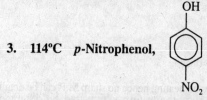

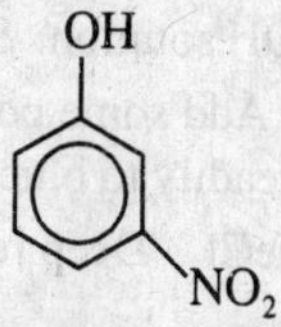

Yellow crystalline solid, soluble in hot water.

(*i*) Dissolve a few crystals of the compound in hot water and add a few drops of $FeCl_3$. A violet red colour is obtained.

(*ii*) The compound dissolves in alkali giving orange-red solution.

Derivatives. Benzoate derivatives, M.P. 95°C.

Bromoderivative, M.P. 91°C.

3. 114°C *p*-Nitrophenol,

Pale yellow crystalline solid, soluble in hot water.

(*i*) Dissolves in an alkali giving yellow solution.

(*ii*) Aqueous solution gives violet colour with $FeCl_3$.

Derivatives. Bromoderivative, M.P. 142°C.

Acetate derivative, M.P. 81°C.

Type [C]. COMPOUNDS CONTAINING NITROGEN AND SULPHUR
[In addition to C, H and O]

When both of these elements are present together, sodium extract of the compound gives a blood red colour with $FeCl_3$ due to the formation of ferric thiocyanate.

$$3NaCNS + FeCl_3 \longrightarrow Fe(CNS)_3 + 3NaCl$$
$$\text{Red colour}$$

Following types of compounds contain both of these elements :

1. Thiourea

(*i*) Simple thiourea upon boiling with NaOH solution gives off ammonia.

(*ii*) Boil a little of the substance with dil. NaOH solution, cool and add lead acetate. A brown or black colour or precipitate indicates that the compound may be a thiourea.

Specific tests for some Thioureas

M.P.

1. 154°C **Phenylthiourea,**

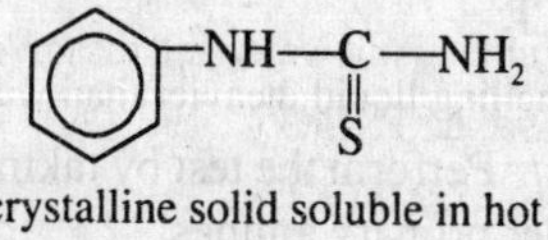

White crystalline solid soluble in hot water.

(*i*) Boil a little of the substance with dil. HCl. A pungent mustard odour is obtained.

(*ii*) On heating with NaOH it gives of ammonia.

2. 180°C **Thiourea,** $H_2N-C-NH_2$

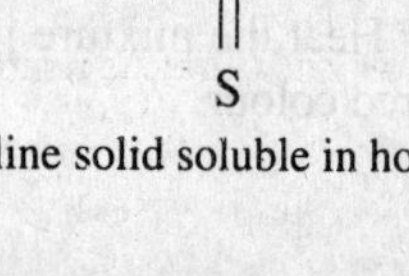

White crystalline solid soluble in hot water.

(i) Boil a little of the substance with conc. NaOH solution. Smell of ammonia is obtained.

(ii) Dissolve a little of the substance in water. Add some potassium ferrocyanide solution and dil. acetic acid. A green colour changing readily to blue is observed.

(iii) Heat the O.S. to melt then add water and $FeCl_3$. Deep red colour

2. Amino Sulphonic Acids

Dissolve a little of the substance in dil. HCl. Cool and add sodium nitrite solution so as to complete the diazotisation. Now add alkaline solution of β-naphthol. Red or orange dye is formed.

Specific tests of some Aminosulphonic Acids

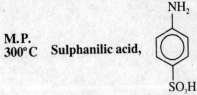

M.P.
300°C   **Sulphanilic acid,**

Light grey solid, soluble in water and decomposes on heating hence no sharp M.P. can be obtained.

(i) Diazotised solution gives orange-coloured dye with alkaline β-naphthol solution.

(ii) To aqueous solution, add bromine water. A precipitate of bromoaniline is obtained.

Type [D]. COMPOUNDS CONTAINING HALOGENS
[In addition to C, H and O]

The halogen present in the organic compounds may generally be chlorine, bromine and iodine. They may generally be classified as follows :

(i) **Aliphatic halogen compounds.** Boil ½ ml of the substance with 5 ml of alcoholic NaOH solution for about 10 minutes. Cool, dilute the solution with water, add excess of dil. HNO_3 and then add $AgNO_3$. A precipitate of silver halide is obtained.

(ii) **Aromatic halogen compounds.** They do not contain labile halogen. The compound containing the halogen may be a simple hydrocarbon or it may contain other functional groups besides halogen. Therefore, in the identification of these compounds, care must be taken to ascertain the presence and nature of the halogen and also the presence of other functional groups by their specific tests and preparation of characteristic derivatives.

1. Specific Tests for some Aliphatic Halogen Compounds

(a) **Chloride containing compounds**

Liquids

B.P.
1. 61°C   Chloroform, $CHCl_3$

Colourless sweet smelling liquid, heavier than water and immiscible with it.

(i) *Gives Carbylamine test.* Perform the test by taking the substance aniline and alcoholic KOH as described under primary amines.

(ii) Dissolve a crystal of resorcinol in 1 ml of dilute sodium hydroxide solution, add 1 ml of chloroform and warm gently. The aqueous layer turns red fading to green.

(iii) *Colour test with pyridine.* Take 2 ml of 20% NaOH solution, 2 ml of pyridine and a drop of the compound in a test tube. Heat the mixture just to boiling and allow it to stand. The pyridine layer assumes a red colour.

(iv) Reduces Fehling solution.

2. 77°C   Carbon tetrachloride, CCl₄

Colourless liquid insoluble in water.

(*i*) Does not reduce Fehling solution.

(*ii*) O.S. + Cu powder + β naphthol + NaOH and heat. Blue colour

(*iii*) Gives isocyanide test only after heating for several minutes.

Solids

M.P.
3.   54°C   Chloral hydrate,   CCl₃CH(OH)₂

Colourless deliquescent solid, characteristic smell (like toddy), soluble in water.

(*i*) Gives colour test with pyridine [test (*iii*) for CHCl₃].

(*ii*) Reduces Fehling solution and Tollen's reagent.

(*iii*) Restores the colour of Schiff's reagent.

(*iv*) On heating with alkaline solution of resorcinol, it gives red colour which changes to violet.

(b) Iodine containing Compounds

M.P.
119°C   Iodoform, CHI₃

Yellow crystals, characteristic smell, insoluble in water but soluble in spirit.

(*i*) Heat a little of the substance in a dry test tube. Violet fumes of iodine are evolved.

(*ii*) To a pinch of the substance, add a little phenol and 1 ml of dilute alcoholic NaOH solution. A red colour is obtained.

(*iii*) Gives positive colour test with pyridine.

(*iv*) Gives carbylamine reaction.

---

2. Aromatic Halogen Compounds

(a) Chlorine containing compounds

Liquids

B.P.
1.   132°C   Chlorobenzene,

Colourless liquid, smells just like benzene. Immisicible with water.

(*i*) Does not give precipitate with AgNO₃.

(*ii*) Undergoes nitration.

In a boiling tube, take 1 ml of substance and add 4 ml of a mixture of conc. HNO₃ and conc. H₂SO₄ (each 2 ml) carefully. Warm the mixture in a water bath for 5-10 minutes. Pour it into cold water. An oily mass begins to separate out which on further cooling and scratching with a glass rod solidfies as yellow crystalline mass. Filter, wash and dry. It gives smell of bitter almonds.

Derivative : 4-Chloronitrobenzene, M.P. 83°C (prepared as above).

2.   179°C   Benzyl chloride,

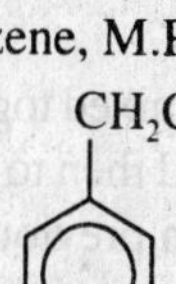

Colourless liquid

(*i*) On warming with AgNO₃ solution, a precipitate of AgCl is obtained.

(ii) To a little amount of the substance add 1 ml of lead nitrate solution and boil. Smell of benzaldehyde (bitter almonds) is obtained.

(iii) When oxidised with $K_2Cr_2O_7$ and dil. H_2SO_4, it gives benzoic acid.

Derivative. Benzoic acid, M.P. 121°C

Solids

3. **M.P.**
   53° C *p*-Dichlorobenzene,

White crystalline solid, characteristic smell.

(i) No precipitate with $AgNO_3$.

(ii) Can be readily nitrated.

Derivative. 2, 5-Dichloronitrobenzene, M.P. 54°C.

Proceed as described in th nitration of chlorobenzene.

(b) Aromatic Bromohydrocarbons

Liquids

1. **M.P.**
   155°C Bromobenzene,

Light yellow liquid, insoluble in water.

(i) Does not give precipitate with $AgNO_3$.

(ii) Can be nitrated by the above method to give *p*-nitrobromobenzene.

Derivative. *p*-Nitrobromobenzene, M.P. 126°C.

Solid

2. **M.P.**
   89° C *p*-Dibromobenzene,

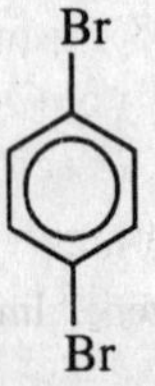

Colourless crystalline compound, insoluble in water.

Upon nitration by above method, it gives 2, 5-dibromonitrobenzene.

Derivative. 2,5-Dibromonitrobenzene, M.P. 84°C.

ORGANIC MIXTURE ANALYSIS

Analysis of Simple Binary Mixture

A binary mixture contains two organic compounds mixed together. The first job in the analysis of the mixture is to separate the individual components and then to proceed with the detailed investigation for the identification of each compound as given in previous sections.

Separation of Binary Mixtures (Solid Components)

Generally the two components present in the mixture are of different nature. They may differ in their solubility in water, acids, alkalies or in some common solvents. They may have similar or different physical state ; one components may be a solid and the other a liquid. A successful separation of the mixture requires general awareness of the chemical behaviour and physical characteristics including solubility etc., of different classes of organic compounds.

An outline of the procedure adopted for the separation of binary mixtures in which both the components are solids is given below :

1. Water Separation

Treat a small quantity of the mixture with cold water, shake well and filter. Evaporate the filtrate in a porcelain dish. If a solid residue is obtained, it indicates that a part of the mixture is soluble in cold water (this will also be indicated by the reduction in the quantity of the undissolved portion of the mixture).

Now, treat the whole mixture with cold water, shake well and filter. Keep the residue on the filter paper as component A. Evaporate the filtrate in a porcelain dish and collect the solid residue as component B. Proceed with the identification of these compounds in the manner given in the previous sections.

(i) The compounds soluble in cold water are generally lower aliphatic acids such as oxalic acids, lower amides such as acetamide and urea, carbohydrates excluding starch, some polyhydroxyphenols and amine salts.

(ii) If acid or phenol is present, the aqueous solution turns blue litmus red. If an acid is present the solution will give effervescence with $NaHCO_3$.

(iii) Amides and carbohydrates will form neutral aqueous solution.

(iv) On evaporating the aqueous solution, a carbohydrate will give syrupy residue.

Higher aliphatic acids such as succinic and adipic acids etc. dissolve in hot water and crystallise out when the solution is cooled. In such cases, treat the mixture with hot water, shake well and filter it quickly. Keep the residue on the filter paper as component A and the solid separated from aqueous solution as component B.

2. Separation with dilute alkali

Almost all aromatic acids and most of the phenols are insoluble in water and hence they are not water separable. In such cases, take a small quantity of the mixture, treat it with dilute aqueous NaOH solution (about 10%), shake well and filter. To the filtrate, add dil. H_2SO_4 until it becomes acidic. If a precipitate is obtained, it indicates the presence of an acid or a phenol.

Now, treat the whole mixture with 10% NaOH solution in excess and shake well. (It is advisable to carry out this process in a conical flask which should be well corked while shaking). Filter and wash the residue with alkali and then thoroughly with water. Press with filter paper, dry and keep it as component A. To the filtrate, add dilute H_2SO_4 until the solution is acidic to litmus, cool, filter, wash the precipitate well with water, dry and test it as component B.

3. Separation with Acid

If one component of the mixture is basic, generally an amine, it can be separated with the help of dilute HCl with which it will form water soluble amine hydrochloride salt. Free amine can now be obtained by neutralisation of the solution with alkali.

Treat the mixture with dilute HCl in excess, shake well and filter. Collect the undissolved residue as component A. To the filtrate, add excess of dil NaOH solution (about 25-30%) until the solution is alkaline to litmus. Cool and scratch the sides of the flask with a glass rod. The component B is gradually precipitated. Filter, dry and proceed for the analysis.

13

Organic Preparations

1. Acetanilide

NH$_2$ $\xrightarrow[\text{CH}_3\text{COOH}]{\text{(CH}_3\text{CO)}_2\text{O}}$ NHCOCH$_3$

Required : Aniline — 10 ml
Acetic anhydride — 10 ml
Glacial Acetic acid — 10 ml

Method. In a 250 ml conical flask, take a mixture of acetic anhydride and acetic acid. Add redistilled aniline gradually with shaking. Now attach the flask to a reflux condenser and heat the reaction mixture for about 30 minutes. Cool the flask in tap water and pour the contents in 150 ml of water containing some ice. The acetanilide is precipitated. Stir the solution well and filter. Wash the precipitate with cold water and recrystallise it from hot water containing a little alcohol (125 ml water and 5 ml alcohol) or from about 60 ml mixture of acetic acid and water (1 : 2). Filter and dry at 50–60°C. Acetanilide is obtained as colourless crystals.

Alternative Method. This method is quicker than previous one. Mix 10 ml of aniline, 10 ml of acetic anhydride, 10 ml of glacial acetic acid and a pinch of zinc dust in a beaker. Heat the mixture slowly on wire gauze for about 50–70 minutes pour hot solution in about 200 ml cold water with stirring. Filter, recrystallise and dry.

Yield ... 8-10 gm
M.P. ... 114°C

2. p-Bromoacetanilide

NHCOCH$_3$ $\xrightarrow{\text{CH}_3\text{COOH}}$ NHCOCH$_3$ (Br) + HBr

Required : Acetanilide — 5 gm
Bromine — 2·1 ml (6·7 gm)
Acetic acid — 50 ml

Method. Dissolve acetanilide in 25 ml cold glacial acetic acid in a 250 ml conical flask. In another flask, prepare a solution of bromine (3 ml) in acetic acid (25 ml) and add this solution gradually to the acetanilide solution with shaking. Allow the reaction mixture to stand at room temperature for about 15 minutes and then pour the solution in about 300 ml of cold water where p-bromoacetanilide is rapidly precipitated. Stir the solution well and filter. Wash the precipitate with cold water and recrystallise from the rectified spirit. The p-bromoacetanilide is obtained as colourless crystals.

Yield ... 7 gm

M.P. ... 167°C

3. Oxalic acid

$$C_{12}H_{22}O_{11} \xrightarrow[\text{(O)}]{\text{Conc. HNO}_3} \begin{array}{c} COOH \\ | \\ COOH \end{array} \ 2H_2O + 3H_2O$$

cane sugar

Oxalic acid

Required : Cane sugar ... 15 gms

Conc. Nitric acid ... 75 ml

Method. Take cane sugar in a 500 ml flat bottomed flask and conc. HNO_3 and heat the flask on a boiling water bath. A vigorous reaction accompanied by copious evolution of NO_2 fumes takes place. Remove the flask from the water bath and place it on a wooden block or a piece of asbestos. When the reaction subsides, transfer the solution into an evaporating dish and evaporate the solution on the water bath until it has a volume of about 15 ml. Now add about 40–50 ml of water to the solution and again evaporate it to about 20 ml. Cool the solution in icy-water where oxalic acid crystallises. Recrystallise it with warm water. Dry it by pressing between pads of drying paper.

Yield ... 5–5·5 gm

M.P ... 101°C

4. Iodoform

$$CH_3COCH_3 + 3I_2 \longrightarrow CI_3COCH_3 + 3HI$$

$$CI_3COCH_3 + NaOH \longrightarrow CHI_3 + CH_3COONa$$

$$3HI + 3NaOH \longrightarrow 3NaI + 3H_2O$$

$$CH_3COCH_3 + 3I_2 + 4NaOH \longrightarrow CHI_3 + 3NaI + 3H_2O + CH_3COONa$$

Required. Acetone ... 2 ml

10% KI soution ... 80 ml

10% NaOH solution ... 30 ml

2N–NaOCl solution ... 70 ml

Method. In a conical flask, take acetone, KI solution and caustic soda solution in the proportion given above. To this mixture, add freshly prepared sodium hypochlorite solution and shake the reaction mixture well. Allow it to stand at room temperature for a few minutes. Filter the yellow precipitate of iodoform, wash and recrystallise from spirit. For this purpose, place the crude material in a 150 ml round bottomed flask fitted with a reflux water condenser. Add a small quantity of methylated spirit and heat to boiling on a water bath and then add more methylated spirit cautiously down the condenser until all the iodoform is dissolved. Filter the hot solution through a fluted filter paper directly into a small beaker or conical flask, and then cool in ice water. The iodoform rapidly crystallises. Filter and dry.

Yield ... 4 gm

M.P. ... 120°C

5. Benzoin

$$C_6H_5CHO + OHC\ C_6H_5 \xrightarrow{CN^-} C_6H_5\ CHOH\ COC_6H_5$$

Required. Benzaldehyde ... 12·5 ml (13 gm)

Powdered sodium cyanide ... 2·5 gm

Rectified spirit ... 25 ml

Water ... 10 ml

Method. Place sodium cyanide, spirit and water in a conical flask and to it, add freshly distilled benzaldehyde. Reflux the mixture gently on a water bath for about 30 minutes to get a clear solution. Now pour the solution into a beaker and cool. A yellow precipitate of benzoin is obtained. Filter, wash well with water and recrystallise from spirit.

$$\text{Yield ... gms}$$
$$\text{M.P. ... 137°C}$$

6. Benzil

$$C_6H_5CHOH\ CO\ C_6H_5 \xrightarrow{\text{oxidation}} C_6H_5-\underset{\underset{O}{\|}}{C}-\underset{\underset{O}{\|}}{C}-C_6H_5 + H_2O$$

Required : Benzoin   ...   5 gm
               Conc. HNO_3 ... 12·5 ml

Method. Reflux the mixture of benzoin and nitric acid in a round-bottom flask on a water bath until the evoution of nitrous fumes ceases. Pour the contents of the flask into 50 ml cold water. The oily product gradually solidifies. Filter, wash with water and recrystallise it from spirit.

$$\text{Yield ... 4 gm}$$
$$\text{M.P. ... 95°C}$$

7. Phenyl Benzoate

$$\langle\bigcirc\rangle-OH + ClCOC_6H_5 \xrightarrow[\substack{\text{Schotten Baumann} \\ \text{reaction}}]{NaOH} \langle\bigcirc\rangle-OCOC_6H_5$$

Required : Phenol   ...   5 gm
               100% NaOH   ... 70 ml
               Benzoyl chloride ...   9 ml

Dissolve phenol in sodium hydroxide solution contained in a conical flask. Add benzoyl chloride gradually with shaking. When the addition is complete, cork the flask and shake the mixture for about 10 minutes more and cool. Filter the solid product, wash it well with cold water and recrystallise from spirit.

$$\text{Yield ... 8 gm}$$
$$\text{M.P. .. 69°C}$$

8. *m*-Dinitrobenzene

$$\langle\bigcirc\rangle\overset{NO_2}{} \xrightarrow[H_2SO_4]{HNO_3} \langle\bigcirc\rangle\overset{NO_2}{}_{NO_2}$$

Required. Nitrobenzene   ... 5 gm
               Fuming HNO_3 ... 5 ml
               Conc. H_2SO_4 ... 7 ml

Method. Take the mixture of nitric and sulphuric acids in a flask and add nitrobenzene gradually with constant shaking. Heat the mixture on the sand bath for 30–40 minutes at a low flame. Now cool the reaction mixture and pour into ice cold water. At first, oily product separates which immediately solidifies. Filter, wash and recrystallise from alcohol.

$$\text{Yield   ... 4–4·5 gm}$$
$$\text{M.P.   ... 90°C}$$

9. Phthalimide

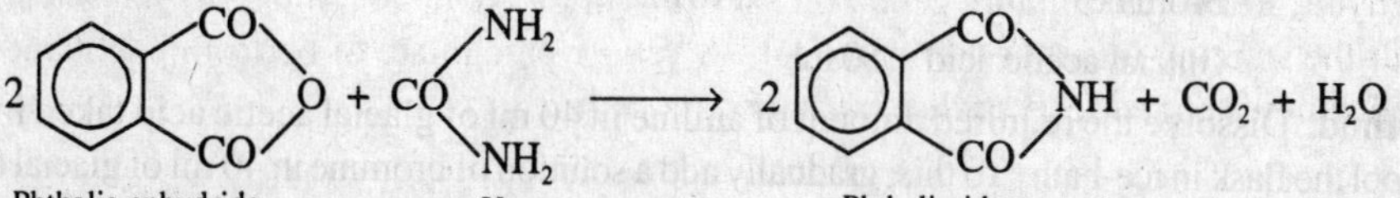

Phthalic anhydride   Urea   Phthalimide

Required :   Phthalic anhydride   ...   15 gm
                 Urea   ...   3·5 gm

Method. Take the mixture of phthalic anhydride and urea in a 250 ml flask and neat it on a sand bath keeping the temperature of the reaction mixture at about 130–135°C. The temperature may be recorded by putting a thermometer in the reaction mixture. After a few minutes, the reaction starts which is indicated by froth of the solid mass and a sharp rise in the temperature.

When the reaction subsides, a spongy solid is obtained. Cool, add some water and transfer the solid over a Buckner funnel. Wash it well and dry. Recrystallise from alcohol.

Yield   ... 12 gm
M.P.   230°C (Crude product)
        233°C (recrystallised).

10. Picric acid

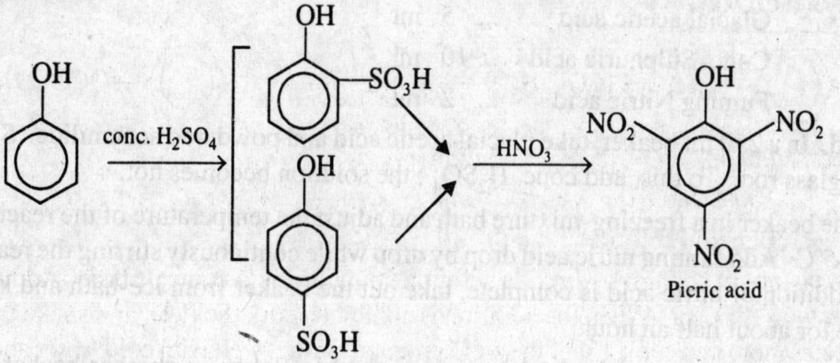

Picric acid

Required :   Phenol   ...   6 gm
                 Conc. Sulphuric acid ... 7·5 ml
                 Conc. Nitric acid   ...   22·5 ml

Method. Take phenol in a dry flask and add conc. H_2SO_4 into it. Shake the mixture and heat over water bath for about half an hour. Now, cool the flask in an ice-bath so that the reaction mixture is cooled to about 10°C. Take out the flask from ice-bath, keep it on the table and gradually add the required amount of conc. HNO_3. Shake the mixture thoroughly and allow it to stand at room temperature. A brisk evolution of brown fumes takes place for a few minutes and then subsides.

Now, heat the flask on a water bath for about 1·5 hours with occasional shaking. Cool and add about 75 ml water. A yellow precipitate soon begins to separate. Filter, wash it with water and recrystallise from dilute alcohol or from water to which a few drops of dilute H_2SO_4 is added.

Yield   ... 7·5 gm
M.P.   ... 122°C

11. Tribromoaniline

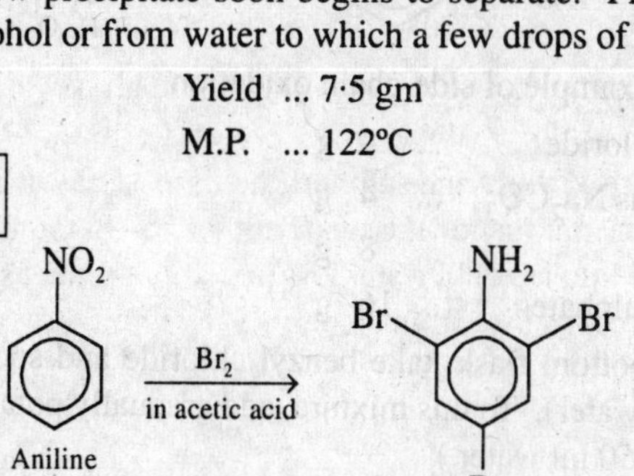

Aniline

2, 4, 6-Tribromoaniline

Required : Aniline   ... 10 ml
                Bromine   ... 16 ml
                Glacial acetic acid ... 80 ml

Method. Dissolve the required amount of aniline in 40 ml of glacial acetic acid taken in a conical flask. Cool the flask in ice-bath. To this, gradually add a solution of bromine in 40 ml of glacial acetic acid drop by drop with constant shaking. (The adding of bromine may be done with a burette.) As bromine is being added, a precipitate is formed which becomes bulkier at the end of the addition.

Pour the contents of the flask in about 150–200 ml of cold water. Filter the precipitated solid, wash it with cold water and recrystallise from rectified spirit.

Yield   ...   22 gm
M.P.   ... 120°C

12. p-Nitroacetanilide

NHCOCH$_3$ $\xrightarrow[\text{H}_2\text{SO}_4]{\text{HNO}_3}$ NHCOCH$_3$ ... NO$_2$

Required : Acetanilide   ... 5 gm
                Glacial acetic acid   ... 5 ml
                Conc. Sulphuric acid   ... 10 ml
                Fuming Nitric acid   ... 2 ml

Method. In a 250 ml beaker, take glacial acetic acid and powdered acetanilide. Stir the mixture well with a glass rod. To this, add conc. H_2SO_4 ; the solution becomes hot.

Place the beaker in a freezing-mixture bath and adjust the temperature of the reaction mixture in the range 0–5°C. Add fuming nitric acid drop by drop while continously stirring the reaction mixture. When the addition of nitric acid is complete, take out the beaker from ice-bath and keep it at room temperature for about half an hour.

Now pour the contents of the beaker over crushed ice taken in another beaker. The product p-nitroacetanilide precipitates out immediately. Filter it at the pump, wash thoroughly with water and recrystallise from methylated spirit.

Yield ... 6-7 gm
M.P.   ... 214°C

13. Benzoic Acid

CH$_2$Cl $\xrightarrow[\text{H}_2\text{O}]{\text{Na}_2\text{CO}_3}$ CH$_2$OH $\xrightarrow[\text{(ii) H}^+_3\text{O}]{\text{(i) KMnO}_4 \ \text{OH}^-}$ COOH

This preparation is an example of side-chain oxidation.

Required : Benzyl chloride   ... 4 g
                Anhydrous Na$_2$CO$_3$ ... 4 g
                KMnO$_4$ ... 8 g
                Sodium sulphate   ... 16 g

Method. In a round bottom flask, take benzyl chloride and sodium carbonate solution (4 g Na$_2$CO$_3$ dissolved in 50 ml water). To this mixture, add gradually potassium permanganate solution (8 g. KMnO$_4$ dissolved in 150 ml water.)

Now, fit the flask with reflux water condenser and boil the reaction mixture gently over the wire gauze for 1½-2 hours until the pink colour of $KMnO_4$ and oily drops of benzyl chloride disappear and a dark brown precipitate is obtained.

Cool the flask and add 35–40 ml of conc. HCl slowly until the mixture becomes acidic and benzoic acid is precipitated along with brown precipitate of manganese dioxide. On adding sodium sulphate solution (16 g Na_2SO_4 dissolved in 80 ml of water) the white precipitate of benzoic acid remains.

Filter the benzoic acid and wash well with water and then recrystallise from boiling water.

Yield ... 4g

M.P. ... 121°C

14. 2, 4, 6-Tribromophenol

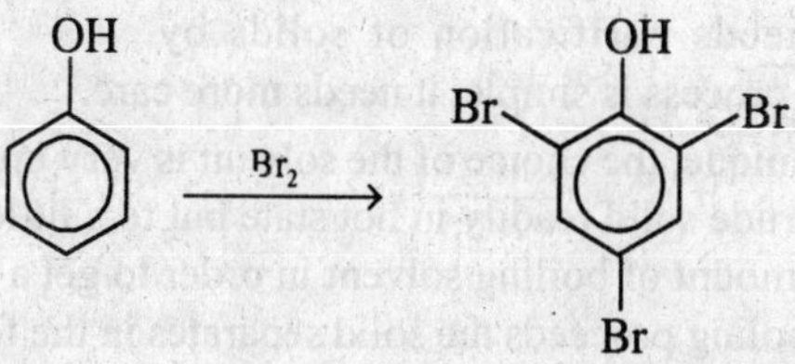

2, 4, 6-Tribromophenol

Required : Phenol ... 5 gm

Bromine ... 9 ml

Method. Make a suspension of phenol in about 100 ml of water. Cool it in ice and then gradually add to it aqueous solution of bromine (9 ml dissolved in about 100 ml water). A white or pale yellow precipitate is obtained. Cool, filter and wash it with water. Recrystallise from the aqueous alcohol.

Yield ... 17 gm

M.P. ... 95°C

15. Preparation of Aspirin (acetyl salicylic acid)

Required :   (i) Salicylic acid 5 gm

       (ii) Acetic Anhydride 10 ml

Method. Mix the contents and add 1 ml conc. H_2SO_4 in a conical flask. Heat it to 50-60°C on water bath for 15 mts. Cool and add ~ 100 ml water and filter the ppt. Recrystallise it with equal volume of alcohol and water yield ~ 5 gm, M.P. 130-135°C.

PURIFICATION OF ORGANIC COMPOUNDS

Purification of Solids. In analytical work, the purity of organic compounds is very important. An impure solid can be ordinarily purified in two ways :

(i) By sublimation and

(ii) By crystallisation.

Sublimation. It is often useful for purification of large number of compounds and can be illustrated as follows :

Purification of Naphthalene

Crude naphthalene is purified by sublimation. *When a solid compound on heating passes readily and directly into vapour state and on cooling changes back to solid state, this phenomenon is known as sublimation.*

The purification of naphthalene by sublimation may be done in a simple apparatus as shown in Fig. 13.1. Crude dry naphthalene is placed in evaporating dish A. This is kept on a wire gauze. Dish A is covered with a perforated filter paper on which a glass funnel is kept in inverted position. The crude compound kept in dish A is now gently heated. Naphthalene vapourises and the vapours pass through the holes of the filter paper in the cold inclosure of the glass funnel. Here, the vapours are cooled and are converted into colourless crystals. These get deposited on the upper surface of the paper and also on the inner walls of the funnel and are finally collected from here.

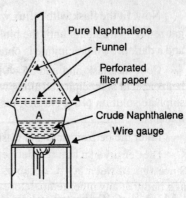

Fig. 13.1

Crystallisation of Solids

During the course of qualitative organic analysis and preparative work, one often needs purification of solids by crystallisation. Although the process is simple, it needs more care.

In the crystallisation technique, the choice of the solvent is very important. A desired solvent is the one which dissolves the crude solid readily in hot state but to a little extent in cold. The solid is then dissolved in minimum amount of boiling solvent in order to get a saturated solution. It is then filtered and cooled. As the cooling proceeds the solid separates in the form of crystals. The crystals are filtered off and are then dried.

There is no strict rule for the selection of a solvent, it is generally chosen by trial. However, a guidance may be obtained from the crude generalisation that "like dissolves like". That is, a solvent will dissolve more readily the compounds similar in constitution to itself. For example, the hydroxy compounds will tend to dissolve in water or alcohol or a mixture of the two while a hydrocarbon will dissolve mostly in liquid hydrocarbon. Those compounds which form hydrogen bond with water are usually soluble in it. Such compounds are alcohols, aldehydes, ketones and amides.

If a substance is soluble in one solvent and insoluble in the other, solvent pairs such as methanol-water, ethanol-water and others are employed. The compound is dissolved in the hot solvent A in which it is very soluble and then hot solvent B in which it is less soluble is added dropwise until a slight turbidity is produced. Then again first solvent A is added to clear the turbidity and then the solution is allowed to cool.

Sometimes, the desired compound may contain coloured impurities. To remove these impurities, a pinch of animal charcoal is added while the solid is being refluxed in the solvent (used for crystallisation) and the solution is filtered hot. The animal charcoal absorbs the coloured impurities and thus the filtered solution now becomes free from them and deposits as pure crystals.

The solvents commonly preferred for crystallisation are water, ethanol, methanol, acetic acid, chloroform, carbon-tetrachloride and acetone. Inflammable solvent like benzene, ether or petroleum ether should normally be avoided or should be used with caution.

Purification of Liquids. In most of the cases, the liquids are purified by simple distillation (for techniques of distillation see the method for boiling point determination of liquids as described in the Chapter).

14

Experiments on Physical Chemistry

Measurement of Density

The density of a liquid is defined as its mass per unit volume. It is also known as specfic gravity. Since 1gm of water at 4°C occupies 1 ml volume, its density at this temperature is one. The density of a liquid at any other temperature is expressed relatively to that of density of water at 4°C.

EXPERIMENT – 1

To measure the density of a given liquid by Pyknometer.

The density of a liquid can either be measured by Pyknometer or by relative density bottle (commonly called R.D. bottle).

Pyknometer is a U-shaped apparatus having a bulb B and two capillary ends (Fig. 14.1). On one arm, there is a mark while the other arm is drawn to a point. In some cases, the two ends are fitted with caps.

Method. The pyknometer is first washed with chromic acid solution and then with distilled water and finally with alcohol and then dried. It is then suspended from the beam of balance with a hook of wire and weighed. Now attaching a rubber tube at one end of pyknometer and placing its other end in a beaker containing distilled water, the water is sucked gently upto mark A. *There should be no air bubble inside the pyknometer.* The pyknometer is then again weighed. After weighing, water is removed from the pyknometer, again washed with alcohol and then dried. The pyknometer is then filled with the liquid whose density is to be determined and weighed again.

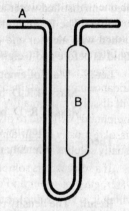

Fig. 14.1. Pyknometer.

Observations.

$$\text{Room temp.} = t\ °C$$
$$\text{Mass of empty pyknometer} = w_1 \text{ gm}$$
$$\text{mass of pyknometer} + \text{water} = w_2 \text{ gm}$$
$$\text{mass of pyknometer} + \text{liquid} = w_3 \text{ gm}$$

Calculations :

$$\text{mass of water} = (w_2 - w_1) \text{ gm}$$
$$\text{mass of liquid} = (w_3 - w_1) \text{ gm}$$
$$\therefore \quad \frac{\text{density of liquid}}{\text{density of water}} = \frac{\text{mass of liquid}}{\text{mass of water}}$$

or
$$\frac{\text{density of liquid } (d_1)}{\text{density of water } (d_2)} = \frac{w_3 - w_1}{w_2 - w_1}$$

$$\therefore \quad d_1 = \frac{w_3 - w_1}{w_2 - w_1} \times d_2$$

Relative density of liquid with respect to water :

$$\frac{d_1}{d_2} = \frac{w_3 - w_1}{w_2 - w_1}$$

Result. The density of a given liquid is .... gm ml^{-1}.

Note. The Pyknometer method is quite convenient and is accurate method for the measurement of density of a liquid. It not only measures the density at room temperature but can do so at other temperatures as well. For this purpose, the pyknometer with sample liquid is kept in thermostat set at required temperature. The measurement of density at various temperature may be required in finding the viscosity at different temperatures.

EXPERIMENT – 2

To measure the density by R.D. bottle.

The R.D. bottle is slightly round bottomed type of glass vessel as shown in Fig. 14.2. It is fitted with a glass cork containing a fine capillary.

The R.D. bottle is first washed with chromic acid solution and then with distilled water and finally with alcohol. It is then dried and weighed. The R.D. bottle is then filled with distilled water and stoppered. There should be no air bubble inside the R. D. bottle. The R. D. bottle is then again weighed. Water is then poured out and washed with alcohol and dried. The R.D. bottle is then filled with experimental liquid as before and weighed again.

25 C. C.
20°C

Fig. 14.2.

Let :   Mass of empty R.D. bottle = w_1 gm

Mass of R.D. bottle + water = w_2 gm

Mass of R. D. bottle + liquid = w_3 gm

Then, $\dfrac{\text{density of liquid } (d_1)}{\text{density of water } (d_2)} = \dfrac{w_3 - w_1}{w_2 - w_1}$

$\therefore$ $d_1 = \dfrac{w_3 - w_1}{w_2 - w_1} \times d_2$

Result. The density of a given liquid is .... gm ml^{-1}.

Surface Tension

Introduction. Surface tension is the characteristic property of every liquid and it is due to intermolecular attractions among molecules of the liquid. A molecule in the interior of the liquid is attracted by surrounding molecules in all directions and hence resultant force on that molecule is zero (Fig. 14.3). The molecule on the surface of the liquid experiences only the resultant downward attraction normal to the surface and, therefore, surface tends to acquire minimum surface area. Thus, the surface of liquid is in a state of tension (called strain or stretching). The force causing this strain is called 'Surface Tension'. It is defined *as the force in dynes acting at right angle to the surface of a liquid along 1 cm length of the surface.* Its unit is dynes cm^{-1}. Its S.I. unit is Newton metre^{-1}. It is denoted by Greek letter gamma (γ).

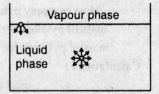

Vapour phase

Liquid phase

Fig. 14.3.

Theory. The measurement of surface tension of liquid is based on the fact that the drop of liquid at the lower end of capillary falls down when weight of drop becomes just equal to the surface tension. Thus -

1.   The force of gravity (wt. of drop) to pull the drop downward = $v\,d\,g$

where v and d are volume and density of the drop respectively and g is the acceleration due to gravity.

2. Force tending to uphold the drop = $2\pi r\gamma$ where $2\pi r$ is circumference of a cirular surface of radius r and γ is its surface tension.

At equilibrium, when the two forces are balanced :

$$2\pi r\gamma = v\,d\,g \qquad \qquad ...(1)$$

If n is the number of drops in volume V of the liquid, the volume of each drop v will be

$$v = \frac{V}{n}$$

Then from eqn. (1) we have $\quad 2\pi r\gamma = \dfrac{V}{n}\,d\,g \qquad \qquad ...(2)$

Surface tension, γ can be obtained from this equation.

Relative Surface Tension. If n_1 and n_2 are the number of drops counted in the same volume of two liquids of densities d_1 and d_2, using the same capillary tube,

$$2\pi r\gamma_1 = \frac{V}{n_1}d_1 g \qquad \qquad ...(3)$$

$$2\pi r\gamma_2 = \frac{V}{n_2}d_2 g \qquad \qquad ...(4)$$

Dividing eqn. (3) to (4), we have

$$\frac{\gamma_1}{\gamma_2} = \frac{n_2.d_1}{n_1.d_2} \qquad \qquad ...(5)$$

If γ_1 = surface tension of the liquid and γ_2 = surface tension of the water then

$\dfrac{\gamma_1}{\gamma_2}$ = Relative surface tension of liquid with respect to water.

Relative surface tension has no unit.

EXPERIMENT – 3

To determine the relative surface tension of a liquid with respect to water at room temperature by stalagmometer.

Apparatus. Stalagmometer, Pyknometer, beaker and rubber tube with a screw pinch cock.

Brief Description of Apparatus. The stalagmometer (Fig. 14.4) consists of a bulbed dropping tube with a capillary. The lower end of capillary is well ground, polished and flattened in order to provide larger dropping surface. There are two marks, C and D above and below the bulb B. The upper end of the stalagmometer is connected with a rubber tube and a screw pinch cock in order to control the rate of flow so that spherical drop may be formed.

Procedure. Clean the stalagmometer first with NaOH then with chromic acid and wash finally with distilled water. Immerse lower end in a beaker containing distilled water and suck up water until it rises above the upper mark C. Control the rate of flow of water with the help of screw pinch cock so that the number of drops falling per minute becomes about 15–20. This adjustment is essential otherwise the drop will not be properly formed (does not become spherical). Start counting the drops when the water meniscus just reaches the upper mark C and stop when the meniscus just passes the lower mark D. Take four readings and then take their mean value.

Now clean the stalagmometer and dry it. Fill it with liquid given until it rises above the upper mark C and count the number of drops as before.

The density of liquid can be determined by pyknometer as done in previous experiment.

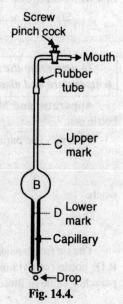

Fig. 14.4.

Observations :

1. Room temperature ....°C
2. Counting of the drops in the volume between C and D

S.No.	With Water		With Liquid	
	No. of drops	Mean (n_2)	No. of drops	Mean (n_1)
1.	....		....	
2.	....		....	
3.	....		....	
4.	....	....	....	....

3. Density measurement of the liquid (d_1) and of the water (d_2)

    Mass of the empty Pyknometer = ...
    Mass of the pyknometer + water = ...
∴ Mass of water filled in = ...
    Mass of the pyknometer + liquid = ...
∴ Mass of the liquid = ...

Calculation :

$$\frac{d_1}{d_2} = \frac{\text{Mass of liquid}}{\text{Mass of water}}$$

$$\text{Relative surface tension of the liquid} = \frac{n_2 \cdot d_1}{n_1 \cdot d_2}$$

By substituting the respective values, the relative surface tension of liquid is calculated.

Result. The relative surface tension of liquid with respect to water at room temperature is ....

Precautions :

1. Number of drops per minute must be in between 15–20.
2. The lower end of stalagmometer should be free from grease hence wash the stalagmometer first with benzene or NaOH and then with chromic acid.
3. Stalagmometer should be kept in a vertical position.

EXPERIMENT – 4

> *To determine the surface tensions of methyl alcohol, ethyl alcohol and n-hexane at room temperature and also calculate the atomic parachors of carbon, hydrogen and oxygen.*

Apparatus and Material. Stalagmometer, distilled water, all the three liquids, beaker and R.D. bottle etc.

Theory. The parachor, P of a liquid is given by

$$P = \frac{M.\gamma^{1/4}}{D}$$

where
γ = S.T. of the liquid,
D = Density of the liquid and
M = Molecular weight of the liquid.

The surface tension of each liquid is determined by means of stalagmometer and the density by R.D. bottle (or pyknometer). Therefore, parachor value of each liquid can be calculated. From parachor value of these liquids, the atomic parachors of C, H and O can be calculated as follows :

1. $$P_{C_2H_5OH} - P_{CH_3OH} = P_{CH_2}$$

$$P_{hexane} - 6 \times P_{CH_2} = 2 \times P_H$$

$$\therefore \qquad P_H = \frac{P_{hexane} - 6 \times P_{C_2H_5OH} + 6 \times P_{CH_3OH}}{2}$$

2. $$P_{CH_2} - 2 \times P_H = P_C$$

$$\therefore \qquad P_C = P_{C_2H_5OH} - P_{CH_3OH} - 2 \times P_H$$

3. $$P_O = P_{CH_3OH} - 4 \times P_H - P_C$$

Procedure. Wash and dry the stalagmometer as given in experiment 6 then fill it with distilled water and count the number of drops between marks C and D. Similarly, wash and fill the stalagmometer with every liquid in turn and count the number of drops as before.

Determine the density of each liquid by pyknometer (or R. D. bottle) as usual. Note the room temperature.

Observations :

1. Room temperature   = ...
2. Counting of drops between marks C and D.

Water		CH_3OH		C_2H_5OH		n-hexane	
No. of drops	Mean (n_w)	No. of drops	Mean (n_m)	No. of drops	Mean (n_e)	No. of drops	Mean (n_h)
...	⎫	...	⎫	...	⎫	...	⎫
...	⎬ ...	...	⎬ ...	...	⎬ ...	...	⎬ ...
...	⎭	...	⎭	...	⎭	...	⎭

3. Density measurement

Mass of empty Pyknometer = ...
Mass of Pyknometer + water = ...
Mass of Pyknometer + CH_3OH = ...
Mass of Pyknometer + C_2H_5OH = ...
Mass of Pyknometer + n-hexane = ...

Calculations :

$$\text{Relative Density of } CH_3OH = \frac{\text{mass of } CH_3OH}{\text{mass of water}} \quad \text{(when volume is same)}$$

$$\text{Relative Density of } C_2H_5OH = \frac{\text{mass of } C_2H_5OH}{\text{mass of water}}$$

$$\text{Relative Density of } n\text{-hexane} = \frac{\text{mass of } n\text{-hexane}}{\text{mass of water}}$$

The surface tension of each liquid is calculated as in previous experiment. Knowing the value of

S.T., density and molecular wt. of the liquid, parachor is calculated by the expression $P = \dfrac{M.\gamma^{1/4}}{D}$.

After calculating parachor values of all three liquids, the atomic parachor of C, H and O are calculated using the expression given in theory of this experiment.

Result. The atomic parachors of C, H and O are ... ..., ... and ...., respectively.

Percautions. The same as in previous experiment.

Viscosity

Introduction. Viscosity is the property of liquids and gases. The resistance offered by one part of the liquid to the flow on the other part is known as viscosity of liquid. It is due to internal friction of molecules and mainly depends on the nature and temperature of the liquid. In the laminar flow, each layer moves parallel to its adjacent layer without intermixing. The unparallel flow of layer is called turbulent flow. In the laminar flow, the resisting force 'F' is directly proportional to the velocity difference between two adjacent layers, v and the area of the surface, 'A' of contact of two layers and is inversely proportional to distance (x) between them. That is

$$F \propto \frac{A.v}{x} \quad \text{or} \quad F = \eta \frac{A.v}{x}$$

where η is constant and is called 'coefficient of viscosity' of the liquid. If in the above equation $A = 1$ sq. cm., $v = 1$ cm./sec., $x = 1$ cm., then $F = \eta$. Hence, η is defined as the force necessary to maintain a velocity difference of unity between two adjacent parallel layers of liquid, 1 cm. apart and having area of surface of contact unity.

Its unit is '*poise*' or '*centipoise*'.

Theory. The measurement of viscosity by viscometer is based on the Poisuille's equation, which is

$$\eta = \frac{\pi r^4 . P t}{8 V l} \qquad \qquad ...(1)$$

where V = volume of the liquid of viscosity η flowing in time t, through the capillary tube of radius r and length l.

P = pressure (hydrostatic pressure) of the liquid.

If t_1 and t_2 are times required to flow for equal volumes of two liquids through the same length of a capillary tube, then from eqn. (1) we have :

$$\frac{\eta_1}{\eta_2} = \frac{P_1 . t_1}{P_2 . t_2} \qquad \qquad ...(2)$$

The pressure of liquid   = $h.d.g.$

Since for the two liquids, h and g are the same, hence

$$\frac{P_1}{P_2} = \frac{d_1}{d_2} \qquad \qquad ...(3)$$

Therefore, $$\frac{\eta_1}{\eta_2} = \frac{d_1 . t_1}{d_2 . t_2} \qquad \qquad ...(4)$$

where d_1 and d_2 are the densities of two liquids and η_1 and η_2 their viscosities

If η_2 = Viscosity of the water

Then $\dfrac{\eta_1}{\eta_2}$ = Relative viscosity of liquid with respect to water.

Relative viscosity has no unit.

EXPERIMENT – 5

To determine the relative viscosity of liquid with respect to water at room temp. by Ostwald's Viscometer.

Apparatus and Material. Stop watch, Ostwald's viscometer, pyknometer and pipette and the liquid.

Description. Ostwald's viscometer is a U-shaped glass tube as shown in the figure 14.5. In one arm, the bulb A is connected with a fine capillary. The lower end of the capillary is connected with a

U-tube provided with a larger bulb B in the second arm. The bulbs are necessary to maintain the hydrostatic pressure during flow of the liquid. Through the capillry tube, the liquid flows with a measurable speed. There are two marks C and D above and below the bulb A. The upper end of the bulb A is attached with a rubber tube. The liquid flows under its own weight.

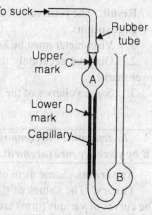

Fig. 14.5. Viscometer

The density of the liquid is determined either by pyknometer or by R.D. bottle, as in earlier experiment.

Procedure. Clean the viscometer with chromic acid and then thoroughly with distilled water. It is finally washed with alcohol or ether and then dried. A sufficient volume of distilled water is introduced by pipette in bulb B so that bend portion of tube and half or a little more than half of bulb B are filled up. Clamp the viscometer in a vertical position. Through the rubber tube attached to upper arm of a bulb A, suck the water until it rises above the upper mark C and allow it to flow under its own weight. The time of flow of water from C to D is measured by starting the stop watch as the meniscus of the liquid just reaches upper mark C and stopping the watch as the meniscus just passes the lower mark D. Take at least 3 to 4 readings with water and then take their mean value.

Now remove the water from viscometer and dry it. Introduce in bulb B almost same volume of liquid and measure the time of flow of liquid as before. Take at least 3 to 4 readings and then take their mean value.

Now, wash and dry the pyknometer (or R.D. bottle) and then weigh it empty and then fill it with distilled water and weigh. Remove the water and dry the pyknometer and then fill it with experimental liquid and weigh. Note the room temperature.

Observations :

1. Room Temperature ....
2. Time of the flow between C and D :

S.No.	Water		Liquid under experiment	
	Time of flow	Mean (t_2)	Time of flow	Mean (t_1)
1.	....		....	
2.	....		....	
3.	....		....	
4.	....	....	....	....

3. Density of liquid (d_1) and of water (d_2)

Mass of the empty R.D. bottle or pyknometer =	.... gm
Mass of the R. D. bottle or pyknometer + water =	.... gm
∴ Mass of the water =	.... gm
Mass of the R. D. bottle or pyknometer + liquid =	.... gm
∴ Mass of the liquid =	.... gm

Calculations :

$$\frac{\text{Density of the liquid } (d_1)}{\text{Density of the water } (d_2)} = \frac{\text{mass of the liquid}}{\text{mass of the water}}$$

$$\text{Relative viscosity of the liquid} = \frac{\eta_1}{\eta_2} = \frac{d_1 t_1}{d_2 t_2}$$

By putting the respective values, the relative viscosity of the liquid is calculated. If relative viscosity is multiplied by viscosity of water at that temperature, it provides absolute viscosity of the liquid.

Result. The relative viscosity of the liquid with respect to water, at room temp. is ....

Precautions.

1.   Viscometer must be kept in vertical position.

2.   Observe accurately the meniscus of liquid and water as they pass through the upper and lower marks.

3.   Same volumes of the water and of the liquid should be taken in viscometer.

EXPERIMENT – 6

> *To determine the composition of the given mixture consisting of two miscible liquids, A and B by viscosity measurement.*

Apparatus. Same as in experiment No. 8.

Theory. The values of η of solutions are plotted against their concentrations (compositions). The curves of various forms are obtained and usually viscosity curves of simple solutions are sagged, *i.e.,* fall below the straight line connecting the viscosities of their components.

Procedure. Perform the experiment as follows :

1.   Prepare a number of solutions, of different compositions by mixing the two liquids A and B in different proportions *e.g.,* 10%, 20% and so on upto 90% of A by volume.

2.   Determine the time of flow for each solution with the help of Ostwald's viscometer as described in the previous experiment.

3.   Find the time of flow for the given (unknown) mixture as usual.

Observations :

1.   Room temperature = ...°C

Time of flow is recorded in tabular form.

S.No.	%age of components		Time of flow in seconds
	A	B	
1.	10	90	—
2.	20	80	—
3.	30	70	—
4.	40	60	—
5.	50	50	—
6.	60	40	—
7.	70	30	—
8.	80	20	—
9.	90	10	—
10.	Unknown solution		—

Calculations. The values of time of flow are plotted against concentration of one component (say A). A straight line is obtained as shown in Fig 14.6. Corresponding to the measured time of flow for unknown solution, the point is located and marked on the straight line. A perpendicular line is drawn on the concentration axis from that point.

The point so obtained on the x-axis gives the percentage of A in the unknown solution. Percentage of B will be 100 minus percentage of A.

Result. The composition of the given mixture is

.... % A

.... % B.

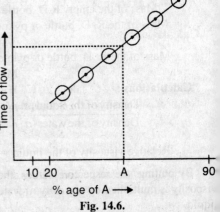

Fig. 14.6.

Solubility of Solid in Liquid

Definition. The solubility of a solid in a liquid (solvent) is defined as the amount of solid dissolved in 100 gms. of a solvent to make the saturated solution at a given temperature.

The determination of solubility of a solid in the liquid involves two steps :

1. Preparation of the saturated solution of solid in specified solvent at the required temperature.
2. Analysis of the saturated solution.

Methods of determination of solubility. There are two methods of determination of solubility.

1. **Gravimetric method.** This method is generally used to determine the solubility of salts. The saturated solution of the salt is prepared at a given temperature. A known volume of it, is taken in a weighed and clean porcelain dish and weighed again. Now the solution is evaporated to dryness and weighed. Thus knowing the amount of solid in the definite quantity of solvent, the amount of solid dissolved in 100 gms. of solvent *i.e.* solubility can be calculated.

2. **Titration Method.** This method is employed to determine the solubility of acids. Here, a known volume of the saturated solution of an organic acid at a given temperature is titrated against the standard solution of an alkali.

EXPERIMENT – 7

> *To determine the solubility of a salt (e.g., KCl) in water at room temperature.*

Apparatus. Beaker, stirring rod and porcelain dish etc.

Method. The solubility of the salt is determined by gravimetric method and involves two steps.

(*a*) **Preparation of saturated solution.** Take distilled water (say 50 ml) in a 250 ml beaker and add successively with a constant stirring, pure KCl until some KCl is left undissolved at the bottom. Now warm the solution very slowly to about 5 °C above the room temperature. More quantity of salt will dissolve and go on dissolving KCl until again some solid remains undissolved at the bottom. Allow it to cool at room temperature. Some solid will be separated out and the solution will become saturated solution at room temperature. Filter it in a clean beaker.

(*b*) **Evaporation.** Take a certain volume (say 25 ml) of the filtered solution in a clean and weighed porcelain dish and evaporate, first on wire gauze and then on water bath to dryness. The residue should become completely dry. Now weigh the dish.

Observations :

1. Room temperature   = ...°C
2. Mass of the empty dish   = w gm.
3. Mass of the dish + soln.   = w_1 gm.
4. ∴ Mass of the solution   = $(w_1 - w)$ gm.
5. Mass of dish + residue .   = w_2 gm.
6. ∴ Mass of residue   = $(w_2 - w)$ gm.
7. Mass of solvent   = $(w_1 - w) - (w_2 - w)$ gm.
                                = $(w_1 - w_2)$ gm.

Calculations :

$$solubility = \frac{100\,(w_2 - w)}{(w_1 - w_2)} \text{ gm per 100 gm of water}$$

Result. Solubility of salt (*e.g.*, KCl) at room temperature is .... gm./100 gm of water.

Note. *If the solubility is to be expressed in gm./litre of solution, calculate it as follows* :

$$Solubility = \frac{1000 \times mass\ of\ residue}{Volume\ of\ soln.\ evaporated} \text{ gm./litre of solution}$$

Precautions :

1 Residue should be completely dry.

2. Repeat at least three times the process of evaporation taking different volumes of same lution (filtrate) and at least two determinations should give same result.

EXPERIMENT – 8

> *To determine the solubility of an organic acid (e.g., oxalic acid) in water at room temperature.*

Material Required. Beaker, pipette, burette, phenolphthalein and standard soln. of NaOH *etc.*,

Method

Preparation of Saturated solution. It is prepared similarly as in previous experiment.

Titration. Take 25 ml. of filtrate in a conical flask and add 2 drops of phenolphthalein, titrate it with standard soln. of NaOH (say N/20) until last two readings coincide.

Observations :

1. Room temp.   = ...°C
2. Normality of NaOH   = ...

Volume of the Filtrate	NaOH readings		Volume of the NaOH used (ml)
	Initial	Final	
25 ml			
25 ml			
25 ml			

Calculations :

Solubility in gm./litre of solution.

$$= \frac{\text{Normality of NaOH} \times \text{Vol. of NaOH} \times \text{Eq. wt. of organic acid}}{\text{Volume of organic acid solution}}$$

Result. Solubility of oxalic acid at room temp. is...gm./litre of solution.

EXPERIMENT – 9

> *To determine the solubility of KNO₃ above room temperature by gravimetric method.*

Method. The details of the method are the same as in experiment No. 9 except that it requires the preparation of saturated solution at higher temperature. The saturated solution of solute at higher temperature is prepared as follows :

Take about 100 ml of distilled water in a beaker with a glass rod and clamp thermometer. Heat it to 10°C above the required temperature and prepare the saturated solution at this temperature. Allow the solution to cool to the required temperature and remove the thermometer without disturbing the solution and then quickly transfer 10 ml of the supernatant solution into a weighed dish. Immediately cover it by a watch glass and weigh again. Evaporate off the solution as in experiment No. 9 and weigh the dish again.

Record the observations as in Experiment No. 1. The calculations are also done as in previous exercise.

EXPERIMENT – 10

> *To determine the solubility of a salt at three different temperatures.*

Apparatus and Material. Three porcelain dishes, thermometer, water bath, 10 ml pipette and the pure substance (here KCl).

Procedure. Prepare approximately 100 ml of saturated solution of KCl at 75°C. (a thermomete: should be dipped and clamped). Allow the solution to cool and then pipette carefully from the top the 10 ml of this solution at 65°, 55° and 45°C and transfer these to already weighed three porcelai: dishes (10 ml to each dish). Now evaporate these solutions on water bath to dryness and calculate th: solubility at each temperature as in the experiment no. 9.

EXPERIMENT – 11

To determine the solubility product of calcium hydroxide using common ion effect of sodium hydroxide or of any other strong alkali.

Principle : $Ca(OH)_2$ is very slightly soluble in water hence remains in following equilibrium in saturated solution.

$$Ca(OH)_2 \rightleftharpoons Ca^{2+} + 2OH^-$$
$$\text{(Insoluble)}$$

The presence of NaOH suppresses the concentration of OH^- ions of $Ca(OH)_2$ and thereby solubility of $Ca(OH)_2$. Knowing this effect quantitatively, K_{Sp} value of $Ca(OH)_2$ can be determined.

Procedure. Prepare 100 ml of each 0.1, 0.05 and 0.025 N NaOH solutions in stoppered flasks from the stock solution of 1 N NaOH. Now to each of this, add about 1gm of solid $Ca(OH)_2$, stopper these flasks tightly and shake vigorously for about one hour and finally allow these to stand for another one hour. Filter these solutions to remove suspended $Ca(OH)_2$ if any and use the filtrate for the titration.

Fill in the burette with 0.05 N HCl and titrate 10 ml of each filtrate using phenolphthalein as an indicator. Note the readings and the temperature as well. Find out the normality of each solution and this determines the normality of $Ca(OH)_2$, OH^- concentration and finally the solubility product.

Calculations with 0.1 N NaOH Solution

We know $N_1V_1 = N_2V_2$

Therefore,   - $N \times 10 = 0.05 \times V_2$

where V_2 is volume of HCl used and N is normality of the filtrate. This normality is the concentration of total OH^- ions in the filtrate so N minus 0.1 will be the OH^- ion concentration of $Ca(OH)_2$. Half of this concentration will be of Ca^{2+} the ions. The K_{sp} of $Ca(OH)_2$ will be :

$$K_{Sp} = [Ca^{2+}][OH^-]^2$$

Repeat it with other two solutions and find K_{Sp} value again. It should be more or less constant.

Result : The K_{Sp} value of $Ca(OH)_2$ at room temperature is ....

EXPERIMENT – 12

To determine the transition temperature of sodium sulphate decahydrate by solubility method.

Apparatus and Material : Six porcelain dishes, thermometer, 500 ml pyrex beaker and pure sodium sulphate decahydrate.

Theory : The decahydrate changes to anhydrous sodium sulphate above 32.4°C. At 32.4°C, both forms are in thermal equilibrium and this temperature is called the transition temperature.

The solubility of $Na_2SO_4.10H_2O$ increases with increase in temperature upto 32.4°C. Above this temperature, it decreases and advantage of it is taken in finding its transition temperature.

If at different temperatures, the solubility of this salt is determined and a curve is plotted, two straight lines are obtained. The point of intersection of these two lines is the transition temperature.

Procedure : Prepare nearly 200 ml of saturated solutions of the compound at 75°C and allow it to stand. Now pipette out 10 ml at each temperature, 60, 50, 45, 35, 30, 25 and at 20°C. Transfer each of 10 ml to a separate already weighed porcelain dish.

Evaporate the solution to dryness in each case and find its solubilities at specified temperatures. Plot these solubilities against the temperature which will provide two lines as shown in the Fig. 14.7. The point of intersection will be the transition temperature as shown in below :

Result : The transition temperature of $Na_2SO_4.10H_2O$ is 32.4°C.

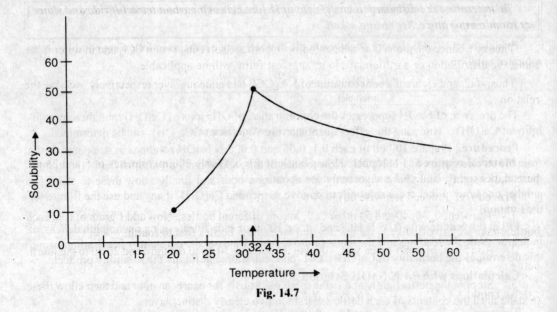

Fig. 14.7

Partition Law

General Information. When a solute is added to two im-miscible liquids in contact with each other, it will distribute itself between them according to the *Partition or Distribution Law* which is defined as follows :

"If a solute distributes itself between two immiscible solvents in contact with each other and its molecular condition is same in both the solvents, the ratio of concentrations of the solute in the two solvents is constant at a constant temperature."

If C_1 and C_2 are the concentrations of solute in two liquids (solvents) 1 and 2 respectively, then

$$\frac{C_1}{C_2} = \text{constant (K)}$$

This constant K is called *Distribution* or *Partition Coefficient* or *Ratio.* This varies, ofcourse with different substances but with a given substance and at a constant temperature, it is independent of total quantity of the solute used.

However, if the *molecular state of the solute changes,* we consider the concentration of the same molecular species only and not its total concentration. Thus when the solute associates, dissociates, chemically combines with solvent or chemically combines with another solute present in the solvent we get different expression for K.

Validity of Partition Law. The *Partition Law* holds good only when the following condition are fulfilled :

1. *Temperature and pressure should remain constant throughout the experiment.*
2. *Equilibrium state should have been attained.*
3. *Neither solution should have reached the saturation state ; greater the concentration larger are the deviations.*

EXPERIMENT – 13

To determine the distribution coefficient of iodine between carbon tetrachloride and water at room temperature.

Theory : Since the molecular state of iodine in both the solvents, *i.e.*, in CCl_4 and in water is the same, the distribution or partition law, in its simplest form, will be applicable.

Thus, if C_1 and C_2 are the concentration of I_2 in CCl_4 and aqueous layer respectively, we have the relation :

$$\frac{C_1}{C_2} = K \text{ (partition coefficient)}$$

Material required : 4 stoppered glass bottles each of nearly 200 ml, pipettes of 5 and 25 ml, burette, beaker, I_2, CCl_4, KI and sodium thiosulphate.

Procedure. Proceed stepwise as follows :

1. Transfer 20, 30, 40 and 50 ml of CCl_4 in four different bottles. Now add 1 gram of I_2 in each bottle, shake well to dissolve I_2 completely and label these bottles first to fourth.

2. Add 50 ml of water to the first, 40 ml to the second, 30 ml to the third and 20 ml to the fourth bottle.

3. Stopper the bottle tightly and shake these vigorously for nearly an hour and then allow these to stand until the contents of each bottle separate in two clearly distinct layers.

4. Prepare 200 ml of 0.1 N solution of sodium thiosulphate (equivalent weight 248.18). Pipette 50 ml from it and dilute it to ten times to make it 0.01N. Thus we have 150 ml of 0.1N and 500 ml of 0.01N thiosulphate solutions.

5. Pipette 5 ml of CCl_4 layer (bottom layer) from each bottle separately and titrate these against 0.1 N sodium thiosulphate taken in burette using starch as an indicator and note the readings.

6. Since concentration of I_2 in aqueous layer is fairly low, pipette 25 ml of aqueous layer from each bottle and titrate it against 0.01 N thiosulphate solution and note the readings.

7. Find the normality of I_2 in each layer of the each bottle using the formula $N_1V_1 = N_2V_2$ and when this normality is multiplied by eq. wt. of I_2, gives the concentration. We can also substitute normality values.

Observations : Room temperature ..... °C.

Bottle number	Volume of 0.1N hypo used with 5 ml of CCl$_4$ layer	Volume of 0.01 N hypo used with 25 ml of aqueous layer	C_1 moles lit^{-1}	C_2 moles lit^{-1}	$\frac{C_1}{C_2} = K$
1	.... ml	... ml	...	....	...
2	... ml	... ml	...	...	...
3	.... ml	... ml	...	....	...
4	... ml	... ml	...	...	...

Result : The partition coefficient of I_2 in CCl_4 and water is .... at ..... °C

Note. If we use the same volume of CCl_4 of the aqueous layer and titrate it with the same hypo solution, we can even substitute volumes of hypo used in place of C_1 and C_2.

Precautions :

1. Temperature and pressure should remain constant throughout the experiment.
2. Equilibrium state should have been attained.
3. Neither solution should have reached the saturation stage : greater the concentrations, larger are the deviations from accuracy.

4. The two liquids should be mutually immiscible. These should remain unaffected by changing the quantity of the solute.

5. The molecular state of solute should not change ; if it changes, consider the concentration of
molecular species common to both solvents and not the total concentration.

EXPERIMENT – 14

> *To determine the partition coefficient of benzoic acid between water and benzene at room*
> *temperature and the molecular state of benzoic acid in benzene.*

Apparatus. 4 bottles (250 ml) with stoppers, conical flasks, pipette (10 ml), burette, beakers,
measuring cylinder and a separating funnel.

Reagents : Benzoic acid, benzene, 0.1 N and 0.01 N NaOH solution and phenolphthalein.

Theory. If the solute associates in one of the solvents, the distribution law expressed is

$$\frac{C_1}{\sqrt[n]{C_2}} = K$$

where n is the number of solute molecules which associate in solvent two.

Benzoic acid exists as normal molecules in water and as a dimer in benzene. Therefore, $\dfrac{C_1}{\sqrt{C_2}}$

should be constant. This can be verified experimentally.

Method : Take 50 ml of distilled water and 50 ml of benzene in each of the four bottles and label
them one to four. Now weigh 1 g, 2 g, 3 g and 4 g of benzoic acid separately and transfer 1 gm to the
first bottle, 2 g to the second bottle and so on. Close these bottles tightly with their stoppers. Shake
them thoroughly for about half an hour and then allow these to stand as such until the two layers of the
liquids are clearly separated. The top layer will be of benzene and the bottom of the water, both
containing dissolved benzoic acid.

Now, transfer slowly the contents of the first bottle to a separating funnel. Extract the bottom
layer in a beaker and take 10 ml of it in a conical flask. Titrate it against 0.01N solution of NaOH
taken in burette using 3-4 ml of phenolphthalein as an indicator and note the end point.

Now from benzene layer, take 10 ml and repeat the titration process with 0.1 N NaOH.

In this way, transfer the contents of second bottle to separating funnel and repeat the process.
Likewise, repeat it with third and fourth bottles too. Note the readings.

Observations :

Titration of aqueous layer against 0.01 N NaOH solution.

Bottle number	Volume of water layer taken	Volume of NaOH used
1	10 ml	V_1 ml
2	10 ml	V_2 ml
3	10 ml	V_3 ml
4	10 ml	V_4 ml

Titration of benzene layer against 0.1 N NaOH solution.

Bottle number	Volume of benzene layer taken	Volume of NaOH used
1	10 ml	V_1' ml
2	10 ml	V_2' ml
3	10 ml	V_3' ml
4	10 ml	V_4' ml

Calculations :

(a) Concentration of benzoic acid in aqueous layer of the first bottle :

Milli equivalent (Meq) of NaOH used $= 0.01\ V_1$. This is equal to Meq of benzoic acid in 10 ml.

Hence concentration of benzoic acid

$$= \frac{0.01\ V_1 \times 122}{10}\ \text{g lit}^{-1}$$

(Equivalent weight of benzoic acid = 122). This is C_1

(b) Concentration of benzoic acid in benzene layer

$$= \frac{0.1\ V_1' \times 122}{10}\ \text{g litre}^{-1}\ \text{or say}\ C_2.$$

In same way, find C_1 and C_2 of aqueous and benzene layer of remaining bottles and tabulate C_1, C_2 and ratios C_1/C_2 and $C_1/\sqrt{C_2}$ as follows :

Bottle No.	C_1	C_2	$\dfrac{C_1}{C_2}$	$\dfrac{C_1}{\sqrt{C_2}}$
1				
2				
3				
4				

It is seen that values of C_1/C_2 are not constant but those of $\dfrac{C_1}{\sqrt{C_2}}$ are found to be constant. Hence, the complexity of molecules of benzoic acid in benzene is 2.

Plot a graph between C_1 and $\sqrt{C_2}$. A straight line is obtained (Fig. 14.8) which verifies that complexity of molecules of benzoic acid in benzene is 2.

Result. 1. The value of partition coefficient between water and benzene is..... at ... °C.

2.   The molecular state of benzoic acid in benzene is dimer.

Precautions :

As in previous exercise.

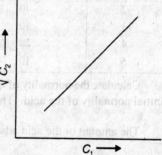

Fig. 14.8.

Adsorption

EXPERIMENT – 15

To determine the adsorption of aqueous acetic acid by activated charcoal and to study the adsorption isotherm.

Theory : Freundlich adsorption isotherm is given by the expression :

$$\frac{x}{m} = KC^{1/n} \qquad\qquad\qquad ...(1)$$

here x and m are the amounts of solute (adsorbate) and adsorbent respectively, C is the equilibrium concentration of adsorbate. K is a constant depends on both, the nature of adsorbent as well as of adsorbate, n is another constant and depends only on the nature of adsorbate.

Taking log, the equation (1) can be written as

$$\log \frac{x}{m} = \log K + \frac{1}{n} \log C$$

when $\log \frac{x}{m}$ is plotted against $\log C$, it gives a straight line whose slope is $\frac{1}{n}$ and the intercept $\log K$.

Material Required. Stoppered bottles, pipette, burette, beakers, powdered activated charcoal, glacial acetic acid and 0.1 N NaOH.

Procedure : Proceed stepwise as follow :

1. Take 50, 40, 30, 20 and 10 ml of 0.5 N acetic acid in stoppered bottles and to these add 0, 10, 20, 30 and 40 ml of distilled water respectively and label them.

2. Transfer 1 gm of activated charcoal to each bottle then shake these vigorously for about one hour and allow them to stand for half an hour to attain the room temperature.

3. Filter off charcoal and while filtering discard initial nearly four ml of filtrate in each case.

4. Take 5 ml of each filtrate and titrate it against standard solution of 0.1 N NaOH using phenolphthalein as an indicator. Note the reading in each case.

Observations :

1. Room temperature = ... °C
2. Amount of charcoal used in each bottle = 1 g
3. Volume of solution taken for titration each time = 5 ml

Bottle No.	Volume of 0.5 N CH_3COOH (ml)	Volume of water added	Initial normality of the acid (normality) C_2	Volume of NaOH used, V
1	50	0	0.50	—
2	40	10	0.40	—
3	30	20	0.30	—
4	20	30	0.20	—
5	10	40	0.10	—

Calculate the normality of the acid after adsorption using $N_1V_1 = N_2V_2$. Subtract this value from initial normality of the acid. This will give you the normality of the acid adsorbed. Let it be N.

The amount of the acid adsorbed, $x = \dfrac{N \times 60}{1000}$ gm

$$m = 1 \text{ g in all cases.}$$

Repeat the calculation of remaining solutions and produce your results in tabular form as below

Bottle number	Initial normality acid	Normality of the acid after adsorption	Normality of acid adsorbed	Amount of acid adsorbed x	x/m	$\log \dfrac{x}{m}$	$\log C$
1	.50	....	....				
2	.40	....	....				
3	.30	....	....				
4	.20	....	....				
5	.10	....	....				

Plot the graph between x/m and log C. The slope is equivalent to $1/n$ and the intercept to log K.

Result. The trend of the graph is in accordance with Freundlich adsorption isotherm. The values of $\frac{1}{n}$ and 'K' are respectively ... and ..... .

Precautions :

1. Shake all the solutions properly and uniformly.

2. Filter the solution before proceeding for titration and discard the initial small volume of the filtrate.

3. Do not use wet filter paper in filtration as it may dilute the solution.

Chemical Kinetics

Chemical kinetics deals with the rate of the reaction as well as mechanism. Rate states, how fast or slow a particular reaction proceeds and is usually expressed in moles lit^{-1} unit time^{-1} which means change in concentration per unit time, usually moles lit^{-1} sec^{-1}. Mechanism discloses the way reaction proceeds.

In kinetics, two terms, order of reaction and molecularity are frequently used.

Order of reaction is defined as "*the number of reacting molecules whose concentration alters as a result of chemical change*". It is equivalent to sum of exponents of all concentration terms in the rate law expression. For instance, if the rate law expression is :

$$R = K[A]^1 [B]^2 \text{ or } K[A][B]^2$$

Its order of reaction will be $1 + 2 = 3$ and, therefore, is a third order reaction. K is rate constant and has a unit which depends upon the order of reaction.

Molecularity is the characteristic of the mechanism part and is defined as "*the actual number of molecules of the reactants taking part in the chemical reaction*". If a reaction takes place is one step, both, the order of reaction and the molecularity will be the same else the two terms differ. The reactions are called bimolecular and termolecular etc. as per their molecularity. A pseudounimolecular reaction is one in which the reaction follows first order but its molecularity is two. Here, one of the reactants is in excess and its concentration remains practically unchanged. Hydrolysis of an ester in acidic medium is a first order reaction. Here, water is in excess. It is a common example of pseudounimolecular reaction.

EXPERIMENT – 16

> *To determine the rate of constant and order of the reaction of the hydrolysis of an ester (methyl acetate) catalyzed by an acid (dilute HCl)*

Theory : The hydrolysis of an ester in acidic medium follows the general equation :

$$RCOOR' + H_2O \underset{}{\overset{H_3O^+}{\rightleftharpoons}} RCOOH + R'OH$$

In case of methyl acetate, we write

$$CH_3COOCH_3 + H_2O \underset{}{\overset{H_3O^+}{\rightleftharpoons}} CH_3COOH + CH_3OH$$

since water is to be taken in large excess, its concentration remains practically constant during the hydrolysis. The rate equation for the above reaction has been found to be :

$$R = K[CH_3COOCH_3]$$

It is a first order reaction but its molecularity is 2, and it is, therefore, is called pseudounimolecular reaction.

Since hydrolysis produces acetic acid, the rate of the reaction can be studied by titrating the acid formed against a standard solution of an alkali, say NaOH. The reaction can be stopped at any instance by lowering temperature of the reaction zone, using the ice. Its rate constant, K is calculated using integrated rate equation of the first order

$$K = \frac{2.303}{t} \log_{10} \frac{V_\infty - V_0}{V_\infty - V_t}$$

where

V_0 = titration reading in the beginning

V_t = titration reading after time 't'

V_∞ = titration reading after infinite time (~ 48 hours).

Material Required : Thermostat or water bath, stoppered conical flasks, burette, 5 ml pipette, stop watch, nearly 250 ml of 0.5 N HCl, 500 ml of 0.1N NaOH and phenolphthalein indicator.

Procedure : (i) Transfer 100 ml of 0.5 N HCl into one ionical flask and 15 ml of methyl acetate into another and place these in a water bath.

(ii) Now fill the burette with 0.1 N NaOH and keep phenolphthalein indicator nearby for ready use.

(iii) Take 3–4 conical flasks containing about 200 ml of icy cold water with some ice pieces in them.

(iv) When the acid and the ester have attained a steady temperature of the bath, pipette 5 ml of the ester and transfer nearly half of it to the acid kept in water bath and start the stop watch. Now transfer the remaining as well and shake well the contents and then immediately, pipette 5 ml of the reaction mixture and transfer into a flask containing icy cold water. This will stop the reaction due to low temperature. Add few drops of phenolphthalein in it and titrate it quickly against NaOH solution. Note the end point (shown by appearance of pink colour). This reading is called V_0 as the time is zero. Titration should be completed within five minutes.

Now when the stop watch shows 10 minutes, take out again 5 ml the from the reaction mixture to another icy cooled conical flask and titrate again quickly. Note the burette reading. It is V_{t_1} as $t = 10$ minutes. Now this is to be repeated after 20, 30 and 40 minutes. These readings will give the values of V_t at different intervals. Now heat the water bath to ~ 60°C for about an hour. This will heat the reaction mixture to complete the hydrolysis. Take 5 ml of it and titrate this also against the NaOH solution. This reading is V_∞. Heating has reduced the time for completion of the reaction.

Observation and Calculations :

Temperature during the experiment = ... °C.

Time in mts.	Volume of NaOH used in titration (ml)	$(V_\infty - V_t)$ ml	$K = \frac{2.303}{t} \log \frac{V_\infty - V_0}{V_\infty - V_t}$
0	V_0	$V_\infty - V_0$	...
10	V_{t_1}	$V_\infty - V_{t_1}$	...
20	V_{t_2}	$V_\infty - V_{t_2}$	...
30	V_{t_3}	$V_\infty - V_{t_3}$	...
40	V_{t_4}	$V_\infty - V_{t_4}$	...
∞	V_∞		....

Result. Since the value of rate constant corresponds to first order integrated rate equation, it i first order reaction.

EXPERIMENT – 17

To determine the rate constant of hydrolysis (saponification) of ethyl acetate with NaOH and to show that the reaction is of second order.

Theory : The alkaline hydrolysis of ethyl acetate takes place according to following equation.

$$CH_3\,COOC_2H_5 + NaOH \longrightarrow CH_3COONa + C_2H_5OH$$

The rate equation for the above reaction is found to be :

$$\frac{dx}{dt} \quad or \quad Rate = K\,[CH_3COOC_2H_5]\,[NaOH]$$

The order of reaction with respect to ester is one and with respect to alkali is also one. The overall order of reaction is two. The integrated rate equation for the second order reaction in above case is expressed as :

$$K = \frac{2.303}{t\,(a-b)} \log \frac{b\,(a-x)}{a\,(b-x)}$$

where

a = initial concentration of NaOH
b = initial concentration of ethyl acetate
x = amount of NaOH or ethyl acetate reacted in time 't'.

Material Required : Stoppered conical flasks of 250 ml, thermostat (or a large water-bath), pipette of 5, 10 and 25 ml, stopwatch, thermometer (0.1°C), N/50 HCl, N/50 NaOH, ethyl acetate, burette and phenolphthalein.

Procedure : Proceed stepwise as follows :

1.   Weigh accurately 0.44 g of ethyl acetate and transfer it in 250 ml of measuring flask and add water to make up it to 250 ml. Now transfer 100 ml of this solution to 250 ml of stoppered conical flask and suspend it in a thermostat, set at room temperature say at 30°C. This is 0.02 molar solution.

2.   Take 100 ml of N/50 NaOH solution in 250 ml conical flask and suspend this too in thermostat. Allow both these solutions to stand for about 25 minutes to acquire the steady temperature of the thermostat.

3.   Pipette 20 ml of N/50 HCl each time and transfer it to seven conical flasks and add few ice pieces in them.

4.   Pipette 50 ml of N/50 NaOH from the thermostat flask and transfer nearly half of it to ester solution. Run the stopwatch now and transfer the remaining NaOH also. Shake the contents of the reaction mixture thoroughly with a glass rod.

5.   After exactly 5 minutes, pipette 20 ml of reaction mixture and transfer it to one of the seven flasks of icy cooled HCl. Shake it well and titrate HCl mixture against sodium hydroxide solution using phenolphthalein indicator. Repeat the same procedure after 10, 15, 20, 25 and 30 minutes. Note the volume of sodium hydroxide used every time in titration. These volumes will be V_{t_1} to V_{t_6}.

6.   Now raise the temperature of the thermostat to about 50°C and reaction mixture is retained for 2 hours. This is done to complete the reaction. Now pipette out 20 ml of reaction mixture, transfer it to the seventh flask containing 20 ml HCl and titrate this also against sodium hydroxide. This reading will be V_∞.

7.   Take 10 ml of N/50 NaOH and transfer it to a flask containing 20 ml of N/50 HCl. Titrate this also against sodium hydroxide solution. The titre value will be V_0.

8.   Take 20 ml of N/50 HCl alone and titrate this also against sodium hydroxide. This reading will be V.

Observations : Temperature of the experiment = …. °C.

$$V = \ldots\ ml$$

$$V_0 = \dots \text{ ml}$$
$$V_\infty = \dots \text{ ml}$$

Time mts.	Titre values using N/20 baryta solution in ml	$V - V_t$ $(a-x)$ ml	$V_\infty - V_t$ $(b-x)$ ml	$V - V_0$ 'a' ml	$V_\infty - V_0$ 'b' ml	K $\dfrac{2.303}{t(a-b)} \log \dfrac{b(a-x)}{a(b-x)}$
0	V_0	...	...	...	...	...
5	V_{t_1}	...	...	...	...	...
10	V_{t_2}	...	...	...	...	...
15	V_{t_3}	...	...	...	...	...
20	V_{t_4}	...	...	...	...	...
25	V_{t_5}	...	...	...	...	...
30	V_{t_6}	...	...	...	...	...
Infinite time.	V_∞	...	...	...	...	...

Calculations :

1.   Initial concentration of sodium hydroxide that is $a - (V - V_0)$ ml of sodium hydroxide solution.

2.   Since NaOH in reaction mixture is kept in excess, small amount of NaOH is left unreacted after the reaction is complete. This excess of NaOH will be equal to $a - b$ or $V - V_\infty$ of sodium hydroxide solution.

3.   Concentration of ethyl acetate in the reaction mixture, b, will be $b = a - (a - b) = (V - V_0)$ $-(V - V_\infty)$ or $(V_\infty - V_0)$ of sodium hydroxide solution.

4.   (i) Total HCl $= V$

(ii) Excess of HCl present after neutralising NaOH of the reaction mixture $= V_t$

So $V - V_t = a - x$ or NaOH still present.

Similarly $V_\infty - V_t = b - x$ or ester still present.

substitute these values in integrated rate equation :

$$K = \frac{2.303}{t(a-b)} \log \frac{b(a-x)}{a(b-x)}$$

If need be, plot $\log\left(\dfrac{a-x}{b-x}\right)$ versus t (abscissa) whose pattern is linear with a slope $\left(\dfrac{a-b}{2.303}\right)K$. This can also give the value of K.

Result :   K values corresponding to rate equation of second order, are found to be constant within experimental limits. It infers that the reaction is of second order.

EXPERIMENT – 18

To study the kinetics of dissolution of magnesium metal in dil. HCl.

Theory :   Mg reacts with dil. HCl according to the equation :

$$Mg + 2\,HCl \longrightarrow MgCl_2 + H_2 \uparrow$$

The reaction is a first order with respect to metal and is second order with respect to the acid. Le 'a' be the amount of the metal taken initially when time, $t = 0$ and out of it, x dissolves in time 't'. Le 'b' be the concentration of the dilute HCl taken.

The rate equation will be :

$$R \quad \text{or} \quad \frac{dx}{dt} = K(a-x)\, b^2.$$

This on integrating and simplifying gives $\log \dfrac{a}{a-x} = K b^2 t$

both 'a' and $(a-x)$ will be kept constant for each working. It, therefore, follows :

$$K b^2 t = \text{constant.}$$
$$b^2 t = \text{another constant}$$

so $$b^2 \propto \dfrac{1}{t}$$

and this relation is to be worked out.

Material Required : 15 inches of Mg ribbon of uniform thickness, conc. HCl, 6 beakers of 250 ml, measuring cylinder of 100 ml and a stopwatch.

Procedure : Proceed stepwise as follows :

1. Take 50 ml of conc. HCl and dilute it by distilled water to 100 ml. From this 100 ml, transfer 30, 25, 20, 15 and 10 ml to five beakers and label them I, II, III, IV and V respectively. Now add 30, 35, 40, 45 and 50 ml of distilled water from I to V beaker respectively to make the volume of each 60 ml.

2. Cut the magnesium ribbon into 5 equal lengths (each of 3" and see that each must weigh same). Drop first piece to I beaker and run the stopwatch. Stop the stopwatch the moment entire piece dissolves and note the time taken. Repeat the experiment by dropping magnesium ribbon from II to V beaker one by one and repeat the experiment.

Observations and Calculations : It can be recorded in a tabular form :

Beaker containing the acid in ml (b)	Volume of HCl squared in ml (b^2)	Time required for dissolution of Mg, t	Reciprocal of of time, 1/t
30	900	...	...
25	625	...	...
20	400	...	...
15	225	...	...
10	100	...	...

Plot the graph using b^2 on y axis and reciprocal of time, $1/t$ on x axis. A graph of the linear pattern is obtained.

Result : Since the graph is a straight line, dissolution of magnesium with respect to HCl acid is second order reaction.

EXPERIMENT – 19

To determine the rate constant of a reaction between acetone and iodine in presence of mineral acid and a catalyst and to show that this reaction with respect to iodine is of zero order.

Theory : Acetone reacts with iodine in presence of mineral acid in following manner :

$$CH_3COCH_3 + I_2 \longrightarrow CH_3COCH_2I + HI$$

Investigations reveal that if concentration of the acid is very small (say 0.005 to 0.05 mole lit^{-1}), the reaction becomes first order with respect to acid but in moderate or higher concentration, the rate remains unaffected by the acid concentration. Further, the rate of reaction is found to be independent of concentration of iodine and hence is of zero order with respect to iodine. With respect to acetone it is of first order. It is, therefore, the rate equation is expressed as :

$$R \quad \text{or} \quad \dfrac{dx}{dt} = K \,[\text{acetone}].$$

but with respect to iodine $$\dfrac{dx}{dt} = K\,[I_2]^0 \qquad \qquad \text{...(i)}$$

Integrating the equation (1), we get

$$x = Kt$$

or

$$K = \frac{x}{t}$$

The constancy of K will establish the zero order reaction with respect to iodine.

Material Required : Hydrochloric acid, iodine, KI, acetone, sodium thiosulphate, starch indicator, 5 beakers of 100 ml, conical flasks, 5 ml pipette, burette and a thermometer.

Procedure : This is to be followed stepwise as below.

1. Take concentrated hydrochloric acid (12 N) and from it prepare 50 ml of N/30 HCl.

2. Dissolve 0.318 g of iodine and 0.415 g of KI in 50 ml of water. This will give you N/20 solution of iodine in 0.1 M KI solution in water. Iodine dissolves in water in presence of KI considerably.

3. Dissolve 0.248 g of sodium thiosulphate (hypo) in 100 ml of water. This gives N/100 solution. Fill this solution in burette for ready use. Keep the starch also nearby.

4. Mix 20 ml of acetone with 30 ml of water. This gives 40% acetone by volume (V/V).

5. Keep 5 beakers of 100 ml capacity each containing nearly 25 ml of icy cooled water with some ice pieces.

6. Mix all the 50 ml of iodine solution with 50 ml of N/30 HCl acid in a conical flask of 250 ml. Start the stopwatch immediately and instantly withdraw 5 ml of reaction mixture and transfer it to beaker containing icy cooled water to arrest the reaction. Titrate it against sodium thiosulphate using starch as an indicator. Note the reading. Now remove 5 ml of the reaction mixture each time after 4, 8, 12, 16 mts and repeat the process.

Observations :

(*i*)   Room temperature during the experiment = ... °C.

(*ii*) Concentration of ....

    (*a*) I_2 solution = N/20

    (*b*) HCl acid = N/30

    (*c*) Acetone solution = 40%.

Time interval (mts)	Volume of N/100 thiosulphate used from burette (ml)	$K = \dfrac{\Delta x}{\Delta t}$
0	V_1	
4	V_2	$K = \dfrac{V_2 - V_1}{4}$
8	V_3	$K = \dfrac{V_3 - V_2}{4}$
12	V_4	$K = \dfrac{V_4 - V_3}{4}$
16	V_5	$K = \dfrac{V_5 - V_4}{4}$

Once the burette is filled with thiosulphate, we run in continuation for V_1 to V_5. So V_5 is inclusive of V_1 to V_4.

The amount of iodine consumed during the reaction is proportional to volume of sodium thiosulphate consumed in titration. Plot a graph between volume of hypo used verses time. This will give a straight line. The slope of line gives the value of velocity constant.

Result : A constant value of K or $\dfrac{\Delta x}{\Delta t}$ indicates the reaction to be of zero order with respect to iodine.

EXPERIMENT – 20

To study the kinetics of decomposition of sodium thiosulphate by a mineral acid.

Theory : The decomposition of sodium thiosulphate in hydrochloric acid takes place according to following equation :

$$Na_2S_2O_3 + 2HCl \longrightarrow 2NaCl + H_2O + SO_2 \uparrow + S \downarrow$$

The precipitation of sulphur is marked by turbidity in solution and this indicates the progress of the reaction. The acid is taken in large excess to keep its concentration more or less constant and solutions of various concentrations of sodium thiosulphate are used to study the kinetics.

Material Required : Sodium thiosulphate, N/2 hydrochloric acid, measuring cylinder, 5–6 beakers of 250 ml, stopwatch and a thermometer.

Procedure : It can be taken stepwise as given below :

1.   Weigh accurately 0.1, 0.3, 0.5 0.7 and 0.9 gm of sodium thiosulphate and transfer it to 5 beakers. Make their solutions in small amount of water and make each upto 100 ml and label them I to V.

2.   Add 20 ml of N/2 HCl in first beaker and run the stopwatch. Make a big cross mark (✗) on a white paper and put the beaker on it. View the cross from above, through the solution. The moment cross mark becomes invisible, stop the stopwatch and note the time taken. Repeat the same experiment with remaining four beakers.

3.   Plot a graph, taking amount of thiosulphate per 100 ml of solution on y-axis and reciprocal of the time required to bring turbidity on x-axis. It gives a linear pattern.

Observations :

Table : Experimental temperature ... °C.

Amount of sodium thiosulphate in gm per 100 ml	Volume of N/2 HCl added (in ml)	Time required for turbidity 't'	Reciprocal of time 1/t
0.1	20	...	....
0.3	20	...	....
0.5	20	...	....
0.7	20	...	....
0.9	20	...	....

The data tell that the reaction is of first order with respect to sodium thiosulphate as reciprocal of time required for turbidity is proportional to initial concentration of the thiosulphate.

Result : Since the graph is linear (straight line), the decomposition of thiosulphate by mineral acid is a first order reaction.

Thermochemistry

Neutralization Equivalent of Acids

Definition. Neutralization equivalent of an acid is defined as the number of parts by its weight which neutralizes 1 gm. equivalent of a base.

EXPERIMENT – 21

To determine the neutralization equivalent of a given acid.

Apparatus and Material. Burette, beaker, supplied acid (standard), NaOH and solution phenolphthalein, *etc.*

Theory. We know that :

1 gm-equivalent of acid neutralizes 1 gm-equivalent of a base.

Let V ml of an alkali of normality $1N$, neutralizes w g of an acid. Hence

$$V \text{ ml of } 1N\text{-NaOH} \equiv w \text{ g of acid}$$

or

$$1000 \text{ ml of } 1N\text{-NaOH} \equiv \frac{w \times 1000}{V} \text{ g of acid}$$

or

$$1000 \text{ ml of } N \text{ normal NaOH} \equiv \frac{w \times 1000}{V \times N} \text{ g of acid.}$$

$$\text{Neutralization equivalent (N.E.) of acid} = \frac{1000 \times w}{V \times N}$$

Method. Weigh accurately 0.2–0.3 g of supplied acid and dissolve it in distilled water (say in about 25 ml or any other volume) in a beaker and add 2 drops of phenolphthalein. Fill the standard NaOH solution in a burette. Titrate the acid solution by adding NaOH from burette until the solution in beaker becomes slight pink. Record the final reading and find the volume of NaOH used. Repeat the titration with atleast two different weights of acid.

Standardization of NaOH solution. Prepare a standard solution of oxalic acid in a measuring flask. Take 25 ml of it in a conical flask and titrate it with supplied NaOH solution (not standard) using phenolphthalein as an indicator. Find the volume of NaOH solution used and hence calculate the normality of NaOH solution

Observation :

1. *Titration with unknown acid :*

S. No.	Mass of unknown acid	Reading of NaOH		Volume of NaOH used (ml)
		Initial	Final	
1	w_1 gm	...	...	V_1
2	w_2 gm	...	...	V_2

2. *Standardization of NaOH solution*

Volume of oxalic acid	Burette Reading		Volume of NaOH used (ml)
	Initial	Final	
say 25 ml			
''			
''			

Calculations : Normality of NaOH :

$$= \frac{\text{Volume of oxalic acid} \times \text{Normality of oxalic acid}}{\text{volume of NaOH used}}$$

$$\text{N.E.} = \frac{1000 \times \text{mass of supplied acid}}{\text{Volume of NaOH} \times \text{normality of NaOH}}$$

Calculate N.E. using different weights of acid. The mean value will be neutralization equivalent of the given acid.

Result. The neutralization equivalent of supplied acid is...

Note. When the given acid does not completely dissolve in water, add to it some alcohol and shake well

EXPERIMENT – 22

To determine the heat of neutralization of sodium hydroxide and hydrochloric acid.

Theory : Heat of neutralization is the amount of heat liberated when 1 gm-equivalent of an acid is neutralized by 1 g-equivalent of a base. When both, acid and base are strong, the heat released is maximum and is 13.7 K cal. In case of HCl and NaOH, it can be shown by an equation.

$$HCl + NaOH \longrightarrow NaCl + H_2O + 13.7 \ K \ cal$$

As both of these are strong, they dissociate cent percent and the reaction is precisely

$$H^+ + OH^- \longrightarrow H_2O + 13.7 \ K \ cal$$

From above, we can offer another definition of heat of neutralization, *i.e.*, the amount of heat released when a mole of water is produced from complete reaction of an acid and a base. It may be noted that if any of the species, acid or the base is a weak electrolyte, heat evolved or heat of neutralization will be less and will be the least when both are weak.

Procedure : The determination of heat of neutralization involves following two steps.

(a) Determination of water equivalent of the calorimeter

The simple calorimeter used in this experiment consists of two glass beakers, one smaller than the other. The small beaker is placed inside the big one and space between them is packed with some insulating material such as cotton to prevent the heat loss by conduction.

To determine the water equivalent, transfer 100 ml of water in the inner beaker, clamp a thermometer in it and allow it to attain a steady temperature and note it accurately. Take a separate beaker and heat in it 100 ml of water to about 60°C and note the temperature of hot water accurately. Quickly transfer the entire hot water to inner beaker of calorimeter and stir it slowly and note the highest temperature attained by the mixed water.

Observations :

Mass of the cold water = 100 g

Mass of the hot water = 100 g

Temperature of the cold water = t_1°C

Temperature of the hot water = t_2°C

Temperature attained after mixing = t_3°C

Calculations :

We know that

heat given by hot water = heat taken by calorimeter + heat taken by cold water.

Therefore, $100 \times (t_2 - t_3) = W \times (t_3 - t_1) + 100 \times (t_3 - t_1)$

where W is water equivalent of calorimeter. Therefore

$$W = \frac{100 \ (t_2 - t_3) - 100 \ (t_3 - t_1)}{(t_3 - t_1)}$$

(b) Determination of heat of neutralization.
Remove the water from calorimeter completely and transfer 100 ml of 0.5 N HCl and note its steady temperature. Add to it, 100 ml of 0.5 N NaOH and stir the mixture thoroughly. Note the final temperature of the mixture.

Observations :

1. Initial temperature of HCl and NaOH = t_4°C

    (It should be room temperature here.)

2. Temperature attained after mixing = t_5°C.

Calculations : The heat liberated, Δh will be

$$\Delta h = - (200 + W) (t_5 - t_4) \text{ cal}.$$

This heat is evolved when 100 ml of each and 0.5 normality of each are mixed. Therefore, heat of neutralization ΔH when 1000 ml of 1 N each is mixed will be :

$$\Delta H = -\Delta h \times 10 \times 2$$
$$= -20 \Delta h.$$

Result : The heat of neutralization of HCl and NaOH is ... K cal.

This value is in negative and should be nearly 13.7 K cal

Precautions : (*i*) Both, HCl and NaOH should be at the same temperature and should have an equal normality.

(*ii*) Heat loss due to conduction, convection and radiation should be negligible.

EXPERIMENT – 23

To determine the heat of solution of KNO_3 by solubility method.

Definition of Heat of Solution. Heat of solution is defined as the heat liberated or absorbed when 1 mole of a solute is dissolved in a large bulk of water.

Apparatus. Beaker, thermometer and pipette etc.

Principle. We simply determine the solubilities of salt at two temperatures say one at room temperature and the other above room temperature and then by using Van't Hoff's equation of solubility which is given below, the heat of solution is determined.

$$\log_{10} S_2 - \log_{10} S_1 = \frac{\Delta H}{2.303 R} \left[\frac{T_2 - T_1}{T_1 T_2} \right]$$

where S_1 and S_2 are solubilities of salt respectively at two temperatures T_1 and T_2 (in degree absolute) ; R is gas constant (*i.e.*, 1.987 cal.) ; ΔH is heat of solution of the solute. The solubilities are expressed in mole per 1000 g solvent.

Method. The methods of determining the solubilities of solute at room temperature and above room temperature have been discussed in the heading of Solubility.

Observation. Room temp. $= T_1 K$

Temperature above room temp. $= T_2 K$

Solubility at $T_1 K$ $= (S_1)$ mole/1000 g H_2O

Solubility at $T_2 K$ $= (S_2)$ mole/1000 g H_2O

Calculations. Heat of solution (ΔH)

$$= \frac{2.303 \times 1.987 \times T_1 . T_2 \times (\log_{10} S_2 - \log_{10} S_1)}{(T_2 - T_1)}$$

By substituting T_1, T_2, S_1 and S_2, we can calculate the heat of solution of the solute.

Result. Heat of solution of $KNO_3 = ...$ cal.

EXPERIMENT – 24

To determine the heat of solution of the given salt in water by calorimeter.

Material. Calorimeter, glass stirrer, thermometer (0.1°C) and salt (say KNO_3).

Principle. Heat of solution is determined by dissolving a small known weight of the given salt in a large measured quantity of water in a calorimeter. The rise or fall in temperature, *i.e.*, Δt is noted Then ΔH is calculated by the formula

$$\Delta H = \frac{(W + m) \times \Delta t \times M}{w}$$

where
$$W = \text{water equivalent of calorimeter}$$
$$m = \text{mass of the water taken}$$
$$w = \text{mass of the solute}$$
$$M = \text{Molecular wt. of solute}$$
$$\Delta t = \text{Change in temperature}$$

Method. 1. To determine the water equivalent of calorimeter, take a known mass of cold water in the inner beaker of calorimeter and note its temperature. Add to it, a known mass of hot water of known temperature and note the highest temperature of the mixture. For calculations, follow previous exercise, heat of neutralization.

2. Remove water from the calorimeter and dry it.

3. Now place 200 ml of distilled water measured by means of graduated cylinder and note the temperature.

4. Weigh accurately about 5 g of the powdered salt (given) and pour it quickly but carefully into the calorimeter. Avoid splashing of water. Immediately, replace the stopper and begin stirring for about 30 minutes or so until the mercury in the thermometer stem is steady at least for 5 minutes. Read the final temperature precisely. Find the difference of the initial and final temperatures, i.e. Δt.

Observations :

Heat of solution of salt :

$$\text{Mass of water taken} = m \text{ g}$$
$$\text{Temp. of water} = t_3 °C$$
$$\text{Mass of salt added} = w \text{ g}$$
$$\text{Final temp. of solution} = t_4 °C$$

Calculation :

Let $\qquad\qquad\qquad\qquad W = \text{water equivalent of calorimeter}$

Since the given salt KNO_3 produces cooling when dissolved in water,

$$\Delta t = t_3 - t_4$$

Heat lost by calorimeter and water in it

$$= (W + m)\,\Delta t$$

$\therefore\qquad$ Heat gained by w g of salt = Heat lost by calorimeter + water

$$= (W + m)\,\Delta t$$

$$\text{Mol. wt. of salt} = M$$

$\therefore\quad$ Heat (ΔH) gained by 1 mole of salt

$$= \frac{(W + m) \times \Delta t \times M}{w}$$

Result. The heat of solution is....cal./mole

Precautions :

1. The salt should be well powdered otherwise it will not dissolve completely.
2. Addition of salt should be quick.
3. Thermometer reading should be taken carefully.

Transition Temperature

Introduction. The transition temperature is defined as the temperature at which one solid form changes into another solid form and at this temperature, both the forms of solid exist in equilibrium with each other. All polymorphous substances show this transition temperature. At all other temperatures, only one form is stable and the other form exists in a metastable state which gradually changes into the stable form. The transition temperature of sulphur is 96°C and equilibrium is expressed as below:

$$\text{Rhombic sulphur} \xrightleftharpoons{96°C} \text{Monoclinic sulphur}$$

The transition temperature is also shown by hydrated salts. During transition phenomenon, there is an absorption or evolution of heat, change in vapour pressure, volume and colour etc. The transition point (temperature) always lies below the melting point of the two forms of solid. The temperature at which one hydrate changes into another hydrate or anhydrous salt is called transition temperature of hydrated salt. For example, at 32.38°C $Na_2SO_4.10H_2O$ and Na_2SO_4 are in equilibrium.

$$Na_2SO_4.10H_2O \xrightleftharpoons{32\cdot38°C} Na_2SO_4 + 10\ H_2O$$

Above 32.38°C, $Na_2SO_4.10H_2O$ changes to anhydrous Na_2SO_4 i.e., favours forward reaction. Below 32.38°C only $Na_2SO_4.10H_2O$ is stable but anhydrous sodium sulphate is unstable and changes into hydrated salt i.e. backward reaction. Hence 32.38°C is called the transition temperature of decahydrate sodium sulphate.

EXPERIMENT – 25

To determine the transition temperature of a hydrated salt e.g., $Na_2SO_4.10H_2O$ by thermometric method.

Material Required. Thermometer (0.1°C), ordinary thermometer, thin walled glass tube, stirrers, hydrated salt, beaker and burner, *etc.*

Principle. When temperature is raised, the hydrated salt becomes partially liquid and a point is reached at which the temperature of the salt remains practically stationary until conversion is complete. It is due to absorption of heat during transition of hydrated form into another or anhydrous form. This point is transition point. During cooling, again a point is reached at which the temperature of salt remains practically stationary for a certain interval. It is due to evolution of heat during the change of anhydrous form to hydrated from. This temperature (point) is also the transition temperature. Hence both the temperatures must coincide theoretically.

Method. Set up the experiment as shown in the Fig. 14.9. Take about 40–50 gms. of powdered $Na_2SO_4.10H_2O$ in a thin walled test tube and insert a thermometer (0.1°C) and stirrer into the salt. The bulb of thermometer should be completely dipped into the salt. Clamp this assembly in a beaker containing water, provided with a stirrer and an ordinary thermometer. Now heat the beaker gently and stir the water regularly for uniform heating. Read the temperature of the salt after every minute by the thermometer (0.1°C) inserted in the test tube. While recording the temperature of the salt, keep the salt stirred. On raising the temperature, the salt becomes liquid. When the temperature has reached about 40°C, stop heating. Then allow it to cool, stirring constantly both salt and water and note the

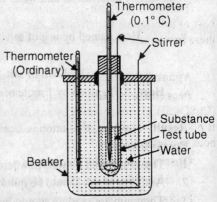

Fig. 14.9. Determination of the transition temperature

temperature of salt after every minute. Heating and cooling should be slow near the transition temperature.

Observations :

S. No.	Timing (minute)	Temperature (°C) While heating	Temperature (°C) While cooling
1	1		
2	2		
3	3		
4	4		
5	5		
6	6		
7	7		
8	8		
9	9		

Calculations. Plot temperature on x-axis and time on y-axis. The curves of heating and cooling as shown in figure are obtained. The curve $ABCDE$ is the heating curve and curve $EDFBA$ is the cooling curve. The two curves should coincide theoretically but as we see in Fig. 14.10, it does not happen so. The reason for this is that the transition of one solid form into another solid form does not take place instantaneously at the transition temperature but at a slightly higher temperature during heating and at a slightly lower temperature during cooling. Therefore, the mean of points (temperatures) C and F will give the transition temperature of $Na_2SO_4.10H_2O$.

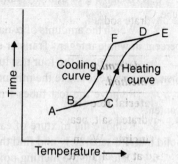

Fig. 14.10. Transition temperature curves.

Result : The transition temperature of $Na_2SO_4.10H_2O$ is...°C.

EXPERIMENT – 26

To determine the heat of hydration of anhydrous copper sulphate.

Theory : The heat of hydration of copper sulphate, ΔH is easily obtained by determining the heat of solution of an anhydrous salt, ΔH, and of hydrated copper sulphate, ΔH_2. The difference in these two values provides the heat of hydration, ΔH of anhydrous salt. Thus

$$\Delta H = \Delta H_1 - \Delta H_2.$$

Apparatus : Calorimeter, glass stirrer and thermometer (0.1°C).

Procedure : Heat of solutions of both, anhydrous copper sulphate and of hydrated one $(CuSO_4.5H_2O)$ are obtained by any of the two methods given in exercise 26 or 27. If anhydrous salt is not readily available, this can be obtained by heating hydrated copper sulphate for about half an hour to a temperature of $\sim 150°C$.

Observations and Calculations :

(a) Heat of solution of anhydrous copper sulphate.

Follow the previous experiment with all observations. Say ΔH_1 is x_1 cal mole^{-1}

(b) Heat of solution of $CuSO_4.5H_2O$. Follow the previous experiment. Say ΔH_2 becomes x_2 cal mole^{-1}.

Result : The heat of hydration of anhydrous copper sulphate is $(x_1 - x_2)$ cals mole^{-1}.

EXPERIMENT – 27

> *To draw the phase diagram of the binary system, diphenyl amine and α-naphthol and find the eutectic temperature.*

Theory : If the substance is not pure, it gives low melting point and the magnitude of their lowering depends upon the two factors.

(*i*) The proportion of each substance present.

(*ii*) Melting points of both substances.

The varying amount of a substance is homogeneously mixed with another substance and melting point of the mixture is determined. A graph is plotted between mole percent of one substance and the melting point of that mixture. The graph gives valuable information.

Material Required : Melting point apparatus, cold water jar, stoppered test tubes, water-bath, diphenyl amine and α-naphthol.

Procedure : It can be taken stepwise :

1. Weigh 1 g of diphenyl amine at four places and transfer each one to separate stoppered test tube.

2. Weigh the amounts of α-naphthol as given in the table below. These amounts make the mole percent in whole integer. Transfer each amount to one test tube and label them carefully.

3. Now put all the four test tubes on water bath and heat them. When the contents melt, shake them for a while to make the mixture uniform.

4. Remove the test tubes from water bath and place these in cold water jar to solidify the contents.

5. Remove the mixture of each test tube separately by breaking the test tubes. Powder each solid mixture separately and find their melting points.

6. Determine the melting points of pure diphenyl amine and α-naphthol separately.

Observations : (*i*) Melting point of α-naphthol = 95°C.

(*ii*) Melting point of diphenylamine = 58°C.

(*iii*) amount of diphenylamine = 1 g

Amount of α-naphthol (in g)	Mole fraction of diphenylamine (dpa)	Mole percent	Melting point ...°C
2·8166	0·2	20	80
1·2780	0·4	40	70
0·5188	0·6	60	45
0·2132	0·8	80	50

Plot a graph between mole percent of diphenylamine in each mixture and its melting point. It will be of the following type (Fig. 14.11).

The cross (×) on the graph represents the thaw points (the temperature at which mixture begins to melt) and the dots represent the melting point (the temperature at which entire mixture melts). The minima in the curve is at *E* and the melting point corresponding to it, is called Eutectic temperature (lowest melting temperature) attained by a mixture) and the corresponding composition is called Eutectic composition.

Result : The eutectic temperature of the mixture is .... and its composition is ....

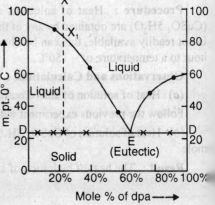

Fig. 14.11.

EXPERIMENT – 28

To determine the pH value of a given solution by indicator method (using buffer solution of known pH).

Procedure. 1. To a small portion of the test solution, add a few drops (5 drops) of the universal indicator and observe the colour change. Then by referring to the colour chart find the approximate pH of the solution.

2. If the pH as indicated by universal indicator lies within 3 and 4.3 then to find out the correct pH value, we must select an appropriate indicator having short working pH range and including this pH limit, *i.e.*, 3 to 4.3. From the inspection of the table, the suitable indicator will then be methyl orange.

3. Prepare a series of buffer solutions (of known pH value differing by 0.1 unit covering the pH range 3 to 4.3) by mixing a salt and a weak acid (or a weak base) in a definite ratio which can be calculated by the equation described earlier.

4. Arrange ten or more test tubes of same dimenstions in a rack. Place in each of these in turn 5 ml of buffer solution of definite pH value *i.e.*, 3, 3.1, 3.2, 3.3, 3.4, 3.5, 3.6, 3.7. 3.8, 3.9, 4.0 4.1, 4.2, 4.3, 4.4. Mark the pH value on the test tubes. To each test tube, add 5 drops of methyl orange solution. Observe the different shades of colour from red to orange.

Place 5 ml of the test solution in a test tube of same shape and size and 5 drops of same indicator (methyl orange). The colour produced in the test solution is then matched with the colours produced in buffer solutions. When a complete match is found, the test solution and corresponding buffer solution have the same pH. If the complete match is not found, but colour lies between those of two successive standard colours, pH value will lie between those of two standards.

Refractivity

When a beam of monochromatic light passes from one medium to another, it suffers change in direction. This phenomenon is called refraction. If it passes from rarer to denser medium (*e.g.*, from air to water) it will bend towards normal and the angle of incidence (*i*) is more than angle of refraction (*r*). According to Snell's law of refraction.

$$\frac{\sin i}{\sin r} = n$$

where *n* is refractive index of the second medium with respect to the first medium and angle *i* and *r* are the angles of incidence and of refraction respectively.

Specific Refraction, *R* (Specific Refractivity)

According to Lorenz and Lorenz, refractive index is related to specific refractive index, *R* by an expression :

$$R = \frac{1}{d}\left(\frac{n^2 - 1}{n^2 + 2}\right)$$

where

R = specific refractivity of medium
d = density of medium
n = refractive index of medium.

Although '*n*' increases with temperature, *R* remains constant as density falls with temperature. In general usage, specific refractivity is called refractivity.

Molar Refraction, R_m (Molar Refractivity) : If the specific refractivity is multiplied by the molecular weight of the liquid, it gives the value of molar refraction, R_m. Thus :

$$R_m = \text{Molecular weight} \times R$$

EXPERIMENT – 29

To determine the refractive index of a given liquid and find its specific and molar refractivities.

Apparatus : Abbe's refractometer, ordinary light lamp, pyknometer and thermometer.

Abbe's Refractometer and method : The instrument is designed in the side figure.

Remove both the prisms, P_1 and P_2 from the spectro-photometer, clean them gently with cotton moistened with alcohol. Put 1-2 drops of the sample liquid on P_1 and press both and clamp them back in the instrument. In doing so, a film of the liquid spreads uniformly between them. Light lamp is kept in such a way that light coming from source L is reflected by mirror M of the instrument and travels towards the prism. This on reaching the surface of P_1 becomes scattered into the film of the liquid. Rays with greater angle of refraction than that of the ray corresponding to grazing incidence do not enter prism P_2. Now view through the telescope. You will find bright and dark portions. Rotate the prism box slowly. Coincide the edge of the bright portion with the cross wire of the telescope, T. Now the scale shows the refractive index of the liquid. Water at controlled temperature is continuously supplied around the prisms in order to maintain constant temperature.

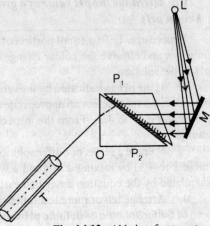

Fig. 14.12. Abbe's refractometer

Repeat the experiment and take at least three readings.

Determine the density of the sample liquid by using density bottle or pyknometer.

Observations : Temperature = ... °C

1.   Refractive index of liquid

(i) ....
(ii) .... Mean value
(iii) .... (n) = ——

2.   Mass of empty pyknometer = w_1 g
Mass of pyknometer + water = w_2 g
Mass of pyknometer + liquid = w_3 g

3.   Molecular weight of the sample liquid = M

Calculations :

$$\frac{\text{Density of liquid } (d)}{\text{Density of water}} = \frac{w_3 - w_1}{w_2 - w_1}$$

∴   Density of liquid (d) = $\dfrac{w_3 - w_1}{w_2 - w_1} \times$ density of water

Calculate the specific refractivity (R) by the formula

$$R = \frac{1}{d}\left(\frac{n^2 - 1}{n^2 + 2}\right)$$

Molar refractivity : R_m = Mol. wt. × R

Results :   1. The refractive index of the given liquid   = ...
2.   Specific refractivity of the liquid   = ...
3.   Molar refractivity of the liquid   = ...

Precautions : 1. Polished surface of the prisms should not be scratched.
2.   Only 1-2 drops of the liquid are placed on the prism.
3.   Light reflected from the upper surface of the upper prism should not enter the telescope.
4.   After putting the given liquid in between the two prisms, press these immediately. Even a small time given will allow some liquid to evaporate obscuring uniform illumination.

Chromatography

Introduction : Chromatography is a laboratory technique which is used to separate and identify the chemical constituents present in a mixture. It was used first of all by botanist T. Swett (1903) and he separated the coloured components of the leaves and from there, it follows its name (chromatus means colour and graphien means writing).

Chromatography involves the flow of a mobile phase, which may be a liquid or a gas over a stationary phase, which may be solid or liquid. It is either based on the principle of adsorption or on partition.

Partition paper chromatography and thin layer chromatography are quite common.

In paper chromatography, we make use of Whatmann no. 1 filter paper. In thin layer chromatography (TLC), in place of filter paper, a fine layer of adsorbent (cellulose or silica gel) is spread on a glass plate and dried. TLC plates can be heated without any damage to them and spot obtained in them are less diffused. They work on the same principle and both identify constituents with the help of R_f values.

EXPERIMENT – 30

To study the separation of amino acids by chromatographic technique.

Material Required : 1. Chromatographic chamber with solvent trough.
2. A sheet of filter paper Whatmann no. 1 of 7×25 cm.
3. Amino acids — glycine, alanine and proline, *etc.*
4. Developing solvent and
5. Locating or visualising reagent.

Procedure : This can be adopted stepwise

(A) **Preparation of Solutions of Amino acids**

1. Weigh approximately 0.5 mg of each amino acid noted above and dissolve separately in 1 ml of water and label them. Now take small amount of each solution and mix them together. This mixture will be a sample mixture to carry out separation.

(B) **Preparation of developing Solvent**

Mix 40 ml of *n*-butanol, 10 ml of glacial acetic acid and 50 ml of water in a separating funnel. Shake the mixture thoroughly and then allow it to separate into distinct layers. The lower aqueous layer is used as developing solvent. Extract it out and fill it in the trough of chromatographic chamber.

(C) **Preparation of Locating Reagent**

Dissolve 0.2 gm of ninhydrin in 10 ml of n-butanol. Fill it in spray bottle for ready use.

(D) **Separation Process**

Draw a pencil line 3–4 cm above one end of the sheet of filter paper. This will be called origin line. Make four cross marks (×) on this line at equal distance and label these with pencil as g, a, P and m. Now with the help of a capillary tube put a small drop of alanine on 'a' cross mark, of glycine on 'g' of proline on 'p' and of mixture on 'm'. The drop must be as compact as possible. Dry these spots by hot air blower. Now suspend this paper in chromatographic chamber with the pencil and dip into the solvent to a depth of only 0.5 cm. This clearly means that the four spots should not touch the solvent of the trough.

Now cover the jar and wait (nearly 3–4 hours) for developing solvent to rise almost upto the top end of the paper. Remove this paper from the jar carefully and draw a pencil line on the height solvent has reached before solvent dries and measure its distance from origin line. This is called solvent front. Dry this paper again and then spray ninhydrin solution on it. The coloured spots will appear. Measure the distance from the origin line to centre of each coloured spot. The pure acid solution will have their spots. The separate spot will also be obtained for the separate amino acid of the mixture. The finished paper is called chromatogram.

Calculate the R_f values of each separated acid and compare with the R_f values of pure acids to ascertain their presence.

Observations :

S. No.	Amino acid	Distance travelled from origin line (cm)		
		Pure amino acid	Amino acid from mixture	by solvent
1.	glycine	...	...	...
2.	alaline	...	...	...
3.	proline	...	...	...

Calculations :

R_f values for amino acids are calculated as follows :

$$R_f \text{ value} = \frac{\text{Distance travelled by an amino acid from origin line}}{\text{Distance travelled by developing solvent from origin line}}$$

Comparison of R_f values

S. No.	Amino acid	R_f value of pure acid	R_f value of the acid separated
1.	glycine	...	...
2.	alanine	...	...
3.	Proline	...	...

Result : The amino acids present in the mixture are glycine, alanine and proline.

EXPERIMENT – 31

> *To separate Cu^{2+}, Pb^{2+} and Cd^{2+} ion by chromatography.*

The process of separation will be the same as in separation of amino acids in previous exercise. Here, we have to use 1% aqueous solutions of cupric chloride, lead nitrate and of cadmium chloride. The developing solvent and locating reagent will ofcourse differ. These are as follows :

Developing solvent

(*i*)   A mixture of 90 ml of ethyl alcohol and 10 ml of 5 N HCl.

(*ii*) A mixture of 45 ml of ethyl alcohol, 45 ml of isopropyl alcohol and 10 ml of 5 N HCl.

EXPERIMENT – 32

> *To separate Pb^{2+}, Ag^+ and Hg^{2+} ions present in a mixture by chromatography.*

Procedure : The nitrate solutions (~ 0.3%) of above ions are made. Distilled water or ethanol can be used as a developing solvent. 0.25 molar solution of potassium chromate serves as a locating reagent.

The entire procedure is the same as explained in exercise 34.

EXPERIMENT – 33

> *To separate Zn^{2+}, Co^{2+} and Ni^{2+} ions present in a mixture by chromatography.*

Procedure : Prepare 1% aqueous solution of their chloride salts. Add few drops of dilute HCl to prevent hydrolysis.

Developing solvent : A mixture containing acetone, ethyl acetate and 6 M HCl in 9 : 9 : 2 by volume as a developing solvent.

Locating reagent : Alcoholic solution of alizarin with 0.1% of salicyldoxime and 0.1% of rubianic acid.

Follow the procedure exactly as in exercise 34.

Colloids (Sols)

EXPERIMENT – 34

1. **To prepare the following sols :**
   (a) **Arsenious sulphide sol**
   (b) **Ferric hydroxide sol**
   (c) **Aluminium hydroxide sol**
   (d) **Sulphur sol**
2. **Purify the sols by dialysis**

(a) **Arsenious sulphide sol.** It is prepared as follows : (Double decomposition method)

Method. Dissolve 0.5–1.0 gm. of arsenious oxide (As_2O_3) in 500 ml of distilled water (free from electrolytes) by boiling. Continue the boiling till a sufficient amount of As_2O_3 has gone into the solution. This will require nearly an hour. Then the hot solution is quickly filtered and cooled. Make the volume of the filtrate to 500 ml. by adding more distilled water. Pure H_2S is passed through an equal volume of distilled water (*i.e.*, in 500 ml.) until it is saturated. Now the two solutions are mixed together. As_2O_3 is converted to As_2S_3 by the action of H_2S as follows

$$As_2O_3 + 3H_2S \longrightarrow As_2S_3 + 3H_2O$$

The fact that sols are coagulated by electrolytes, can be used to test the complete conversion of As_2O_3 to As_2S_3. Take a small quantity of above sol in a separate test tube and add to it a little $BaCl_2$ solution, filter to pass into the filterate. If an yellow ppt. is now obtained, it shows the incomplete conversion of As_2O_3 to As_2S_3 sol. Hence pass excess of H_2S to the whole sol until it becomes saturated or add more H_2S water to the whole sol. Since As_2S_3 sol is not very unstable to boiling, large excess of H_2S is removed by boiling the sol and removal of H_2S is tested with lead acetate paper. If H_2S is not removed, it may cause coagulation of sol to some extent. *Only a small excess of H_2S is permissible.* Filter the sol and the deep yellow coloured filtrate is arsenious sulphide sol. The sol prepared this way is slightly opalescent owing to the Tyndall effect. For further purification, the sol may be dialysed.

(b) **Ferric hydroxide sol**

Hydrolysis Method : Boil $FeCl_3$ soln. with water. Stop boiling when colour becomes reddish brown. This reddish brown solution is the ferric hydroxide sol which may be purified by dialysis to remove HCl formed during hydrolysis.

(c) **Aluminium hydroxide sol** It is also prepared by hydrolysis of $AlCl_3$ solution. The details of the method are the same as in case of ferric hydroxide sol.

(d) **Sulphur sol**

Reduction method : When SO_2 solution is reduced by H_2S or acid is added to thiosulphate solution, colloidal soln. of sulphur is formed. Hence sulphur sol is prepared as under :

(i) *From reduction of SO_2.* The solution of SO_2 is reduced by passing H_2S. On filtration, the colloidal sulphur passes into the filtrate. To this filtrate, add saturated solution of NaCl to precipitate colloidal sulphur. Wash this precipitate of sulphur on filter paper with distilled water until all the chloride ions are removed. When water is further added, the sulphur changes to colloidal solution, *i.e.*, sulphur sol.

(ii) *From thiosulphate solution.* Prepare a concentrated solution of sodium thiosulphate (about 3–4 N). About 100 ml of this solution is added gradually to 10–12 ml of conc. H_2SO_4 with shaking. This will give colloidal sulphur which is precipitated and impurities are removed as above. This precipitate is converted to colloidal sulphur as above.

2. **Purification of sol by dialysis.** In dialysis, the sol is kept in a parchment bag. The parchment bag allows the passage of only solvent molecules and ions of electrolytic impurities but not the

...lloidal particles. This bag containing sol is immersed into a vessel containing water. The water is being removed continuously. This process is very slow and takes days for purification. The above process of diffusion is accelerated by applying a certain voltage through two metal electrodes dipped into the vessel in which parchment bag is immersed. The ions move faster towards oppositely charged electrodes and in this way the electrolytic impurities are quickly removed.

Precipitation (coagulation) of Lyophobic Colloids. This lyophobic colloids are very much sensitive to electrolytes. It has been found that in some cases even very small quantities of electrolyte cause the coagulation of sol. *The minimum concentration of an electrolyte necessary to produce precipitation of a sol is called precipitation (coagulation or flocculation) power of that electrolyte for a sol.* The precipitation value of an electrolyte for a sol depends on the nature of colloid, on the method of its preparation, concentration, nature and valency of the electrolyte ion, etc. The higher the valency, the smaller is the precipitation value. The precipitation value of an electrolyte is expressed in *milli-moles of electrolyte per litre of the mixed solution of the sol and electrolyte.*

EXPERIMENT – 35

To determine the precipitation values of NaCl, CaCl$_2$ and AlCl$_3$ for As$_2$S$_3$ sol. Test the validity of Hardy-Ehulze law.

Theory. As already discussed before. According to Hardy-Schulze law, the precipitation value should decrease with increasing valency of active ions.

Method. Prepare arsenious sulphide sol using 0.5% solution of arsenious oxide (see the method of preparation in Expt. No. 1).

[A] NaCl

Prepare 0.2 M NaCl solution (*i.e.*, 11.7 gram NaCl per litre).

Take ten clean and dry test tubes, number them from 1 to 10 and then arrange them serially in a test tube rack.

Add the following volumes of 0.2 M NaCl soln. and distilled water in each of these test tubes keeping total volume in each case to 10 ml.

Test tube No.	Vol. of 0.2 M NaCl (ml.)	Vol. of distilled water (ml.)	Observations
1	1	9	
2	2	8	
3	3	7	
4	4	6	Note against each test
5	5	5	tube whether solution
6	6	4	remain clear or turbid
7	7	3	after adding As$_2$S$_3$ sol.
8	8	2	
9	9	1	
10	10	0	

Add to each test tube in turn 10 ml of As$_2$S$_3$ sol. After each addition, cork the test tube, mix well by inverting the tube two or three times (do not shake) and then allow them to stand in the rack for one or two hours.

Note in which test tube the solution becomes just turbid. Record this observation in the above table.

Suppose the solution remains clear up to third test tube but becomes turbid in each of the next test tubes. This means that precipitation occurs between 3 and 4 ml. of 0.2 M NaCl. In the next set, prepare the following solutions :

Test tube No.	Vol. of 0.2 M NaCl (ml.)	Vol. of distilled water (ml.)	Observations
11	4·0	6·0	
12	3·9	6·1	
13	3·8	6·2	
14	3·7	6·3	
15	3·6	6·4	
16	3·5	6·5	Note the same ob-
17	3·4	6·6	servations as in pre-
18	3·3	6·7	vious table.
19	3·2	6·8	
20	3·1	6·9	
21	3·0	7·0	

Add to each test tube in turn 10 ml of As_2S_3 ml. After each addition, cork the test tube, mix well by inverting the tube two to three times (do not shake) place in the rack for a definite time, say, one or two hours.

Note from which test tube the precipitation just occurs, i.e., solution becomes just turbid. That will give the precipitation value for NaCl.

[B] CaCl₂

Prepare 0·005 M solution of $CaCl_2$ (0·555 g/litre) and then proceed in the same way as for NaCl.

[C] AlCl₃

Prepare 0.005 M solution of $AlCl_3$ (0.668 gm. per litre) and for stock solution dilute this 10 times and the dilute in the same proportions as in case of NaCl.

Note. *Similar experiments can be carried out with ferric hydroxide sol In this case KCl, K_2SO $K_4[Fe(CN)_6]$ and $K_3[Fe(CN)_6]$ are the precipitating electrolytes.*

Calculations : The concentrations of the electrolyte can be calculated as follows :

Let.

Strength of eletrolyte solution $= x$ M

Volume of electrolyte solution for precipitation

$$= v \text{ ml}$$

Volume of mixed solution of sol and electrolyte, which is kept constant

$$= V \text{ ml}$$

Then :

Concentration of electrolyte in mixed solution

$$= \frac{x.v.\,1000}{V} \text{ millimoles per litre.}$$

NaCl

Conc. of original solution = 0·2 M

Vol. of mixed solution = 20 ml

∴   Conc. of NaCl in first test tube

$$= \frac{0.2 \times 1 \times 1000}{20}$$

= 10 milli-moles/litre

Similarly, the concentrations of NaCl in mixed solution in other test tubes will be 20, 30, 40, 50, 60, 70, 80, 90 and 100 milli-moles per litre respectively.

Similarly, the concentrations of NaCl in mixed solution in test tubes 11 to 21 can be calculated. The concentration in milli-moles per litre can thus be found from the second set of test tubes, which will give precipitation value of NaCl.

$CaCl_2$ and $AlCl_3$. Make similar calculations for $CaCl_2$ and $AlCl_3$ and thus find out their precipitation values.

Result : The precipitation values of Na^+, Ca^{2+} and Al^{3+} ions for As_2S_3 sol. are ... ... and ... respectively.

Since in the three electrolytes, the active ions are Na^+, Ca^{2+} and Al^{2+} respectively, we observe that the precipitation value decreases in the order Na^+, Ca^{2+} and Al^{3+} which proves Schulze-Hardy law.

Precautions : The turbidity in each test tube should be observed very carefully.

15

Conductometric Titrations

Introduction

The electrical conductivity of an aqueous solution is the measure of its ability to conduct electricity. Current is carried by ions (very rarely by solvated elections) in solution and conductance of a solution is infact, sum of the conductances all the ions available in the solution. The conductance of an individual ion will depend upon:

1. Size of the ion: Smaller the size of ion, less strongly it is hydrated and faster it can move and can carry the current and finally can provide greater conductivity. The conductivity of certain ions are given below which carry the same magnitude of charge.

Relative conductance of few ions with respect to acetate ion

$$CH_3COO^- = 1$$
$$Na^+ = 1.2$$
$$Cl^- = 1.9$$
$$OH^- = 4.8$$
$$H^+ = 8.5$$

H^+ being smallest in size has the greatest conductance.

2. Charge on the ion. As the magnitude of charge increases on any ion, the amount of electricity carried by it becomes more and thus contributes more to the conductivity. Fe^{3+} ion is more conducting than Fe^{2+} ions.

As the concentration of ions increases in solution, number of current carriers increases and hence conductivity too. In extreme cases, where concentration becomes very high, each ion resists the movement of its neighbour thus will not allow to increase conductivity in proportion to the concentration.

It is now clear that the conductance of solution is reciprocal to its resistance likewise it will have the unit ohm^{-1} or mho.

Following terms are used to express conductivity of a solution.

Specific conductivity. As the conductivity is reciprocal of resistivity, specific conductance is reciprocal of specific resistance. By definition, it is the conductance of cubic centimeter of a solution.

It is denoted by symbol k (Kappa). Thus sps. conductivity, $k = \dfrac{1}{\rho}$

Its unit is Ohm^{-1} cm^{-1} or siemens per meter. Since mobility of ions changes with temperature, it is generally taken at 25° C.

Equivalent Conductivity. It is the product of specific conductivity and the volume of solution in ml which contains one gram equivalent of solute.

Alternatively, it is defined as the conductivity of a solution obtained by dividing its specific conductance with gram equivalent of solute per cubic centimeter of the solvent. It is denoted by symbol, $\wedge$ (Capital Lambda)

Molar Conductivity. It is the product of specific conductance and the volume in ml of a solution which contains 1 mole of solute.

It is alternatively expressed as the specific conductance of a solution divided by number of moles of solute per cubic centimeter. It is denoted by the symbol $\wedge_m$ and at infinite dilution by $\wedge_\infty$.

Effect of dilution. The value of specific conductance, on dilution, decreases as the number of current carriers per unit volume drops. However both, equivalent conductivity and molar conductivity increase on increasing the dilution of the solution and attains a limiting value at infinite dilution.

Conductivity Cell. To determine the conductance of electrolytic solution, a special type of cell is employed called electrolytic cell. It is made up of the pyrex glass, fitted with two platinum electrodes. In an ideal cell, two 1 cubic cm platinum surfaces are kept 1cm apart. The position of the electrodes are rigidly fixed. This cell has a cell constant 1. This rigid condition is not followed in practice; for highly conducting solutions, the space between two electrodes is increased and for low conducting ones, it is decreased. Each such cell has its own characteristic cell constant which is defined as the ratio of the space between the two electrodes to the area of electrode.

The cell constant, K having the value 0.1 are more common. The cell is dipped into the solution under investigation and is used as one arm of the Wheatstone bridge which evaluates its resistance. The reciprocal of the resistance is its conductance. The observed conductance is then multiplied by cell constant provides the conductivity value of the solution.

Wheat Stone Bridge. It is device for measuring the electrical resistance of a substance and here of the electrolytic solution. It has circuit diagram as shown below

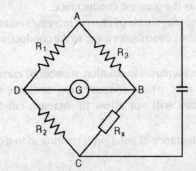

Fig. 15.1 The Circuit diagram of Wheatstone Bridge.

R_1, R_2, and R_3 are known resistances whereas R_x is the resistance of the conductivity cell under investigation and G is galvanometer.

At the point of Balance, we have

$$\frac{R_2}{R_1} = \frac{R_x}{R_3}$$

The value of R_x can be determined. In conductivity experiments, it is also called conductivity bridge.

EXPERIMENT – 1

Object: *To determine the strength of a given hydrochloric acid conductometrically by titrating it against a standard solution of sodium hydroxide solution.*

Theory : HCl is a strong acid as it ionises completely into H^+ and Cl^- ions. The conductivity of this solution is very high. When sodium hydroxide is added to it, OH^- ions of the latter combine with

H^+ ions of the acid producing non-electrolytic water. So during the progress of the titration, H^+ ions are replaced by Na^+ ions. The relative conductance of H^+ ions are nearly seven times more than that of Na^+ ions. Consequently the conductivity of the solution decreases until end point is reached.

After the end point, no H^+ ions remain and any addition of sodium hydroxide solution furnishes OH^- ions and increases the conductivity (OH^- ions here are second most conducting). So end point has minimum value of conductivity.

Trend of the graph. If the graph is plotted by taking conductivity on Y-axis and volume of NaOH added on X-axis, a V-shape curve is obtained. The end point of the titration lies at the intersection of two linear lines of the titration curves as shown below.

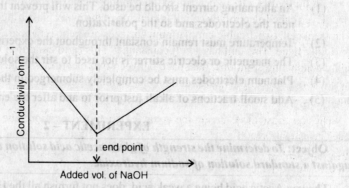

Fig. 15.2 Conductometric titration graph of NaOH and HCl

The end volume is used to calculate the strength of hydrochloric acid.

Apparatus and Reagents. Conductivity bridge, standard sodium hydroxide solution burette, pipette, measuring flask and hydrochloric acid of unknown strength.

Experimental Procedure

Take about 10 ml of given HCl acid in a beaker and dip the conductivity cell in it in a vertical position and clamp it. Use mechanical stirrer. Fill in the burette the given standard solution of NaOH, say 0.1N. You can also prepare the solution at your own. Make the connections of the conductivity bridge.

Add NaOH in 1 ml increments and find the resistance of the solution each time. We will see that in every addition of sodium hydroxide solution, the sliding contact is moved to the left to get balance point indicating decrease in conductivity. After adding sufficient amount of alkali further addition the balance point is obtained by sliding the contact to the right. Repeat the experiment with fresh 10 ml of HCl and now before end point, add only in drops to get the precise value.

Titration against 10 ml of unknown HCl acid.

S.No	Added Vol. of NaOH solution (ml)	Observed Resistance (ohm)	Conductivity (mho) or ohm^{-1}
1	-- -- -- ml	-- -- -- ohm	-- -- --
2	-- -- -- ml	-- -- -- ohm	-- -- --
3	-- -- -- ml	-- -- -- ohm	-- -- --
4	-- -- -- ml	-- -- -- ohm	-- -- --
5	-- -- -- ml	-- -- -- ohm	-- -- --

Calculation: Plot the graph as stated earlier an get the end point volume, say V ml.

Normally of HCl solution: $10 \times N = 1 \times V$

$$N = \frac{.1 \times V}{10}$$

Strength of given HCl acid $= \dfrac{36.5 \times .1 \times V}{10}$ gm per litre

Precautions

(1) An alternating current should be used. This will prevent the build up of reaction products near the electrodes and so the polarization.

(2) Temperature must remain constant throughout the experiment.

(3) The magnetic or electric stirrer is not used to stir the solution.

(4) Platinum electrodes must be completely submerged in the acid solution.

(5) Add small fractions of alkali just prior to and after the end point.

EXPERIMENT – 2

Object: *To determine the strength of given acetic acid solution conductometrically by titrating against a standard solution of sodium hydroxide.*

Theory: Acetic acid being a weak acid, does not furnish all the H^+ ions it has so in the beginning of the experiment, its conductivity is small. As the solution of sodium hydroxide is added, OH^- ions react with H^+ ions of the acid available to form water. To make up the loss of H^+ ions, acetic acid ionises further to give H^+ ions acetate ions as shown below.

$$CH_3COO^- + H^+ \xrightarrow{\quad Na^+ OH^- \quad} H_2O + Na^+ + CH_3COO^-$$
$$\updownarrow$$
$$CH_3COOH$$

The Na^+ ions and CH_3COO^- ions increase the conductivity of the solution gradually until the end point is reached. After the end point, further addition of NaOH solution increases the conductivity sharply conductance of OH^- ions is nearly five times to that of acetate ions.

Trend of the Graph. If conductance of the solution is plotted on Y-axis and the volume of the alkali on X-axis added, a following type of the curve is obtained

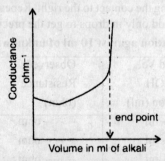

Fig.15.3 Conductometric titration curve of CH_3COOH and NaOH solution.

As the graph shows that there is slight decrease in the conductance in the beginning of the reaction. It is due to suppression in the ionisation of acetic acid due to common ion effect (CH_3COONa CH_3COOH) and buffering action.

The end point corresponds to sudden inflection in the conductance.

Experimental Procedure- Apparatus and reagents required are the same but for acetic acid in place of hydrochloric acid. NaOH solution will be taken in burette and the acetic acid in the titration flask. The conductivity cell will be dipped in acetic acid solution. Follow each and everything as in the HCl and NaOH titration.

Advantage of conductometric titration

(1)   It works well even in highly coloured or turbid solutions where observation of end point by the use of an indicator is difficult.

(2)   This method is also suitable in titrations where proper indicator is not available.

(3)   This is applicable to even highly diluted solution.

(4)   It can be used for all kinds of titrations acid-base, redox, precipitation and complex titrations.

16

Potentiometric Titrations

Introduction: Potentiometric titrations like conductivity titrations, are volumetric titrations but here the end point is obtained from the variation in electrode potential with the change in concentitration of potential determining ions. These titrations provide an accurate and precise results and are quite suitable to coloured, turbid or opaque solutions where visual indicators fail to accomplish the experiment.

To measure the change in electrode potential, e.m.f. is measured using two electrodes; the reference electrode and the indicator electrode. A brief description of both the types is given below.

(*a*) **Reference Electrodes:** The purpose of the reference electrode is to complete the circuit for measuring e.m.f. It should provide a stable and reproducible potential so as to measure potential difference due to indicator electrode. The reference electrodes are mainly of the following types.

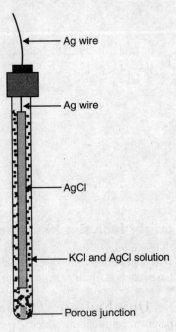

Fig 16.1 Silver electrode

(*i*) **Hydrogen electrode:** This electrode is made up of rectangular platinum foil further coated with platinum black to enhance the adsorption. This is placed in a standard solution of an acid. Hydrogen at a known pressure is constantly bubbled through this acid solution.

Standard hydrogen electrode (SHE) or so called normal hydrogen electrode (NHE) is quite specific where hydrogen gas at 1 atmosphere is bubbled through the acid solution whose H^+ activity is unity. The magnitude of this electrode potential is considered to be 0.00 volt.

This electrode is seldom used in laboratory practice as to maintain the pressure of 1 atmosphere of hydrogen gas is quite difficult. Further, it is not workable is presence of any oxidising agent.

(*ii*) **Silver Electrode:** A silver wire is used which is coated with silver chloride. To do so, wire is dipped in molten silver chloride solution. This wire is now placed in a saturated solution of potassium chloride. The electrode potential of it largely depends upon the concentration of chloride ions and so provides a constant potential as long the concentration of chloride ions remains constant. It has a potential of 0.199V compared with standard hydrogen electrode. This electrode is cheap, simple and can work well even upto a temperature of 130°C. It suffers from a disadvantage that it is easily contaminated (Fig. 16.1).

(*iii*) **Saturated Calomel Electrode:** This is based in the reaction between mercury and mercurous chloride (Calomel). The mercury remains in contact with the aqueous phase and mercurous chloride is in contact with saturated solution of potassium chloride. (Fig. 16.2).

The cell operates with a redox reaction:

$$Hg_2^{2+} + 2e^- \rightleftharpoons 2Hg \text{ (liquid)}$$

and its electrode potential is calculated as

$$E = E°_{Hg_2^{2+}/Hg} - \frac{RT}{2F} \log \frac{1}{\left[Hg_2^{2+}\right]}$$

Where E° is its potential at standard conditions.

The potential of this electrode is 0.2415 compared with standard hydrogen electrode.

Many people discourage the use of this electrode owing to toxic nature of the mercury. If glass tube containing mercury is broken and mercury spills, it is hazardous.

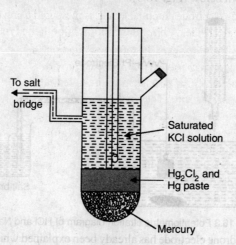

Fig 16.2 Calomel electrode

(b) Indicator Electrodes: These provide a voltage which is proportional to the concentration of potential determining ions. These should be nonpolarized and should give a constant potential with time. Indicator electrodes are selected as per the requirement of the experiment. For instance, in acid-base reaction, quinhydrone electrode is used.

Quinhydrone Electrode: When equimolar proportions of hydroquinone and parabenzoquinone are added to a solution under experiment and an inert platinum wire (inert electrode) is dipped into it, a redox potential is developed which largely depends upon H+ ion concentration. The electrode half reaction for this is

$$\text{Hydroquinone} \rightleftharpoons \text{Quinone} + 2e^-$$

This electrode works well in the pH range (1 to 8 but does not serve the purpose in presence of any strong oxidising or reducing agents.

Antimony-antimony oxide electorde is another alternative to it.

TYPE OF POTENTIOMETRIC TITRATIONS

Following types of titrations are performed under this head.

(1)   acid- base titrations

(2)   redox titrations and

(3)   Precipitation titrations

For examples, follow the experiments

EXPERIMENT – 1

Object: *To find out the strength of a given HCl solution potentiometrically by titrating against standard solution of sodium hydroxide.*

Apparatus and Reagents. Calomel electrode, Quinhydrone electrode, salt bridge, standard sodium hydroxide solution, burette and pipette, *etc*.

Experimental Procedure: Take 25 ml of HCl solution in a beaker and add equimolar amounts of quinhydrone and parabenzoquinone and clamp inert platinum electrode. Clamp a burette filled with standard sodium hydroxide solution. Put a magnetic bar in it for sirring and connect it with a calomel electrode through a salt bridge. The two end wires of electrodes are connected to potentiometer. The setting diagram is shown below (Fig. 16.3).

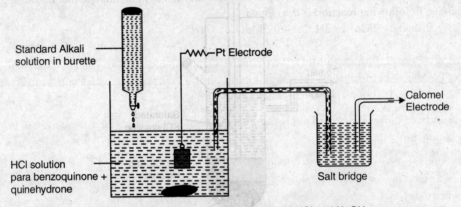

Standard Alkali
solution in burette

Pt Electrode

Calomel
Electrode

HCl solution
para benzoquinone +
quinehydrone

Salt bridge

Fig 16.3 Potentiometric titration diagram of HCl and NaOH

The function of quinhydrone electrode has already been explained while discussing the electrodes.

Add sodium hydroxide solution in 1 ml increamnts and measure the e.m.f. each time. The sudden change in e.m.f. is your end point and this volume is end volume. Repeat the experiment with fresh lot and this time add drop-wise sodium hydroxide before 2 ml of end point

Plot the graph using e.m.f. on *Y*-axis and the volume of alkali on *X*-axis. The end point is indicated by maximum slope in the graph shown below (Fig. 16.4).

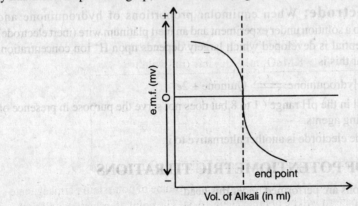

e.m.f. (mv)

end point

Vol. of Alkali (in ml)

Fig 16.4 Potentiometric titration curve of HCl and NaOH

Precautions

(1)   Take the e.m.f. after every addition of alkali when the reaction is complete.

(2)   Add sodium hydroxide solution slowly and in drops before the end point

EXPERIMENT – 2

Object: *To determine the strength of given ferrous ammonium sulphate solution potentiometrically by titrating it against a standard solution of potassium permanganate.*

Theory: Ferrous ions are oxidised by acidic potassium permanganate. As the reaction advances, ferrous ions are gradually oxidised and the ratio of Fe^{3+} ions to ferrous ions increases. This brings the change in the electrode potential of the cell which is sharp in the beginning and slow until end point is reached. Once all the ferrous ions are oxidised further addition of potassium permanganate, the cell potential due to following reaction is developed.

$$2MnO_4^- + 3Mn^{2+} + 2H_2O \longrightarrow 5MnO_2\downarrow + 4H^+$$
$$\text{brown}$$

This cell potential is due to MnO_4^-/Mn^{2+} and it again provides a sharp change. Due to intense colour of potassium permanganate the solution turns pink at the end point which subsequently turns brown due to formation of manganese dioxide.

Apparatus and Reagents: Potentiometer, calomel electrode, platinum inert electrode, salt bridge, beaker, burette and pipette, *etc.*

Experimental Procedure: Take 20 ml of supplied ferrous ammonium sulphate solution in a beaker and add sufficient dilute sulphuric acid (it should not fall short in oxidising all the ferrous ions). Clamp a platinum electrode and burette filled with a standard solution of potassuim permanganate. Connect this to calomel electrode through a salt bridge. Now connect the terminals of both cells to potentiometer. Design the apparatus as shown in the earlier experiment.

Add potassium permanganate solution from the burette in 1 ml increments and after every addition take the *e.m.f.* of the cell by potentiometer. When solution acquires just pink colour, note the end volume and *e.m.f.* carefully.

Repeat the experiment with fresh lot and now add only dropwise potassium permanganate just before end point and after the end point to about 1 ml volume. Take these as final readings.

Observation.

It can be written in tabular form

S.No	Vol. of KMnO$_4$ ml	e.m.f. observed mv (milivolts)
1	-- -- --	-- -- --
2	-- -- --	-- -- --
3	-- -- --	-- -- --
4	-- -- --	-- -- --
5	-- -- --	-- -- --

Graph: Plot the graph taking e.m.f. of the cell on Y-axis and volume of potassium permanganate on X-axis. A graph of following pattern will be obtained (Fig 16.5). The point of inflection will be the end point.

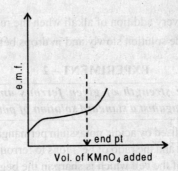

Fig 16.5 Titration curve of ferrous ammonium sulphate and $KMnO_4$

Precautions

1. Temperature should be maintained constant throughout the experiment.
2. After every addition of potassium permanganate, solution should be stirred to get stable reading.
3. $KMnO_4$ is toxic so should be handled carefully.
4. Pt-electrode should be cleaned after the experiment.

EXPERIMENT – 3

Object: *To determine the strength of given ferrous ammonium sulphate solution potentiometrically by titrating against the standard solution of potassium dichromate.*

This titration is exactly similar to the previous experiment with following change.

(1) Take potassuim dichromate solution in burette in place of potassium permanganate

(2) The light green colour due to ferrous ions is discharged at the end point and solution may turn slightly yellow.

Calculation

Once the end point volume is obtained, calculation part is simple. Let the normality of ferrous ammonium sulphate be N. Its volume used is 20 ml

End volume is V_1 and normality of standard solution is N_1. Therefore,

$$N \times 20 = N_1 V_1$$

$$N = \frac{N_1 V_1}{20}$$

$$\text{Strength} = \frac{N_1 V_1}{20} \times 392.12 \text{ gms / litre}$$

EXPERIMENT – 4

Object: *To find out the strength of a solution containing mixture of potassium halides potentiometrically by titrating it against a standard solution of silver nitrate.*

Theory: Both chloride as well as iodide are precipitated by silver nitrate solution. Since silver chloride and silver iodide are not isomorphous, they are precipitated independently. Silver iodide is precipitated first followed by silver chloride. The first inflection in the graph curve will correspond to iodide ion concentration where as the second to chloride ion concentration.

Apparatus and Reagents: Potentiometer, Calomel electrode, silver electrode, standard silver nitrate solution, burette and pipette, *etc.*

Experimental Procedure. Take about 5 ml of supplied solution containing chloride and iodide ions and make up its volume by distilled water to 100 ml. Make the connections as per the diagram (Fig. 16.6) given here.

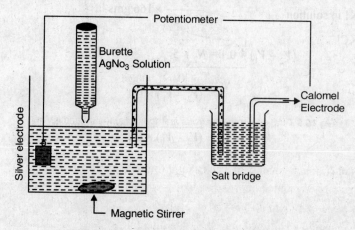

Fig 16.6 Potentiometric titration of silver halides.

Now add silver nitrate solution in 1 ml aliquots and note the e.m.f. of the cell everytime with potentiometer. A sudden inflex in e.m.f. corresponds to iodide ion concentration. Note this volume carefully. Continue adding silver nitrate solution in 1 ml aliquots by noting the e.m.f. at various additions. A further sharp increase in e.m.f. will correspond to second end point, *i.e.,* chloride ion concentration.

Plot the graph using e.m.f. in millivolts on *Y*-axis and volume of silver nitrate added on *X*-axis. The trend of the graph will follow the pattern (Fig. 16.7).

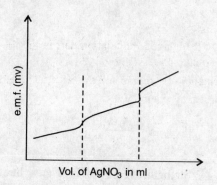

Fig 16.7 Titration curve of mixture of halides

Calculation

Let V_1 = first end point

V_2 = second end point

0.1 = normality of silver nitrate solution.

5 ml = volume of supplied solution taken

N_1 and N_2 be the normality of KI and KCl.

Therefore.

$$V_1 \times 0.1 = N_1 \times 5$$

$$N_1 = \frac{5}{V_1 \times 0.1}$$

Strength of KI in solution $= \dfrac{5}{V_1 \times 0.1} \times 166 \text{ gms lit}^{-1}$

Strength of KCl

$$(V_2 - V_1) \times 0.1 = N_2 \times 5$$

$$N_2 = \frac{5}{(V_2 - V_1) \times 0.1}$$

$$\text{Strength} = \frac{5}{(V_2 - V_1) \times 0.1} \times 74.5 \text{ gms lit}^{-1}$$

Molecular Weight Determination

To determine the molecular weight of a volatile substance by Victor Meyer's method.

Principle : In this method, the volume of vapours of a known amount of volatile substance is measured at room temperature and atmospheric pressure and its corresponding volume at N.T.P. is determined. Vapour density (V.D.) of the substance is thus calculated from the formula given below and double of its value is the molecular weight.

$$\text{V.D.} = \frac{\text{mass of substance taken}}{\text{its volume at NTP} \times 0\cdot000089}$$

(0·000089 = density of H_2 at N.T.P.)

Apparatus : (Fig 17.1) The Victor Meyer's (V.M.) tube consists of a long glass tube having a cylindrical bulb at its lower end and open cylindrical funnel at the top end which can be closed with a rubber stopper 'S_2'. It has a narrow side tube 'A', quite near to top whose mouth is dipped few cm deep into water contained in a trough. An inverted graduated cylindrical tube called Eudiometer tube filled with water is placed over the mouth of the side tube.

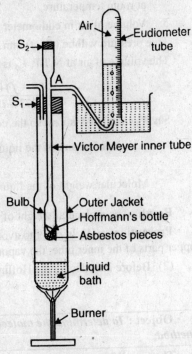

The V.M. tube is inserted into an outer jacket, usually made of metal by removing the cork 'S_2'. A glass tube is also inserted in the S_1 cork.

Hoffmann's bottle is a small bottle with a glass stopper which will carry the substance under investigation into bottom of Victor Meyer's tube.

Material Required. Thermometer, large jar to dip eudiometer and Victor Meyer's assembly.

Procedure. Fill the bulb of the outer jacket more than half with a liquid whose boiling point is 30–50°C higher than the boiling point of the liquid under investigation. If the sample liquid is acetone (b. pt. 56°C), methyl alcohol (b. pt. 65°C) or carbon tetrachloride (b. pt. 77°C), water can be filled in the outer jacket shown as liquid bath. Put some pumic stone in the liquid bath to avoid bumping during boiling. Clean the V.M. tube and dry it completely. Place small amount

Fig. 17.1.

of sand or asbestos pieces on the bottom of this tube and place it back into the apparatus.

Weigh accurately about 0·1 gm of the liquid under investigation in the Hoffmann's bottle and stopper the tube tightly to avoid evaporation.

Start heating the outer jacket to make the water boil. During the course of heating, the air in V.M. tube expands and comes out in the form of bubbles through the mouth of the side tube dipped in water. When the bubbles stop coming, fill the eudiometer tube with water and put it over the mouth of the side tube as shown in the figure 17.1.

Loosen the stopper of the Hoffmann's bottle and drop it at once to the bottom of V.M. tube and tighten the stopper S_2 of V.M. tube instantly.

The liquid evaporates spontaneously and displaces equal volume of air which is collected in the eudiometer. When bubbles cease to come, open the stopper 'S_2' and stop heating.

Now close the eudiometer firmly with your thumb and put it inverted into a large jar full of water. Now shift the eudiometer up and down to make the level of the water inside equal to level of water in the large jar. Allow it to stand for sometime to acquire room temperature. Now this volume of the air in the eudiometer tube is at atmospheric pressure and at room temperature. Note the volume carefully.

Observations and Calculations :

1. Room temperature $t°C$ = $(273 + t)°$ K
   (use thermometer)

2. Atmospheric pressure   = P mm of Hg.
   (use barometer)

3. Mass of liquid taken   = w g

4. Aqueous tension (if water)   = f mm.
   at room temperature

5. Volume of air in eudiometer tube = V ml

True pressure will be $(P - f)$ mm.

The volume of air at N.T.P, V_0 is calculated as

$$\frac{(P - f)V}{273 + t} = \frac{760\ V_0}{273}$$

Say V_0 calculated is x ml so the corresponding mass of the H_2 will be $x \times 0.000089$.

$$\text{Vapour Density of the liquid} = \frac{w}{x \times 0.000089}$$

$$\text{Molecular weight of the liquid} = \frac{2 \times w}{x \times 0.000089}$$

Result. The molecular weight of the given substance is ....

Precautions : (1) In order to avoid the diffusion and condensation of vapours on the colder and upper parts of the inner tube, the vaporization of substance must take place as rapidly as possible.

(2) Before introducing the Hoffmann's bottle, its stopper should be loosened.

EXPERIMENT-2

Object : To determine the molecular weight of a nonvolatile solute by Beckmann's freezing method.

Theory : When a nonvolatile solute is dissolved in a solvent the freezing point of this solution is always less than that of pure solvent. This is because of the inter-action between solute and solvent particles. The following expression works well in molecular weight determination.

If w = mass of the solute dissolved

W = mass of the solvent

m = molecular weight of the solute.

Following expression holds:

$$m = \frac{1000\ K_f w}{W\ \Delta T}$$

Where ΔT is the depression in freezing point and K_f is molal depression constant.

This depression in freezing point is to be measured precisely and accurately and for this purpose, a special type of thermometer called Beckmann's thermometer is used.

Beckmann's Thermometer

This is unusually a large size thermometer invented by Otto Beckmann and does not measure the absolute temperature instead it does the small differences in temperatures.

In its brief description, it has a large reservoir of mercury at the bottom end of the thermometer and from which stems a capillary which serves as a scale. It has the scale of 5-6° and each degree is divided into 100 division thus enables us the measure the temperature change of even 0.01°. At the upper end of the capillary, there is S-shaped loop which is a small reservoir for mercury. This thermometer is quite suitable for measuring the colligative properties such as depression in freezing point and elevation in boiling point *etc.* For each experiment, thermometer needs setting.

Setting: First of all invert the thermometer gently so as to bring sufficient of mercury in the bottom reservoir. Now heat slowly at the bottom so that the mercury rises in the capillary and joins the small reservoir at the top Now cool the thermometer $1-2°C$ higher than the freezing point of the solvent used in experiment.

Now tap the upper end of the thermometer gently so that mercury in column is broken from S-bond. It is now set for the depression in freezing point experiment.

Apparatus: Beckmann's thermometer, Beckmann's freezing point apparatus, two manual stirrers, ice and NaCl.

Experimental Procedure: Take ice in a beaker and hold the thermometer vertically and see that its bulb is dipped in ice. If the mercury in capillary scale is almost full and is disconnected from top reservoir, thermometer is set. If not, do it by hit and trial method as given earlier. Now remove the thermometer. Take 600 ml or a large beaker and make an inch of bed of crushed ice on its bottom. Sprinkle a small layer of salt. Continue layers of salt and crushed ice until the beaker fills two third (Fig. 17.2).

Take 25gm of deionized water in an inner test tube (a test tube of nearly 40-50 ml). Put one stirrer in this test tube and another in large beaker. Clamp the thermometer vertically such as its bulb remains dipped in water. Stir the salt and ice slowly and watch the point in thermometer where water freezes. Note this point.

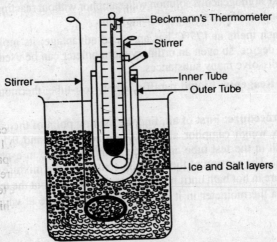

Fig. 17.2. Beckmann's apparatus for determination of molecular weight.

remove the inner test tube and warm it gently to melt the frozen water and dissolve accurately 0.20 gm of solute whose molecular weight is to be determined. Make clear with no residue and place back the inner test tube in ice salt bath again. Stir the ice-salt bath the time, the solution freezes once again. Note this reading Repeat the experiment with varying of solute with increments of 0.02 gm and note the temperatures.

servation: This can be shown in tabular form:

S.No.	Mass of Solute (w)	Freezing pt. of solution °C	ΔT
1	0.20 gm	.....	$(T - T_l)$
2	0.22 gm	.....	.....
3	0.24 m	.....	.....

Freezing point of the water = T °C

Molal depression constant of water = 1.86.

Mass of the water taken (W) = 25 gm

Calculation: The molecular weight m of the solute is calculated from the following expression.

$$m = \frac{1000 \, K_f w}{W \, \Delta T}$$

where various terms have their usual meanings.

Result: The molecular weight of the given substance is ....

Precautions:

1. Water should be pure and deionized.
2. Solute should not dissociate in solution.
3. Stirring should be uniform and slow.

EXPERIMENT – 3

Object: *To determine the molecular weight of a nonvolatile solute by Rast's camphor method.*

Theory: Rast developed a simple method to determine the molecular weight of nonvolatile solutes which can make homogeneous solution with camphor without reacting with it. Like Beckmann's method, it is also based on depression in freezing point.

Camphor which melts at 177°C has an added advantage; its molal depression constant is quite high, *i.e.,* 39.7 degree so even an ordinary thermometer can be used. It can be easily purified and has an ability to dissolve many substances.

Apparatus and Reagents: Liquid paraffin, empty test-tube, thermometer, beakers and Rast's apparatus.

Experimental Procedure: First of all, find the melting point of the camphor in use by melting point apparatus. Now weigh camphor about 1.0 gm accurately and 0.1 gm of the solute under investigation. Put both in the test tube and immerse this test tube in a paraffin-wax bath, already heated to about 180°C. The portion of the test tube containing the mixture should remain dipped in the bath. Keep a mixture in hot bath until it melts completely. Take out the test tube from the bath and allow it to cool. Dip a thermometer in it and note the temperature at which the camphor solution freezes (Fig. 17.3).

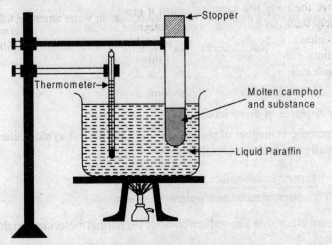

Fig. 17.3 Rast comphor diagram

Calculation: Substitute the values in the following expression

$$m = \frac{1000\,K_f w}{\Delta T\,W}$$

$$K_f = 39.7°$$

ΔT = is the difference of two temperatures, *i.e.*,

m.pt. of camphor and melting point of the mixture.

$$w = 0.1\ gm$$

$$W = 10.0\ gm$$

Substituting the values, obtain the molecular weight of the given solute.

Precautions:

1. The mixture, camphor and solute should not remain more than 1 minute in the hot bath as this may allow some of the camphor to sublime changing the mass of the solvent.

2. Reading of the freezing point should be taken at once, freezing starts.

EXPERIMENT – 4

Object: *To determine the apparent degree of dissociation of an electrolytes (e.g., NaCl) in aqueous solution at different concentrations by ebullioscopy.*

Theory: The boiling point of a solvent is increased by addition of a nonvolatile solute. This is called elevation in boiling point. The same expression holds good with certain modification as was used for depression in freezing point:

$$m = \frac{1000\,K_b w}{W\,\Delta T}$$

Where K_b is molal elevation constant and is also called ebullioscopic (Greek meaning boiling viewing) constant.

Elevation in boiling point is smaller than depression in freezing point for the same solvent for water, it is 0.512 degree and for naphthalene, it is as high as 5.8 degree.

When an electrolyte is used as a nonvolatile solute elevation in boiling point becomes higher as it dissociates into smaller particles increasing their number.

ake the case of sodium chloride which dissociates in water attaining following equilibrium.

$$NaCl \rightleftharpoons Na^+ + Cl^-$$

Before dissociation	1	0	0
After dissociation	$(1-x)$	x	x

Where x is degree of dissociation.

Due to increase in number of particles $(1-x+x+x=1+x)$ molecular weight gives lower value experimentally and therefore, it follows.

$$\frac{\text{experimental molecular weight}}{\text{Normal molecular weight}} = \frac{1}{1+x}$$

For a known electrolyte like sodium chloride the normal molecular weight is 58.5. Now X can be easily calculated.

Apparatus: Landsberger's apparatus, Beckmann's thermometer and steam generating round bottomed flask.

Experimental Procedure: The boiling point of pure solvent and its boiling point elevation can be measured accurately by an instrument called ebullioscope. This was developed by Brossard which measured the absolute temperatures only. Later on, Beckmann's thermometer was used as a modification.

It is in normal practice to find the boiling point elevation by Landsberger's method.

Beckmann's thermometer is set for 2-3 higher than boiling point of water and it is fitted in an assembly shown in the figure 17.4.

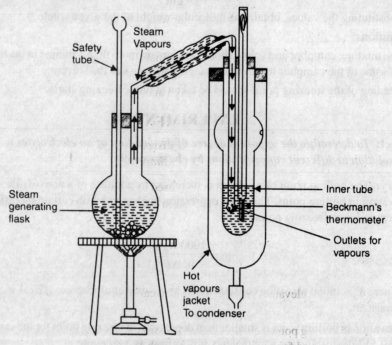

Fig. 17.4. Lands berger's Apparatus

Take about 200 ml of water in steam generating round bottomed flask. Take 20 ml of pure and deionized water in inner tube and attach Beckmann's thermometer in it.

Now steam generating flask is heated and its vapours are allowed to pass through the inner tube which is graduated and will boil the water under experiment. Note the temperature at which water boils. This is the boiling point of water in the experiment.

Now remove the inner tube and pour off the water and make it only 10 ml. Add 0.1 gm of sodium chloride in it and make its clear solution. Replace this in original position and note the boiling point of solution and its volume as well. Determine the density mass of this solution to get the total mass of the solution. This mass when subtracted by mass of sodium chloride, gives the mass of solvent, *i.e.*, water.

Observation and Calculation

If m = molecular weight of the sodium chloride

      w = mass of sodium chloride

      W = mass of water

      ΔT = Difference in boiling points

      $K_b = 0.52°$

$$m = \frac{1000\, K_b\, w}{W\, \Delta T}$$

This is experimental molecular weight. If degree of dissociation is x

$$\frac{m}{58.5} = \frac{1}{1+x}$$

$$x = \frac{58.5}{m} - 1$$

This is degree of dissociation if needed percentage of dissociation x is multiplied by 100.

Viva -Voce

CONDUCTOMETRIC AND POTENTIMETRIC TITRATIONS

1. What is the titration ?

Ans. It is method used in the laboratory to determine the unknown concentration of a known reagent. The solution to be titrated is called analyte and the solution used to react with it is called titrant.

2. What is conductometric titration ?

Ans. It is volumetric titration in which the end point is located by measuring the electrical conductance of a solution.

3. What is conductivity ?

Ans. The conductivity is reciprocal of electrical resistivity. In fact, it is a measure of the ability of a solution to conduct electricity.

4. What is specific conductance ?

Ans. It is the conductance of cubic centimeter of a solution. It is denoted by the symbol Kappa, k. Its unit is $ohm^{-1} cm^{-1}$. Its S.I. unit is siemens $meter^{-1}$.

5. What is equivalent conductance ?

Ans. It is the product of specific conductance and the volume of solution which contains 1 gm equivalent of a solute. It is denoted by symbol capital lambda, λ and equivalent conductivity at infinite dilution is denoted by $\lambda\infty$.

6. What is molar conductance ?

Ans. It is the product of specific conductance and the volume of a solution which contains 1 mole of solute. It is denoted by symbol λm.

7. What is cell constant ?

Ans. It is defined as the ratio of the space between two electrodes to the area of the electrode. Its unit is cm^{-1}.

8. What is the effect of dilution on the conductivity of a solution.

Ans. The specific conductance decreases on dilution of the solution. This is because number of current carriers (ions) decreases per unit area. However both, equivalent and molar conductance increase on dilution and attain a limiting value at infinite dilution.

9. Explain why in potentiometric titration of acetic acid against sodium hydroxide, the graph pattern shows slight decline initially when conductance taken on y-axis.

Ans. It is due to suppression in the ionisation of acetic acid due to common ion effect.

10. What is a Potentiometric titration ?

Ans. It is volumetric titration in which the end point is obtained from the variation in electrode potential with the change in concentration of potential determining ions.

11. What is Nerst equation ?

Ans. This equation is used to determine the equilibrium reduction potential or oxidation potential of a half-cell in a electrochemical cell. It is expressed as:

$$E_{red} = E^{\circ}_{red} - \frac{RT}{nF} \log_e \frac{[\text{Reduced form}]}{[\text{Oxidised form}]}$$

Where $R = 8.314 \text{ JK}^{-1} \text{ mole}^{-1}$.

12. What is the redox potential for ferric and ferrous ions ?

Ans. The expression will be :

$$E_{red} = E^{\circ}_{red} - \frac{RT}{F} \frac{[Fe^{2+}]}{[Fe^{3+}]} \qquad (n = 1)$$

13. Is it necessary to continue the titration for a while even after the end point has reached in conductometric and potentiometric titrations ?

Ans. Yes, it is because graph is to be made to locate the end point.

14. What is the function of a reference electrode ?

Ans. The purpose of the reference electrode is to complete the circuit for measuring the e.m.f. of the cell. Hydrogen electrode, silver electrode and calomel electrode are the examples of reference electrodes.

15. What are indicator electrodes ?

Ans. These provide voltage which is proportional to the concentration of potential determining ions. Quinhydrogen electrode is an example of this type.

16. Is any indicator used in these titrations?

Ans. No.

17. What is the purpose of salt bridge in a electrochemical cell.

Ans. The function of the salt bridge is to maintain electroneutrality and permit the current to flow. If the salt bridge is not used, electrons will accumulate and the current would stop.

Molecular Weight Determination

18. What is a melting point and a freezing point ?

Ans. The melting point of a solid is the temperature at which its vapour pressure equals to the vapour pressure of its liquid. The two phases exist in equilibrium. Likewise, the freezing point of a liquid is the temperature at which its vapour pressure of liquid equals to that of its solid. Mostly these two are equal.

19. What is freezing point depression ?

Ans. The freezing point of a liquid (a solvent) is depressed when a nonvolatile solute is added to make its solution. It is, therefore, solution will freeze at a lower temperature than pure solvent. The difference in the two temperatures is called freezing point depression. It is a colligative property.

20. What is the relation between depression in freezing point of a solution and the molecular weight of the solute dissolved in it ?

Ans. It is expressed as :

$$m = \frac{1000 \, K_f \, w}{\Delta T \, W}$$

where the terms have their usual meanings.

21. What is Beckmann's thermometer ?

Ans. It is large size thermometer, inverted by Otto Beckmann. It does not measure the absolute temperature but does the very small deference in temperatures.

22. What is boiling point of a liquid ?

Ans. The boiling point of a liquid is that temperature at which its vapour pressure is equal to the environmental pressure surrounding the liquid.

23. What is boiling point elevation ?

Ans. Whenever a nonvolatile solute is added to a liquid (solvent), the boiling point of the resulting solution is elevated. This phenomenon is called boiling point elevation. It is a colligative property and is called ebullioscopy. This depends upon number of particles present and not on their identity.

24. What is molal elevation constant ?

Ans. The elevation in the boiling point produced when 1 mole of a nonvolatile solute is dissolved in 1000 gms of the solvent.

25. What is the relation between molal elevation, constant and molecular weight of the solute dissolved ?

Ans. The following expression holds.

$$m = \frac{1000 \, K_b \, w}{\Delta T \, W}$$

Where terms have their usual meanings.

26. What is the relation between boiling point elevation and lowering of vapour pressure of a solution ?

Ans. The lowering of vapour pressure of a solution is directly proportional to its elevation in boiling point.

REFRACTOMETRY

27. What do you mean by refractive index ?

Ans. The refractive index or index of refraction is the measure of the angle at which light is bent when it passes from one material to another. This phenomenon takes place because the velocity of light changes from one medium to another.

If i is the angle of incidence of a ray in vacuum, perpendicular to the surface of medium and r is the angle of refraction, the refractive index n, will be:

$$n = \frac{\text{sine } i}{\text{sine } r}$$

It may be mentioned that light travels fastest in vacuum and is taken to be a physical constant.

28. On what factors, does the refractive index of a liquid depend ?

Ans. The refractive index of a liquid depends upon the temperature of the liquid as well as the wave length of the incident light.

29. What is specific refractivity ?

Ans. If the refractive index of a medium is n, its refractivity will be $n - 1$. Specific refractivity R is given by

$$\text{Specific refractivity, } R = \frac{n-1}{d}$$

Where d is the density of the medium

According to Lorentz and Lorentz:

$$R = \frac{n^2 - 1}{n^2 + 2} \cdot \frac{1}{d}$$

30. What is molar refractivity ?

Ans. When specific refractivity is multiplied by molar mass, gives the value of molar refractivity.

APPENDICES

1. Preparation of the chemical reagents used in laboratory

These have been given in brief :

1. *Acetic acid (dilute)*. Take 285 ml of conc. acetic acid and dilute it with 1000 ml of water : 5 N.

2. *Aluminon Reagent*. Dissolve 1g of the substance in one litre of water.

3. *Alcoholic Caustic Soda solution*. Dissolve 20 g of caustic soda in 1 litre of rectified spirit : N/2.

4. *Alcoholic Caustic Potash solution*. Dissolve 28 g of caustic potash in one litre of rectified spirit : N/2.

5. *Ammoniacal Silver Nitrate (Miller's reagent)*. Take about 1·7 g of $AgNO_3$ and 17ml. of conc. NH_4OH. To it, add 25 g of KNO_3 and add water to make it one litre.

6. *Ammoniacal Acetate (2 N solution)*. Dissolve 154 g of substance in one litre of distilled water.

7. *Ammonium Carbonate*. Dissolve 160 g of substance in one litre of a mixture made by adding 140 c.c. of conc. ammonia solution in 860 ml water : 4 N.

8. *Ammonium Chloride*. Dissolve 270 g of substance in one litre of distilled water. Shake it well and filter out the undissolved substance and use the solution : 5 N.

9. *Ammonium Hydroxide (dilute)*. 340 ml of conc. ammonia (sp.gr.0.88) in one litre of distilled water : 5 N.

10. *Ammonium thiocyanate (alcoholic solution)*. A saturated solution of NH_4SCN in 95% ethyl alcohol.

11. *Ammonium Nitrate*. Dissolve 80 g of substance in one litre of distilled water : 1 N.

12. *Ammonium oxalate*. Disslove 35 g of substance in one litre of distilled water : 0·5 N.

13. *Ammonium Sulphate*. Dissolve 132 g of salt to one litre of distilled water : 2 N.

14. *Ammonium Thiocyanate*. Dissolve 38 g of substance in one litre of distilled water : 0·5 N.

15. *Barium Chloride*. Dissolve 61 g of substance in one litre of distilled water. 0·5 N

16. *Baryta Water [$Ba(OH)_2$]*. Dissolve 70 g of hydrated Barium hydroxide in one litre of distilled water and filter : 0·4 N.

17. *Benedicts Solution*. Dissolve 1·75 g of crystalline $CuSO_4$ in 100 ml of distilled water and 17·5 g of sodium nitrate and 10 g of anhydrous sodium carbonate, prepared in 75 ml of water. Dilute it again by 100 ml of distilled water.

18. *Bismuth Nitrate Solution*. Dissolve 81 g of the substance in one litre of distilled water (containing 100 ml of dil. HNO_3).

19. *Bromine Water*. Dilute 10–15 ml. liquid Br_2 by one litre of distilled water carefully.

20. *Calcium Chloride (dilute)*. Dissolve 55 g of hydrated substance in 1 litre of distilled water : 0·5 N.

21. *Calcium Hydroxide (Lime water)*. Dissolve 1·7 g of CaO or 2–3 g $Ca(OH)_2$ in one litre of water. Decant the solution after an hour, 0·04 N.

22. *Caustic Potash*. Dissolve nearly 31·0 g of potassium hydroxide in distilled water and make up to one litre : 0·5 N.

23. *Caustic Soda.*

    (a) *Concentrated.* Dissolve 225 g of NaOH in distilled water and make up to the volume to one litre : 5 N.

    (b) *Dilute.* Dissolve only 80 g similarly : 2 N.

24. *Cerric Ammonium Nitrate Solution.* Dissolve 100 g of the substance to 250 ml of hot dil. HNO_3 (2 N).

25. *Chlorine Water.* Saturate the desired amount of distilled water by passing chlorine gas : 0·2 N.

26. *Diphenyl Amine.* Dissolve 10 g of substance in 1litre of conc. H_2SO_4.

27. *Diphenyl Carbazide.* Dissolve 10 g of substance in one litre of 95% alcohol.

28. *Diphenyl Glyoxime.* Dissolve 10 g of substance in one litre of alcohol (95%).

29. *Dithiozone.* Dissolve 1 g of substance in 1 litre of $CHCl_3$ or CCl_4.

30. *2,4 Dinitrophenylhydrazine.* Dissolve 5 g of substance in 50 ml of conc. H_2SO_4. To it, add one litre of rectified spirit carefully. Allow the solution to stand for a while and then filter.

31. *Dimethyl Glyoxime.* Dissolve 1g of substance in 100 ml of 95 % alcohol (rectified spirit).

32. *Fusion Mixture.* Mix equimolar proportions of anhydrous sod. carbonate and pot. carbonate .

33. *Ferric chloride.* Dissolve 81 gms. of the substance in one litre of distilled water containing 23 ml of conc. HCl : 0·5 M.

34. *Ferric Alum.* Dissolve 40 g of ferric ammonium sulphate in 100 ml of water and acidify with a few drops of dil. HNO_3.

35. *Iodine Solution.* Dissolve 20 g of KI and 13 g of iodine in 30 ml of distilled water and make up the solution to one litre: 0·1 N.

36. *Lead Acetate.* Dissolve 95 g of substance in small amount of distilled water and make up the solution to one litre: 0·5 N.

37. *Lime Water.* See calcium hydroxide preparation.

38. *Mercuric Chloride.* Dissolve 27·2 g of the salt in a litre of distilled water : 0·2 N.

39. *Magneson.* Dissolve 5 gms. of substance in 1litre of 0.25 NaOH.

40. *Methyl Orange.* Dissolve 2 g of methyl orange in one litre of hot water. Allow to cool and filter if necessary.

41. *Magnesium nitrate reagent.* Dissolve 130 g $Mg(NO_3)_2$ and 240 g NH_4NO_3 in water ; add 15-28 ml conc. NH_4OH and dilute to one litre with water.

42. *β-Naphthol* (alkaline). Dissolve 10 g of the substance in one litre of caustic soda solution (10%).

43. *Phenolphthalein.* Dissolve 10 g of substance in one litre of 50% alcohol. Filter if necessary.

44. *Potassium Chromate.* Dissolve 98 g of the substance in 2 litres of distilled water to use it as precipitant : 0·5 N.

45. *Potassium dichromate.* Dissolve 75 g of the substance in one litre of distilled water.

46. *Potassium Ferricyanide.* Dissolve 55 g of substance in one litre of distilled water : 0·5 N.

47. *Potassium ferrocyanide.* Dissolve 53 g of the substance in one litre of distilled water : 0·5 N.

48. *Potassium Cyanide .* Dissolve 33 g of substance in 1 litre of water very carefully. It is extra poisonous (should not be touched by hand) : 0·5 N.

49. *Potssium Iodide.* Dissolve 166 g of substance in two litres of distilled water : 0·5 N.

50. *Potassium - permanganate.* Dissolve 3 g of substance in 1 litre of water and filter : 0·1 N.

51. *Phenyl Hydrazine Reagent.* Dissolve 25 g of substance in one litre of water .

52. *Silver Nitrate.* Dissolve 17 g of substance in one litre of distilled water : 0·1N.

53. *Disodium hydrogen phosphate.* Dissolve 120 g $Na_2HPO_4.12H_2O$ in one litre of water : 1 N.

54. *Sodium Acetate Solution.* Dissolve 340 g of hydrated sodium acetate in one litre of distilled water 2·5 N.

55. *Sodium Nitroprusside.* Dissolve 3 g of substance in one litre of water.

56. *Sodium Hydroxide.* See caustic soda preparation.

57. *Starch Solution.* Take 1 g of starch and make its paste in water and add to it 100 ml. of boiling water. The contents are further boiled for 5 minutes. Cool it. The solution stored over 20 hours is not used and requires fresh preparation.

58. *Yellow ammonium Sulphide.* Digest one litre of 6 N ammonium sulphide with 25 g flowers of sulphur for few hours and filter : 6 N.

Or

Saturate 150 ml of conc. NH_4OH with H_2S, keeping the solution cold ; and 10 g flowers of sulphur and 250 ml of conc. NH_4OH. Shake until sulphur has dissolved and then dilute to one litre : 6 N.

59. *Uranyl Acetate.* Dissolve 300 g of the substance in one litre of distilled water.

60. *Zinc Uranyl Acetate.* Dissolve 20 g of the substance and 12 g of acetic acid (30%) in 100 ml of distilled water.

2. Preparation of some special Regents

1. *Brady's reagent.* Mix 4 g 2, 4-dinitrophenyl hydrazine with 8 ml conc. H_2SO_4. Add gradually with shaking and cooling 70 ml methyl alcohol, warm and make up the solution to 100 ml by adding water.

2. *Barford's reagent.* Dissolve 6·5 g copper acetate in 100 ml 1% acetic acid.

3. *Cerric ammonium nitrate.* This is obtained by dissolving 40 g cerric ammonium nitrate per 100 ml of 2 N nitric acid.

4. *Fehling's soultion.* Consists of two parts, A and B.

   (a) *Solution A* : Dissolve 35 g crystalline copper sulphate $CuSO_4.5H_2O$ in 500 ml water. Add few drops of conc. H_2SO_4.

   (b) *Solution B* : Dissolve 173 g sodium Potassium-tartarate (Rochelle salt) and 60 g of caustic soda in 500 ml water.

   Equal volumes are mixed before use.

5. *Tollen's reagent.* Add 2-3 drops of dilute NaOH (1%) solution to 5 ml $AgNO_3$ solution. A white ppt. is obtained. Now add ammonium hydroxide drop by drop till the ppt. just dissolves. When required, the reagent should be prepared fresh.

6. *Molisch's reagent.* Dissolve 10 g α-naphthol in 100 ml alcohol or 10 g. β-naphthol in 100 ml chloroform.

7. *Schiff's reagent.* Dissolve 1 g *p*-rosaniline in 50 ml water with gentle warming. Cool, saturate with SO_2 and filter.

Or

Dissolve 5 g of *p*-rosaniline hydrochloride (dye) in 500 ml distilled water. Add 500 ml saturated with SO_2 distilled water to this solution. If sodium bisulphate is available, add 5 g of it in place of SO_2 saturated water.

8. *Phenyl hydrazine solution.* Dissolve 25 g of phenyl hydrazine in 100 ml glacial acetic acid.

9. *Ammonium molybdate.* Dissolve 45 g of the commercial salt in a mixture of 40 ml conc. NH_4OH, 60 ml water and 120 g NH_4NO_3 and dilute to one litre with water.

Or

Dissolve 45 g of the commercial salt or 40 g of MoO_3 in a mixture of 70 ml conc. NH_4OH and 140 ml water and then add with vigorous stirring to mixture of 250 ml of conc. HNO_3 and 500 ml of water, now dilute it to one litre. Allow it to stand for 1-2 days and decant the clear solution.

10. *KSCN solution.* Dissolve 49 g salt in one litre water : 0·5 N.

11. *Mercuric chloride (saturated).* Add 70 g of salt in one litre water.

12. *Magnesium Uranyl acetate.* Treat separately 8·5 g of uranyl acetate and 50 g of magnesium acetate, each with 6 ml of glacial acetic acid and 95 ml of water, heating at 70° C until clear solution are obtained. Mix at this temperature and allow to cool. Filter after standing for two hours at 20° C.

13. *Carminic acid.* 1 g of acid in one litre conc. H_2SO_4 .

14. *p-nitrobenzeneazo resorcinol.* 5 g of dye dissolved in 1 litre of 0·25 N-NaOH. It is also called *Magneson I.*

15. *α-Naphthylamine.* 3 gm. of reagent in one litre of water, filter if necessary.

16. *Rhodamine reagent.* Dissolve 0·03 g of Rhodamine in 100 ml acetone or alcohol.

17. *Rhodamine-B.* 0·1 g of reagent in one litre of water.

18. *Nessler's reagent.* Dissolve 10 g of KI in 10 ml water. Add to it saturated solution of $HgCl_2$ until a slight permanent reddish ppt. is formed. Dissolve separately 40 g of KOH in 80 ml of water. Cool and add this KOH solution to the previous solution. Allow to stand overnight. Decant it next day. Store in dark place.

Or

Dissolve 23 g of HgI_2 and 16 g of KI in ammonia free water and make up the volume to 100 ml add 100 ml of 6 N-NaOH. Keep it for 24 hours and then decant the solution. Keep in dark place.

19. *Magnesia mixture.* 100 g of $MgCl_2.6H_2O$ are dissolved in one litre of water, 250 g of NH_4Cl and then 300 ml of conc. ammonia solution are added. After allowing it to stand for several days the clear liquid is used.

20. *Methyl violet indicator.* Dissolve one g of indicator in 25 ml of 95% alcohol and make up volume to 100 ml with water.

21. *Methyl red indicator.* Dissolve one g of solid indicator in a mixture of 600 ml alcohol and 400 ml distilled water with constant stirring. Filter if necessary.

22. *Stannous chloride (1N).* Dissolve 56 g of $SnCl_2.6H_2O$ in 100 ml conc. HCl. Allow it to stand until solution becomes clear and then dilute it to one litre. keep a few pieces of tin in bottle to prevent oxidation.

23. *Methylene blue indicator.* Dissolve 1 g of powdered indicator in water and make up to 500 ml.

24. *Litmus solution (blue).* Dissolve one g of solid in 100 ml of water.

25. *Litmus solution (red).* To the above blue litmus solution, add a drop or two of dil. HCl in order to change its colour to red.

26. *Indigo solution.* Warm gently 1g of indigo with 12 ml of conc. H_2SO_4. Allow it to stand for two days and then pour it into 240 ml of water and then filter.

27. *Picric Acid.* Prepare 2 % solution in rectified spirit.

28. *Oxine reagent.* Dissolve 2 g of oxine in 100 ml of $2NCH_3COOH$.

3. Atomic weights and other constants

Elements	Symbol	At. No.	At. wt.	Density (gm . ml)	Specific Heat
Aluminium	Al	13	26. 98	2· 703	0·214
Arsenic	As	33	74. 92	5· 73	0·083
Barium	Ba	56	137. 34	3· 78	0·068
Bromine	Br	35	79. 909	3· 12	0·107
Calcium	Ca	20	40. 08	1· 548	0·149
Carbon	C	6	12. 011	3· 51	0·17
Chlorine	Cl	17	35. 453	0· 00322	—
Chromium	Cr	24	51. 996	6· 22	0·112
Copper	Cu	29	63. 24	8· 95	0·093
Fluorine	F	9	18. 998	0· 001694	—
Gold	Au	79	196. 967	19· 3	0·031
Hydrogen	H	1	1· 008	0· 00008987	—
Iodine	I	53	126· 904	4· 94	0·054
Iron	Fe	26	55· 847	7· 86	0·113
Lead	Pb	82	207· 19	11· 4	0·0305
Magnesium	Mg	12	24· 312	1· 74	0·248
Manganese	Mn	25	54· 938	7· 06	0·122
Mercury	Hg	80	200· 59	13· 515	0·033
Nitrogen	N	7	14· 007	0· 0012507	—
Phosphorus	P	15	30· 974	1· 83	0·202
Potassium	K	19	39· 102	0· 875	0·166
Silver	Ag	47	107· 870	10· 5	0·058
Sodium	Na	11	22· 990	0· 971	0·293
Sulphur	S	16	32· 064	1· 96	0·174
Tin	Sn	50	118· 69	7· 2	0·055
Zinc	Zn	30	65· 37	7· 1	0·093

4. List of Equivalent weights of substances commonly used in Volumetric Analysis

Titration and Substances used	Formula	Eq. Weights
Acid - Alkali Titration :		
Sodium carbonate	Na_2CO_3	M.W./2 = 53·06
Sodium bicarbonate	$NaHCO_3$	M.W./1 = 84·00
Potassium carbonate	K_2CO_3	M.W./2 = 69·00
Potassium bicarbonate	$KHCO_3$	M.W./1 = 100·00
Sodium hydroxide	$NaOH$	M.W./1 = 40·00
Barium hydroxide	$Ba(OH)_2.8H_2O$	M.W./2 = 157·75
Oxalic acid (hydrated)	$H_2C_2O_4.2H_2O$	M.W./2 = 63·03
Hydrochloric acid	HCl	M.W./1 = 35·46
Sulphuric acid	H_2SO_4	M.W./2 = 49·04
Redox Titrations :		
Pot. permanganate	$KMnO_4$	*M.W./5 = 31·606
Pot. dichromate	$K_2Cr_2O_7$	M.W./6 = 49·03
Ferrous sulphate (hydrated)	$FeSO_4.7H_2O$	M.W./1 = 278·00
Ferrous sulphate (anhydrous)	$FeSO_4$	M.W./1 = 152·00
Ferrous amm. sulphate	$FeSO_4.(NH_4)_2SO_4.6H_2O$	M.W./1 = 392·12
Oxalic acid (hydrated)	$H_2C_2O_4.2H_2O$	M.W./2 = 63·04
Iron (ferrous)	Fe	At.W/1 = 55·84
Iodine Titrations :		
Iodine	I_2	M.W./2 = 126.92
Hypo (Sod.thiosulphate		
Hydrated)	$Na_2S_2O_3.5H_2O$	M.W./1 = 248.19
Anhydrous	$Na_2S_2O_3$	M.W./1 = 158.09
Arsenious Oxide	As_2O_3	M.W./4 = 49.45
Copper sulphate (hydrated)	$CuSO_4.5H_2O$	M.W./1 = 249.68
Argentometric Titrations :		
Silver nitrate	$AgNO_3$	M.W./1 = 169.89
Silver	Ag	At.W/1 = 107.88
Sodium chloride	$NaCl$	M.W./1 = 58.45
Chloride ion	Cl^-	At.W./1 = 35.46
Pot. sulphocyanide	$KCNS$	M.W./1 = 97.17
Amm. sulphocyanide	NH_4CNS	M.W./1 = 76.12

* in acidic medium

5. Data on the specific gravity, percentage composition, normality etc., of some reagents

(a) Concentrated reagents

Reagent	Sp.gravity	Percentage by weights	Approx. normality	Weight of anhydrous reagent (gm.) in 1 ml
H_2SO_4	1·84	96	36 N	1·80
HCl	1·19	37·9	12 N	0·366
HNO_3	1·42	69·8	16 N	0·991
NH_4OH	0·882	34·95	15 N	0·308
Glacial-CH_3COOH	1·055	99·5	17 N	1·80

(b) Dilute reagents

Reagent	Sp. gravity	Normality (approx)	Anhydrous reagent (gm.) in 1 ml	Preparation Add
CH_3COOH	1.0256	3 N	—	180 vol.conc. acid to 820 vol. H_2O
H_2SO_4	1.15	5 N	0.240	1 vol. conc. acid to 7 vol. H_2O
HCl	1.05	3 N to 5 N	0.104	3 vol. conc. acid to 7 vol. H_2O
HNO_3	1.165	5 N	—	3 vol. conc. acid to 10 vol. H_2O
NH_4OH	0.96	5 N	0.089	4 vol. conc. soln. to 10 vol. H_2O
NaOH	1.113	3 N	0.110	10 g NaOH to 90 ml water
$Na_2CO_3.10H_2O$	1.079	1·5 N	0.080	20 g decahydrate 80 ml water
$BaCl_2$	1.095	N	0.109	10 g $BaCl_2.2H_2O$ to 80 ml water

6. Ionizaton Constant of weak Bases at Room Temperature(25°C)

Name of Base	Formula	Ionization Constant
Ammonium hydroxide	NH_4OH	$1·8 \times 10^{-5}$
Aniline	$C_6H_5NH_2$	$3·8 \times 10^{-10}$
Dimethyl amine	$(CH_3)_2NH$	$5·12 \times 10^{-4}$
Ethanol amine	$C_2H_5ONH_2$	$2·77 \times 10^{-5}$
Ethyl amine	$C_2H_5NH_2$	$5·6 \times 10^{-4}$
Hydrazine	$(NH_2)_2$	3×10^{-6}
Lead hydroxide	$Pb(OH)_2$	$9·6 \times 10^{-4}$
Methyl amine	CH_3NH_2	$4·38 \times 10^{-4}$
Pyridine	C_5H_5N	$1·4 \times 10^{-9}$
Trimethyl amine	$(CH_3)_3N$	$5·27 \times 10^{-5}$

7. Ionization Constants of Weak Acids at Room Temperature

(In case of dibasic acids, K_1 signifies the ionization constant for the first replaceable H atom, K_2 is the ionization constant for the second replaceable H atom.)

Acid	Formula	Ionization Constant		
		K_1	K_2	K_3
Acetic	CH_3COOH	1.75×10^{-5}		
Arsenic	H_3AsO_4	5×10^{-3}	4×10^{-5}	6×10^{-10}
Arsenious	H_3AsO_3	6×10^{-10}		
Benzoic	$HC_7H_5O_2$	6.3×10^{-5}		
Boric	H_3BO_3	5.8×10^{-10}		
Carbonic	H_2CO_3	4.3×10^{-7}	5.6×10^{-11}	
Chloroacetic	$HC_2H_2O_2Cl$	1.4×10^{-3}		
Chromic	H_2CrO_4	2×10^{-1}	3.2×10^{-7}	
Citric	$H_3C_6H_5O_7$	8.7×10^{-4}	1.8×10^{-5}	4×10^{-6}
Cyanic	$HCNO$	2×10^{-4}		
Dichloroacetic	$CHCl_2COOH$	5×10^{-2}		
Formic	$HCOOH$	1.77×10^{-4}		
Hydrazoic	HN_3	2.6×10^{-5}		
Hydrocyanic	HCN	7.2×10^{-10}		
Hydrofluroic	HF	7.2×10^{-4}		
Hydrogen sulphide	H_2S	5.7×10^{-3}	1.2×10^{-15}	
Hypochlorous	$HOCl$	3.5×10^{-8}		
Iodic	HIO_3	1.6×10^{-1}		
Lactic	$HC_3H_5O_3$	1.39×10^{-4}		
Nitrous	HNO_2	4×10^{-4}		
Oxalic	$H_2C_2O_4$	6.5×10^{-2}	6.1×10^{-5}	4.8×10^{-12}
Phenol	C_6H_5OH	1.3×10^{-10}		
Phosphoric	H_3PO_4	7.5×10^{-3}	6.3×10^{-8}	
Phosphorous	H_3PO_3	1.6×10^{-2}	7.0×10^{-7}	
Phthallic	$H_2C_3H_4O_4$	1.3×10^{-3}	3.9×10^{-6}	
Propionic	$HC_2H_5O_2$	1.34×10^{-3}		
Succinic	$H_2C_4H_4O_4$	6.4×10^{-5}	2.7×10^{-6}	
Sulphurous	H_2SO_3	1.7×10^{-2}	6.24×10^{-3}	
Tartaric	$H_2C_4H_4O_6$	9.6×10^{-4}	2.9×10^{-5}	

8. Ionization (Instability) Constant of Complex Ions at Room Temperature

Dissociation Equilibrium			Instability Constant
$[Al(OH)_4]^-$	$\rightleftharpoons$	$Al(OH)_3 + OH^-$	$2 \cdot 5 \times 10^{-2}$
$[Cd(NH_3)_4]^{2+}$	$\rightleftharpoons$	$Cd^{2+} + 4NH_3$	$2 \cdot 5 \times 10^{-7}$
$[Cd(CN)_4]^{2-}$	$\rightleftharpoons$	$Cd^{2+} + 4CN^-$	$1 \cdot 4 \times 10^{-17}$
$[CdI_4]^{2-}$	$\rightleftharpoons$	$Cd^{2+} + 4I^-$	5×10^{-7}
$[Co(NH_3)_6]^{3+}$	$\rightleftharpoons$	$Co^{3+} + 6NH_3$	$1 \cdot 25 \times 10^{-5}$
$[Co(NH_3)_6]^{3-}$	$\rightleftharpoons$	$Co^{3+} + 6NH_3$	$2 \cdot 2 \times 10^{-34}$
$[Cr(OH)_4]^{2-}$	$\rightleftharpoons$	$Cr(OH)_3 + OH^-$	1×10^{-2}
$[Cu(NH_3)_4]^{2+}$	$\rightleftharpoons$	$Cu^{2+} + 4NH_3$	$4 \cdot 6 \times 10^{-14}$
$[Cu(CN_3)_6]^{2-}$	$\rightleftharpoons$	$Cu^+ + 3NC^-$	5×10^{-28}
$[Cu(CN)_4^{3-}]$	$\rightleftharpoons$	$Cu^+ + 4CN^-$	5×10^{-28}
$[I_3]^-$	$\rightleftharpoons$	$I^- + I_2$	$1 \cdot 4 \times 10^{-3}$
$[Fe(SCN)_6]^{3-}$	$\rightleftharpoons$	$Fe^{3+} + 6SCN^-$	8×10^{-10}
$[Fe(SCN)]^{2+}$	$\rightleftharpoons$	$Fe^{3+} + SCN^-$	$3 \cdot 3 \times 10^{-2}$
$[Pb(OH)_3]^-$	$\rightleftharpoons$	$Pb(OH)_2 + OH^-$	$50 \cdot 0$
$[HgBr_4]^{2-}$	$\rightleftharpoons$	$Hg^{2+} + 4Br^-$	$2 \cdot 2 \times 10^{-22}$
$[HgCl_4]^{2-}$	$\rightleftharpoons$	$Hg^{2+} + 4Cl^-$	$1 \cdot 1 \times 10^{-16}$
$[HgI_4]^{2-}$	$\rightleftharpoons$	$Hg^{2+} + 4I^-$	5×10^{-31}
$[Hg(CN)_4]^{2-}$	$\rightleftharpoons$	$Hg^{2+} + 4CN^-$	4×10^{-42}
$[Hg(SCN)_4]^{2-}$	$\rightleftharpoons$	$Hg^{2+} + 4SCN^-$	$1 \cdot 0 \times 10^{-22}$
$[Ni(CN)_4]^{2-}$	$\rightleftharpoons$	$Ni^{2+} + 4CN^-$	$1 \cdot 0 \times 10^{-22}$
$[Ag(NH_3)_2]^+$	$\rightleftharpoons$	$Ag^+ + 2NH_3$	$6 \cdot 8 \times 10^{-8}$
$[Ag(CN)_2]^-$	$\rightleftharpoons$	$Ag^+ + 2CN^-$	$1 \cdot 0 \times 10^{-21}$
$[Ag(S_2O_3)_2]^{3-}$	$\rightleftharpoons$	$Ag^+ + 2S_2O_3^{2-}$	$1 \cdot 0 \times 10^{-13}$
$[Sn(OH)_3]^-$	$\rightleftharpoons$	$Sn(OH)_2 + OH^-$	2×10^3
$[Sn(OH)_6]^{2-}$	$\rightleftharpoons$	$Sn(OH)_4 + 2OH^-$	5×10^3
$[Zn(NH_3)_4]^{2+}$	$\rightleftharpoons$	$Zn^{2+} + 4NH_3$	$2 \cdot 6 \times 10^{-10}$
$[Zn(CN)_4]^{2-}$	$\rightleftharpoons$	$Zn^{2+} + 4CN^-$	2×10^{-17}
$[Zn(OH)_4]^{2-}$	$\rightleftharpoons$	$Zn^{2+} + 4OH^-$	$10 \cdot 0$

[**Note.** *The more stable the complex, the smaller is the instablility constant.*]

9. Solubility Products at Room Temperature

Substance	Solubility Product	Substance	Solubility Product
$Al(OH)_3$	1.9×10^{-33}	CaC_2O_4	2×10^{-9}
$AgBr$	3.5×10^{-33}	$Ca_3(PO_4)_2$	1×10^{-25}
$AgBrO_2$	5×10^{-5}	$Ca(C_4H_4O_6)_2$	7.7×10^{-7}
$AgCN$	2.2×10^{-12}	$Ce(IO_3)_4$	3.5×10^{-10}
$AgCl$	1.56×10^{-10}	$Ce(C_2O_4)_2$	2.6×10^{-29}
Ag_2CO_3	5×10^{-12}	$Ce(C_4H_4O_6)_2$	9.7×10^{-29}
$Ag_2C_2O_3$	1.1×10^{-11}	$CdCO_3$	2.5×10^{-14}
Ag_2CrO_4	1.1×10^{-12}	CdC_2O_4	1.5×10^{-8}
$Ag_2Cr_2O_7$	2×10^{-7}	$Cd(OH)_2$	1.0×10^{-14}
$AgSCN$	1.2×10^{-12}	CdS	3.6×10^{-29}
AgI	1.5×10^{-16}	$Ca(OH)_2$	1×10^{-30}
$AgIO_3$	2.3×10^{-8}	CoS	3×10^{-26}
Ag_2O	1.9×10^{-8}	$Cu(OH)_2$	5.6×10^{-20}
$AgOH$	2×10^{-8}	CuC_2O_4	2.9×10^{-8}
Ag_3PO_4	3.4×10^{-14}	$Cu(IO_3)_2$	1.4×10^{-7}
Ag_2SO_3	1.9×10^{-11}	CuS	8.5×10^{-45}
Ag_2SO_4	1.2×10^{-5}	Cu_2S	2.5×10^{-50}
Ag_2AsO_3	1.9×10^{-11}	$CuBr$	4.1×10^{-8}
Ag_2AsO_4	5.6×10^{-14}	$CuCl$	1.8×10^{-7}
Ag_2S_3	1.6×10^{-49}	CuI	2.6×10^{-12}
$Ag(CH_3COO)$	1.8×10^{-3}	$CuSCN$	1.6×10^{-11}
$Ag_2(H_5COO)$	9.3×10^{-5}	FeC_2O_4	2.1×10^{-7}
$Ag_4[Fe(CN)_6]$	1.6×10^{-41}	$FeCO_3$	2.1×10^{-11}
$BaCO_3$	7×10^{-9}	FeS	1.5×10^{-19}
$BaCrO_4$	23×10^{-11}	$Fe(OH)_2$	1.6×10^{-14}
BaF_2	1.7×10^{-6}	$Fe(OH)_3$	4×10^{-38}
$Ba(IO_3)$	6.5×10	Fe_2S_3	1×10^{-38}
BaC_2O_4	1.6×10^{-7}	Hg_2Cl_2	2×10^{-18}
$BaSO_4$	9.9×10^{-11}	Hg_2Br_2	1.3×10^{-21}
$BiOCl$	2×10^{-9}	Hg_2I_2	1.2×10^{-28}
Bi_2S_2	1.6×10^{-72}	$Hg_2(SCN)_2$	3×10^{-20}
$Ba_3(PO_4)$	1.3×10^{-29}	$Hg_2(CN)_2$	5×10^{-40}
$CaCO_3$	4.8×10^{-9}	Hg_2CrO_4	2×10^{-9}
CaF_2	3.9×10^{-11}	$Hg_2C_2O_4$	1×10^{-13}
$Ca(IO_3)_2$	6.5×10^{-7}	Hg_2S	1×10^{-45}
$CaSO_4$	1.3×10^{-8}	$Hg_2(CH_3COO)_2$	2×10^{-15}
$CaSO_4$	6.1×10^{-5}	HgS	3×10^{-53}

Substance	Solubility Product	Substance	Solubility Product
HgI_2	$2 \cdot 5 \times 10^{-26}$	PbF_2	$3 \cdot 7 \times 10^{-8}$
HgO	$1 \cdot 4 \times 10^{-26}$	PbI_2	$1 \cdot 4 \times 10^{-8}$
$MgCO_3$	$2 \cdot 6 \times 10^{-5}$	$Pb(IO_3)_2$	3×10^{-13}
MgC_2O_4	$8 \cdot 6 \times 10^{-5}$	$Pb_3(PO_4)_2$	3×10^{-44}
MgF_2	$6 \cdot 4 \times 10^{-9}$	PbS	$3 \cdot 4 \times 10^{-26}$
$MgNH_4PO_4$	$2 \cdot 5 \times 10^{-13}$	$PbSO_4$	$1 \cdot 1 \times 10^{-8}$
$Mg(OH)_2$	$1 \cdot 2 \times 10^{-11}$	$Pb(OH)_2$	$2 \cdot 5 \times 10^{-16}$
$MnCO_3$	$8 \cdot 8 \times 10^{-11}$	SrF_2	$2 \cdot 8 \times 10^{-9}$
$Mn(OH)_2$	$4 \cdot 5 \times 10^{-14}$	$SrCrO_4$	$3 \cdot 6 \times 10^{-5}$
MgS	6×10^{-16}	$SrSO_4$	$2 \cdot 9 \times 10^{-7}$
$Ni(OH)_2$	7×10^{-14}	SrC_2O_4	$5 \cdot 6 \times 10^{-8}$
NiS	$1 \cdot 4 \times 10^{-24}$	$SrCO_3$	$1 \cdot 6 \times 10^{-9}$
		$Sr(OH)_2$	$3 \cdot 2 \times 10^{-4}$
$PbBr_2$	$6 \cdot 3 \times 10^{-6}$		
$PbCl_2$	$2 \cdot 4 \times 10^{-4}$	$ZnCO_3$	6×10^{-11}
$PbCO_3$	$4 \cdot 0 \times 10^{-14}$	ZnC_2O_4	$1 \cdot 5 \times 10^{-9}$
PbC_2O_4	$2 \cdot 8 \times 10^{-11}$	$Zn(OH)_2$	$4 \cdot 5 \times 10^{-17}$
$PbCrO_4$	$2 \cdot 0 \times 10^{-14}$	ZnS	$1 \cdot 2 \times 10^{-23}$

10. Transition Temperatures

(a) Hydrated salts

	Equilibrium	Transition temperature (°C)
1.	$Na_2CO_3.10H_2O \rightleftharpoons Na_2CO_3.7H_2O$	32·02
2.	$Na_2SO_4.10H_2O \rightleftharpoons Na_2SO_4$	32·38
3.	$NaBr.2H_2O \rightleftharpoons NaBr$	50·67
4.	$SrCl_2.6H_2O \rightleftharpoons SrCl_2.2H_2O$	61·33
5.	$Na_2CrO_4.10H_2O \rightleftharpoons Na_2CrO_4.6H_2O$	19·53
6.	$CuCl_2.2KCl.2H_2O \rightleftharpoons KCl + KCl.CuCl_2 + 2H_2O$	92·0

(b) Polymorphic substances

		Transition temperature (°C)
1.	Rhombic S $\rightleftharpoons$ Monoclinic S	96
2.	Rhombic KN_3 $\rightleftharpoons$ Rhombohedral	128·5
3.	Hexagonal AgI $\rightleftharpoons$ Cubic AgI	146·5
4.	Rhombic $AgNO_3$ $\rightleftharpoons$ Rhombohedral $AgNO_3$	159·5
5.	Vermilion HgI_2 $\rightleftharpoons$ Yellow HgI_2	127·0

11. Viscosity of Liquids in (centipoise) at different Temperatures

Note—One centipoise = 10^{-2} poise

Liquid	Viscosity		Liquid	Viscosity	
1. Acetaldehyde	0·255	(10°C)	14. Ethylene glycol	19·9	(20°C)
	0·22	(20°C)		9·13	(40°C)
2. Acetic Acid	1·31	(15°C)	15. Formic acid	1·804	(20°C)
	1·555	(25°C)		1·465	(30°C)
	1·04	(30°C)		1·219	(40°C)
3. Acetone	0·337	(15°C)	16. Glycerine	1·490	(20°C)
	0·316	(25°C)		0·954	(25°C)
	0·295	(30°C)		0·629	(30°C)
4. Aniline	3·71	(25°C)	17. Heptane	0·409	(20°C)
	3·16	(30°C)		0·386	(25°C)
5. Benzene	0·652	(20°C)		0·341	(40°C)
	0·564	(30°C)	18. Hexane	0·326	(20°C)
	0·504	(40°C)		0·294	(25°C)
6. Benzaldehyde	1·39	(25°C)	19. Iso-Butyl alcohol	4·703	(15°C)
7. Butyl Alcohol	2·948	(20°C)	20. Iso-Propyl alcohol	2·86	(15°C)
	2·30	(30°C)		1·77	(30°C)
	1·782	(40°C)	21. Methyl alcohol	0·597	(20°C)
8. Carbon tetra-chloride	0·969	(20°C)		0·547	(25°C)
	0·843	(30°C)		0·510	(30°C)
	0·739	(40°C)		0·458	(40°C)
9. Chlorobenzene	0·799	(20°C)	22. Methyl acetate	0·381	(20°C)
	0·631	(40°C)		0·320	(40°C)
10. Chloroform	0·58	(20°C)	23. Nitrobenzene	2·03	(20°C)
	0·512	(25°C)	24. Toluene	0·590	(20°C)
	0·514	(30°C)		0·526	(30°C)
11. Cyclohexane	1·02	(17°C)		0·471	(40°C)
12. Ethyl acetate	0·455	(20°C)	25. Turpentine	1·487	(20°C)
	0·441	(25°C)		1·272	(30°C)
	0·400	(30°C)		1·070	(40°C)
13. Ethyl alcohol	1·200	(20°C)	26. Water	1·005	(20°C)
	1·003	(30°C)		0·8007	(30°C)
	0·834	(40°C)		0·656	(40°C)
				0·469	(60°C)

12. Surface Tension of Liquids (in dynes/cm).

Liquid	Surface Tension		Liquid	Surface Tension	
1. Acetaldehyde	22·4	(10°C)	10. Ethyl Alcohol	24·05	(0°C)
	21·2	(20°C)		22·27	(20°C)
	17·0	(50°C)		18·22	(70°C)
2. Acetic Acid	28·8	(10°C)	11. Ethyl Bromide	25·5	(10°C)
(vapours)	27·8	(20°C)		24·2	(20°C)
	22·3	(75°C)		21·5	(40°C)
3. Acetone	26·2	(0°C)	12. Formic Acid	38·7	(10°C)
	23·7	(20°C)		37·6	(20°C)
	18·6	(60°C)		29·0	(100°C)
4. Aniline	44·0	(10°C)	13. Glycerol	63	(20°C)
	42·9	(20°C)		59	(90°C)
5. Benzene	31·6	(0°C)	14. Methyl Alcohol	24·5	(10°C)
	28·9	(20°C)		22·6	(20°C)
	21·3	(80°C)		15·7	(100°C)
6. Carbon tetra-	28·0	(10°C)	15. Nitrobenzene	46·4	(0°C)
chloride	26·8	(20°C)		43·9	(20°C)
	20·2	(75°C)		34·4	(100°C)
7. Chloroform	28·5	(10°C)	16. Nitroethane	33·4	(10°C)
	27·1	(20°C)		32·3	(20°C)
	21·7	(60°C)		22·5	(100°C)
8. Cyclohexane	27	(10°C)	17. Phenol	40·9	(20°C)
	25·3	(20°C)		34·4	(80°C)
	15·7	(80°C)	18. Pyridine	40·8	(0°C)
9. Ethyl Acetate	26·5	(0°C)		38·0	(20°C)
	23·9	(20°C)		26·4	(100°C)
	17·4	(75°C)	19. Toluene	30·74	(0°C)
				28·43	(20°C)
				19·39	(100°C)

13. Surface Tension of Water (in dynes/cm.)

Temperature	Surface tension	Temperature	Surface tension
0°C	75·64	40°C	69.56
5	74·92	45	68
10	74·22	50	67·91
15	73·49	60	66·18
18	73·05	70	64·42
20	72·75	80	62·61
25	71·97	90	60·75
30	71·18	100	58·85
35	70·38		

14. Densities at 25°C (g/ml)

Liquid	Density	Liquid	Density
Acetone	1·0524	Ethyl acetate	0·8941
Benzene	0·8734	Methyl acetate	0·9277
Carbon tetrachloride	1·5844	Toluene	0·8725
Chloroform	1·4797	Water	0·9971
Ethyl alcohol	0·7850		

15. Density and Specific Volume of Water at Different Temperatures

Temperature (°C)	Density (g/ml)	Volume (ml) of 1 g water
0	0·999878	1·000122
10	0·999739	1·000261
15	0·999154	1·000847
20	0·998272	1·001731
22	0·997859	1·002158
25	0·997140	1·002868
30	0·99577	1·00425
35	0·99417	1·00586

16. Solubilities

S.No.	Substance	Solubilities in gm. per 100 ml. water at									
		20°C	25°C	30°C	35°C	40°C	45°C	50°C	60°C	75°C	100°C
1.	$Al_2(SO_4)_3.18H_2O$	107·35	—	127·6	—	167·6	—	201·4	263	—	1132
2.	Potassium alum	15·13	—	—	—	—	—	44·1	—	—	357·5
3.	NH_4Cl	37·28	39·3	41·4	43·6	45·8	48	50·4	55·2	62·8	77·3
4.	$(NH_4)_2SO_4$	75·4	76·7	78	79·5	81	82·7	84·4	·88	93·4	103·3
5.	Ferrous amm. sulphate	21·6	—	28·1	31·6	—	36·2	38·2	44·6	56·7	—
6.	KCl	34·5	36	37·5	38·9	40·3	41·7	43·1	45·9	49·6	56·08
7.	KNO_3	33·0	39·7	46·7	55·6	65·5	75·1	84·1	101·1	153·9	248·6
8.	NaCl	35·9	36·1	36·2	36·34	36·48	36·6	36·83	37·3	37·81	39·24
9.	$NaNO_3$	88·1	92·13	96·15	101	105	110	115	124	142	177
10.	$NH_4CO_3.10H_2O$	92·8	—	—	—	—	—	142	—	—	540
11.	$CuSO_4.5H_2O$	42·31	—	—	—	—	—	65·8	—	—	203
12.	Ferrous sulphate	26·6	29·8	32·9	36·4	40·2	44·1	48·5	55	47·5	36
13.	Calcium chloride	74	85·9	100·8	107·5	115·3	130·4	132·6	136·8	141·6	159·1
14.	K_2CrO_4	62·2	63·7	65·1	66·5	67·9	69·5	70·9	74·6	78·7	86·8
15.	$K_2Cr_2O_7$	12·1	14·8	18·1	21·7	25·4	29·33	33·4	45·4	59·9	98·9
16.	$KMnO_4$	6·25	—	—	—	—	—	35	—	—	94
17.	$BaCl_2$	35·7	37	38·2	39·5	40·7	42·2	43·6	46·4	50·9	58
18.	KI	144·2	148·4	152·3	156	160	164	168	176	188	209
19.	$ZnSO_4.7H_2O$	161·5	—	—	—	—	—	264	—	—	653·6
20.	Cane sugar	204	211·4	219·5	228·4	238	248·7	260·4	287	340	487·2
21.	NH_4NO_3	92	214	242	265	297	—	344	421	—	871
22.	C_6H_5COOH	·29	·345	·41	48	·555	·65	·77	1·55	2·2	—
23.	$C_6H_4(OH)COOH$ (Salicylic acid)	·27	·32	·39	46	·555	·66	·80	1·22	2·55	—

17. Viscosity (Centipoise) of Different Aqueous Solutions

Substance	(% by Wt.)	Density (g/ml)	η (Centipoise) at temp. °C					
			20	25	30	35	40	45
C₂H₅OH	10	0·9839	1·538	1·323	1·160	1·006	0·907	0·812
	20	0·9715	2·183	1·815	1·553	1·332	1·160	1·015
	30	0·9577	2·71	2·180	1·860	1·580	1·368	1·189
	40	0·9396	2·91	2·35	2·02	1·72	1·492	1·284
Glycerol density for pure (1·2644 g/ml)	2	1·00475	1·055	0·935	0·836			
	5	1·012	1·143	1·010	0·900			
	6	1·01425	1·175	1·037	0·924			
	10	1·02370	1·311	1·153	1·024			
	15	1·03605	1·517	1·331	1·174			
	20	1·04840	1·769	1·542	1·360			
Sugar	1	1·0021	1·03		0·80			
	5	1·0179	1·27		0·889			
	10	1·0381	1·322	1·20	1·044		0·836	
	20	1·0810	1·967	1·710	1·510	1·336	1·197	1·074
	40	1·1764	6·223	5·206	4·398	3·776	2·261	2·858
	60	—	56·70	44·20	26·6	1·3	17·24	14·06
Sucrose	1		1·03	—	0·80		—	
	5		—		—		—	
	10		1·322	—	1·044		0·836	
	20		1·945	1·695	—			
	25		2·447	2·118	—			
	30		3·187	2·735	—			
	40		6·167	5·164	—			
	50		15·43	12·40	—			
	60		58·49	40·03	—			
Pure water	—	—	1·005	0·8937	0·8007	0·7225	0·6560	0·5980

18. Surface Tension of Different Aqueous solns.

Substance	Percent-age (W/W)	Density (g/)	(S.T. dynes/cm.) at				
			20°C	25°C	30°C	40°C	50°C
Acetic acid density for pure (1·05 gm./ml)	1· 00	0· 9996	—		67· 98		
	2· 475	—	—		64· 4		
	5· 00	1· 0055	61· 5		60· 1		
	10· 01	1· 0125	56· 0		54· 6		
	20· 00	1· 0263	49· 6		47· 7		
	30· 09	1· 0384	—		43· 6		
	50· 00	1· 0585	—		38· 4		
Acetone (density = 0·799 g/ ml)	5· 00	0· 987		55· 50			
	10· 00	0· 974		48· 9			
	20· 00	0· 948	—	41· 1			
	25· 00			38· 3			
	50· 00	—		30· 4			
C_2H_5OH (den.= 0·7893 (g/ml)	5	0· 9913		54· 2			
	10	0· 98361		45· 9			
	25	—		34· 1			
C_2H_5OH	%(V/V)						
	5	—	—			54· 9	53· 35
	10		—			48· 25	40· 77
	20		—			35· 5	34· 32
	24		38· 34			31· 6	30· 70
	48		30· 10			28· 2	28· 24
Pure water	—	—	72· 75	71· 91	71· 18	69· 59	67· 91
CH_3OH	%(V/V)						
	7· 5	—	60· 9		59· 33		56· 2
	10		59· 04		57· 27		55· 01
	25		46· 38		45· 30		43· 24
	50		35· 31		34· 5		32· 95

19. Hints for Preparing solution for gravimetric estimations

Exercise	Substance taken in the form of solution	Amount to be dissolved in one litre			
		For ·25 gm of ppt. in		For ·3 gm. of ppt. in	
		10 ml.	25 ml.	10 ml.	25 ml.
1. Estimation of Ba as $BaSO_4$. $Ba = 137\cdot34$ $BaSO_4 = 233\cdot42$	$BaCl_2.2H_2O$ M. wt. $= 244\cdot302$	26·18 gm.	10·47 gm.	31·4 gm.	12·56 gm.
2. Estimation of SO_4^{--} as $BaSO_4$.	K_2SO_4 M. wt. $= 172\cdot26$	18·75 gm.	7.5 gm.	22·5 gm.	9·0 gm.
3. Estimation of Pb as $PbSO_4$. $Pb = 207\cdot19$ $PbSO_4 = 333\cdot27$	$Pb(NO_3)_2$ M. wt. $= 331\cdot23$	27·3 gm.	10·92 gm.	32·8 gm.	13·11 gm.
4. Estimation of Ag as AgCl $Ag = 107\cdot87$ $AgCl = 143\cdot34$	$AgNO_3$ M. wt. $= 169\cdot88$	29·65 gm.	11·86 gm.	35·6 gm.	14·23 gm.
5. Estimation of Cl^- as AgCl	NaCl M. wt. $= 58\cdot45$	10·2 gm.	4·08 gm.	12·25 gm.	4·90 gm.
6. Estimation of Zn as ZnO $Zn = 63\cdot37$ $ZnO = 81\cdot38$	$ZnSO_4.7H_2O$ M. wt. $= 287\cdot546$	88·18 gm.	35·27 gm.	158 gm.	43·32 gm.
7. Estimation of Fe as FeO $Fe = 55\cdot847$ $FeO = 159\cdot68$	$FeSO_4.(NH_4)_2SO_4.6H_2O$ M. wt. $= 391\cdot60$	122·8 gm.	49·12 gm.	143·3 gm.	58·92 gm.
8. Cu as CuO $Cu = 63\cdot54$ $CuO = 79\cdot54$	$CuSO_4.5H_2O$ M. wt. $= 249\cdot68$	78·53 gm.	31·41 gm.	94·18 gm.	37·67 gm.
9. Al as Al_2O_3 $Al = 26\cdot98$ $Al_2O_3 = 101\cdot96$	$(NH_4)_2SO_4.Al_2(SO_4)_3.24H_2O$ M. wt. $= 906.69$	222·5 gm.	89·0 gm.	266·85 gm.	106·74 gm.

20. Laboratory First Aid, Safety and Treatment of Fires etc.

The laboratory work may be subject to hazards of many kinds, of which every person should be aware while working. Most of these hazards occur due to the ignorance and carelessness of the students.

A First-aid Box should be kept in a readily accessible place in the laboratory and should contain the following articles. *All bottles and packages should be clearly labelled.*

Bandages, lint, gauze, cotton-wool, adhesive plaster, and sling.

Delicate forceps, needles, thread, safety pins and scissors.

Fine glass dropping tube. Glass eye-bath.

Vaseline, salt, mustard (powder). Castor oil. Olive oil. Sal volatile, zinc oxide ointment. Boric acid (powder). Chloramine-T (fine hydrated crystals, *not* pellets).

Bottles of the following :

Acriflavine Emulsion* (in quantity)

Saturated aqueous picric acid solution (in quantity).

Lime-water (in quantity).

2% Iodine solution. 1% Boric acid. 1% Acetic acid.

8% Sodium bicarbonate solution (*i.e.*, saturated in the cold).

1% Sodium bicarbonate solution.

BURNS

(1) *Caused by Dry Heat* (*i.e.*, by Flames, Hot Metal, etc.)

    (*a*) For very small burns, hold the burn in cold saturated (8%) sodium bicarbonate solution for sometime then cover freely with zinc oxide ointment (or failing this, with vaseline) and bandage to exclude air.

    (*b*) For large burns, *do not put on oil or ointment*, but always apply the acriflavine emulsion freely and without delay. If the burn is on the hand or arm, after applying the emulsion, cover the burn *lightly* with a layer of cotton-wool also soaked in the acriflavine emulsion.

    Alternatively, a saturated aqueous solution of picric acid can be used. This should be poured freely on to the burn and also on to a pad of cotton-wool which should be applied *lightly* to the burn.

    Of the two treatments, the acriflavine treatment is the more efficacious. Either treatment, however, should be regarded so preliminary that it needs qualified medical attention.

(2) *Scalds* (by boiling water). Apply at once the acriflavine or picric acid treatment.

(3) *Acid Burns.* Wash immediately and thoroughly with cold water and then with dilute (8%) sodium bicarbonate solution.

    If burn is severe, wash again with water and then apply the acriflavine or picric acid treatment.

(4) *Caustic Alkali Burns.* The same as for acid burns, except that after *thorough* washing with water, wash with very dilute (*e.g.*, 1%) acetic acid solution in place of sodium bicarbonate. Then continue as before.

(5) *Bromine Burns.* When experiments involving the use of liquid bromine are being performed, small bottles of petrol (b.p. 80—100°) should be available on the bench. If bromine is spilt on the hands, *immediately* wash it off with an ample supply of petrol and the bromine is to be completely removed from the skin, (If subsequently, the skin which has been in contact with the petrol feels tender and "smarts" owing to the removal of the normal film of grease, cover gently with olive oil.)

*This emulsion can be purchased from most of the pharmacists.

If petrol is not immediately available, wash under a steady stream of water for sometime and then wash with dilute (8%) sodium bicarbinate solution. Water followed by bicarbonate is not, however, as affective as petrol for removing the bromine which has already started to penetrate the skin.

(6) *Phosphorus Burns*. Wash in cool water, and then immerse in aqueous silver nitrate solution. If serious, wash again with water and apply the acriflavine treatment.

(7) *Sodium Burns*. Most frequently caused by small molten pellets flying out of heated tubes. If a small solidified pellet of sodium can still be seen, remove carefully with forceps and then wash the burn thoroughly in water, then in dilute (1%) acetic acid and then cover with gauze soaked in olive oil. For serious burns, apply the acriflavine treatment.

(8) *Methyl sulphate*. If spilt on the hands : wash immediately with an ample supply of concentrated ammonia solution. Then rub *gently* with a piece of cotton wool soaked in the ammonia solution.

EYE ACCIDENTS

In all cases, the patient should see a doctor. If the accident appears serious, a doctor should be summoned immediately, *while* first-aid is applied. If the accident appears to be of a minor character, the patient should be sent to a doctor after first-aid has been applied.

(1) *Acid in Eye*. If acid is dilute, wash the eye repeatedly with 1% sodium bicarbonate solution. If the acid is concentrated, first wash the eye well with water, and then continue with the bicarbonate solution.

(2) *Caustic Alkali in Eye*. Proceed just as for acid in the eye, but was with 1% boric acid solution in place of bicarbonate solution.

(3) *Bromine in Eye*. Wash well with water, and then *at once* with 1% sodium bicarbonate solution.

(4) *Glass in Eye*. Remove loose glass either with forceps, or by washing with water in an eye-bath.

If glass has penetrated the eye, remove the fragments with forceps *only*. If the glass has entered in the white part of the eye away from the immediate neighbourhood of the iris and pupil. If glass is not removed, place patient on his back and hold eye gently open until doctor arrives.

Soreness which may follow very minor accidents to the eye may be relieved by placing 1 drop of castor oil in the corner of the eye.

CUTS

If only a minor cut, wash well in 1% aqueous Chloramine-T solution or in 2% iodine solution. Remove dirt or glass, wash again and apply sterilised dressing firmly bandaged to check bleeding.

For serious cuts, send at once for a doctor, and meanwhile endeavour to check bleeding whenever possible.

Most cuts which occur in the laboratory are caused either by glass tubing, condensers, etc., snapping while they are being forced through perforated corks, with the result that the broken jagged end cuts the hand holding the cork, or by test tubes and boiling tubes breaking whilst being too forcibly corked with similar results. Such accidents in either case can be avoided by careful working

POISONS

Solid and Liquids.

(a) *In the mouth but not swallowed* : Spit out atonce and wash the mouth cut repeatedly with water.

(b) *If swallowed* :

(i) **Acids (including oxalic) or alkalis** : Dilute their effect by drinking much water. For acids follow by drinking much lime water. Milk may be given, but no emetics.

(ii) **Salts of heavy metals** : Give milk or white of egg.

(*iii*) **Arsenic or Mercury Compounds** : Give emetic without delay, *e.g.*, one teaspoonful of mustard, or 1 tablespoonful of salt, in a tumbler of *warm* water.

[(*iv*) **Potassium Cyanide** : Immediately fatal.]

GAS POISONING

Remove patient to fresh air, and loosen clothing at neck. If breathing has stopped, give artificial respiration until the doctor arrives.

To counteract chlorine or bromine fumes if inhaled in only small amounts, inhale ammonia vapours, or gargle with sodium bicarbonate solution. Afterwards such eucalyptus pastilles, or drink warm dilute peppermint or cinnamom essence, to soothe the throat and lungs.

ELECTRIC SHOCK

Switch off and treat for burns and shock.

TREATMENT OF FIRES

Clothes. laboratories should be equipped with a sufficient number of fireproof blankets, so that a blanket available at any point of the laboratory at a few seconds' notice. Each blanket should be kept in a clearly-labelled box, the lid of which is closed by *its own weight* and not by any mechanical fastening, which might delay removal of the blanket. The box itself should be kept in some open and unencumbered position in the laboratory.

The blanket when required should be atonce wrapped firmly around the person whose clothes are on fire, the person then placed in a prone position on the floor with the ignited portion upwards, and the blanket pressed firmly over the ignited clothes until the fire is extinguished.

BENCH FIRES

Most of the available fire extinguishers are unsuitable for chemical laboratory use. Those which give a stream of water are useless for extinguishing burning ether, benzene, petrol, etc., and exceedingly dangerous if metallic sodium or potassium is present. Those which give a vigorous and fine stream of carbon tetrachloride, although of greater use than the water type, frequently serve merely to fling the burning material (and particularly burning solvents) along the surface of the bench without extinguishing the fire, the area of which is thus actually increased.

The following methods should, therefore, generally be used :

(1) *Sand*. Buckets of dry sand for fire extinguishing should be available in the laboratory and should be *strictly reserved* for this purpose and not encumbered with sand-baths, waste-paper, etc. Most fires on the bench may be quickly smothered by the ample use of sand. Sand once used for this purpose should always be thrown away afterwards, and *not* returned to the buckets, as it may contain appreciable quantities of inflammable, non-volatile materials (*e.g.*, nitrobenzene), and be dangerous if used a second time.

(2) *Carbon tetrachloride*. Although sand is of great value for extinguishing fires, it has the disadvantage that any apparatus around which the fire centres is usually smashed under the weight of the sand. Alternatively, therefore, for *small* fires carbon tetrachloride may be poured in a copious stream from a Winchester bottle on to the fire, when the "blanketing" effect of the heavy carbon tetrachloride vapour will quickly extinguish the fire. In such cases it should be remembered, however, (*a*) not to use carbon tetrachloride if metallic sodium or potassium is present, as violent explosions may result, (*b*) to ventilate the laboratory immediately after extinguishing the fire, in order to disperse the phosgene vapour which is always formed when carbon tetrachloride is used in this way.

(3) If a liquid which is being heated in a beaker or a conical flask catches fire, turn off the gas (or other source of heating) and then stretch a clean duster tightly over the mouth of the vessel. The fire

quickly dies out from lack of air and the (probably valuable) solution is recovered unharmed.

Student should bear in mind that the majority of bench fires arise from one of the two causes, both of which result from careless manipulation by the student himself. These causes are (1) the cracking of glass vessels which are being heated (usually for distillation purposes) while containing inflammable liquids. This cracking may occasionally be due to faulty apparatus but is almost invariably caused by an unsuitable method of heating, the latter furthermore being often hastily applied. (2) the addition of unglazed porcelain to a heated liquid which bumps badly—and the superheated liquid suddenly froths over and catches fire. Porcelain should never be added to a "bumping" liquid until the latter has been allowed to cool for a few minute.

The most dangerous solvent in the laboratory is carbon disulphide, the flash-point of which is so low that its vapour is ignited, e.g., by a gas-ring 3-4 minutes after the gas has been turned off. Carbon disulphide should, therefore, never be used in the laboratory unless an adequate substitute as a solvent can not be found. Probably the next most dangerous liquid for general manipulation is ether, which, however, has to be frequently employed. If the precautions described are *always* followed. the manipulation of ether should, however, be quite safe.

EXPLOSIONS

Gaseous explosions also rank among those accidents which are almost invariably due to careless work. They are usually caused either by :

(*i*) Faulty condensation of a heavy inflammable varpours, such as ether. The precautions mentioned in the previous paragraph if observed will prevent this occurrence.

(*ii*) Igniting an inflammable gas before all air has been removed from the containing vessel. Whenever an inflammable gas is collected, a sample in a small test-tube should first be ignited at a safe distance from the main experiment. If it burns quietly, without any sign of even a gentle explosion, the main body of the gas be safely ignited. Even then, this should be done with the smallest volume of gas suitable for the purpose concerned.

(*iii*) Experiments in which metallic sodium has been used and in which the product has subsequently to be treated with water, great care should be taken to ensure that no unchanged sodium remains when the water is added.

21. Consolidated List of Organic Compounds

Solids with their melting points (°C)

42	Phenyl Salicylate	122	Benzoic acid
44	o-nitro phenol	122	β-naphthol
45	p-toluidine	125	Succinimide
48	Benzophenone	127	Benzidine
50	α-napthylamine	128	Benzamide
53	p-dichlorobenzene	131	Malonic Acid
54	o-nitrotoluene	132	Urea
54	Chloral hydrate	132	Pyrogallol
54	Methyl oxalate	133	Cinnamic acid
57	p-anisidine	139	Salicylamide
61	α-nitronaphthalene	140	p-phenylene diamine
63	Metaphenylene diamine	146	Glucose
70	Diphenyl	150	Adipic Acid
80	Naphthalene	153	Citric Acid
82	Acetamide	154	Phenyl thiourea
89	p-dibromobenzene	155	Salicylic acid

90	m-dinitrobenzene
95	Acenaphthene
95	α-naphthol
95	Benzyl
97	m-nitrophenol
97	Glutaric acid
100	Maltose
100	Phenanthrene
101	Oxalic acid
102	o-phenylenediamine
105	Fructose
111	β-naphthylamine
114	Acetanilide
114	p-nitrophenol
118	Resorcinol
119	Iodoform

157	Bromobenzene
160	Sucrose (decomposes)
161	Benzanilide
166	Galactose (anhy.)
169	Quinol
169	Tartaric acid
180	Thiourea
185	Succinic acid
195	Phthalic acid (decomposes)
203	Lactose
216	Anthracene
233	Phthalamide (decomposes)
242	Succinamide
300	Sulphanilic acid decomposes—starch.

Liquids with their boiling points (°C)

49	n-propyl amine
56	Acetone
57	Methyl acetate
61	Chloroform
65	Methyl alcohol
77	n-butyl amine
77	Ethyl acetate
78	Carbon tetrachloride
78	Ethyl alcohol
80	Benzene
80	Ethyl methyl ketone
82	Isopropyl alcohol
97	Acetaldehyde
97	n-propyl alcohol
98	Formaldehyde
99	Tert. butyl alcohol
101	Formic acid
102	Diethyl ketone
108	Iso-butyl alcohol
110	Toluene
116	n-butyl alcohol
118	Acetic acid
132	Chlorobenzene
136	Ethylbenzene
137	p-xylene
138	n-amyl alcohol
139	m-xylene
140	Propionic acid

142	o-xylene
155	Cyclohexanone
160	Cyclohexanol
163	n-Butyric acid
179	Benzyl chloride
179	Benzaldehyde
182	Phenol
183	Aniline
185	Benzyl amine
186	Diethyl oxalate
191	o-Cresol
193	Methyl aniline
193	Dimethyl aniline
196	Salicyldehyde
197	o-toluidine
197	Ethylene glycol
198	Methyl aniline
199	m-Toluidine
201	p-cresol
202	m-cresol
202	Acetophenone
205	Benzyl alcohol
210	Nitrobenezene
213	Ethyl benzoate
219	o-nitrotoluidine
224	Methyl salicylate
290	Glycerol

LOGARITHMS TABLES
COMMON LOGARITHMS

	0	1	2	3	4	5	6	7	8	9	Mean Differences								
											1	2	3	4	5	6	7	8	9
10	0000	0043	0086	0128	0170	0212	0253	0294	0334	0374	4	8	12	17	21	25	29	33	37
11	0414	0453	0492	0531	0569	0607	0645	0682	0719	0755	4	8	11	15	19	23	26	30	34
12	0792	0828	0864	0899	0934	0969	1004	1038	1072	1106	3	7	10	14	17	21	24	28	31
13	1139	1173	1206	1239	1271	1303	1335	1367	1399	1430	3	6	10	13	16	19	23	26	29
14	1461	1492	1523	1553	1584	1614	1644	1673	1703	1732	3	6	9	12	15	18	21	24	27
15	1761	1790	1818	1847	1875	1903	1931	1959	1987	2014	3	6	8	11	14	17	20	22	25
16	2041	2068	2095	2122	2148	2175	2201	2227	2253	2279	3	5	8	11	13	16	18	21	24
17	2304	2330	2355	2380	2405	2430	2455	2480	2504	2529	2	5	7	10	12	15	17	20	22
18	2553	2577	2601	2625	2648	2672	2695	2718	2742	2765	2	5	7	9	12	14	16	19	21
19	2788	2810	2833	2856	2878	2900	2923	2945	2967	2989	2	4	7	9	11	13	16	18	20
20	3010	3032	3054	3075	3096	3118	3139	3160	3181	3201	2	4	6	8	11	13	15	17	19
21	3222	3243	3263	3284	3304	3324	3345	3365	3385	3404	2	4	6	8	10	12	14	16	18
22	3424	3444	3464	3483	3502	3522	3541	3560	3579	3598	2	4	6	8	10	12	14	15	17
23	3617	3636	3655	3674	3692	3711	3729	3747	3766	3784	2	4	6	7	9	11	13	15	17
24	3802	3820	3838	3856	3874	3892	3909	3927	3945	3962	2	4	5	7	9	11	12	14	16
25	3979	3997	4014	4031	4048	4065	4082	4099	4116	4133	2	3	5	7	9	10	12	14	15
26	4150	4166	4183	4200	4216	4232	4249	4265	4281	4298	2	3	5	7	8	10	11	13	15
27	4314	4330	4346	4362	4378	4393	4409	4425	4440	4456	2	3	5	6	8	9	11	13	14
28	4472	4487	4502	4518	4533	4548	4564	4579	4594	4609	2	3	5	6	8	9	11	12	14
29	4624	4639	4654	4669	4683	4698	4713	4728	4742	4757	1	3	4	6	7	9	10	12	13
30	4771	4786	4800	4814	4829	4843	4857	4871	4886	4900	1	3	4	6	7	9	10	11	13
31	4914	4928	4942	4955	4969	4983	4997	5011	5024	5038	1	3	4	6	7	8	10	11	12
32	5051	5065	5079	5092	5105	5119	5132	5145	5159	5172	1	3	4	5	7	8	9	11	12
33	5185	5198	5211	5224	5237	5250	5263	5276	5289	5302	1	3	4	5	6	8	9	10	12
34	5315	5328	5340	5353	5366	5378	5391	5403	5416	5428	1	3	4	5	6	8	9	10	11
35	5441	5453	5465	5478	5490	5502	5514	5527	5539	5551	1	2	4	5	6	7	9	10	11
36	5563	5575	5587	5599	5611	5623	5635	5647	5658	5670	1	2	4	5	6	7	8	10	11
37	5682	5694	5705	5717	5729	5740	5752	5763	5775	5786	1	2	3	5	6	7	8	9	10
38	5798	5809	5821	5832	5843	5855	5866	5877	5888	5899	1	2	3	5	6	7	8	9	10
39	5911	5922	5933	5944	5955	5966	5977	5988	5999	6010	1	2	3	4	5	7	8	9	10
40	6021	6031	6042	6053	6064	6075	6085	6096	6107	6117	1	2	3	4	5	6	8	9	10
41	6128	6138	6149	6160	6170	6180	6191	6201	6212	6222	1	2	3	4	5	6	7	8	9
42	6232	6243	6253	6263	6274	6284	6294	6304	6314	6325	1	2	3	4	5	6	7	8	9
43	6335	6345	6355	6365	6375	6385	6395	6405	6415	6425	1	2	3	4	5	6	7	8	9
44	6435	6444	6454	6464	6474	6484	6493	6503	6513	6522	1	2	3	4	5	6	7	8	9
45	6532	6542	6551	6561	6571	6580	6590	6599	6609	6618	1	2	3	4	5	6	7	8	9
46	6628	6637	6646	6656	6665	6675	6684	6693	6702	6712	1	2	3	4	5	6	7	7	8
47	6721	6730	6739	6749	6758	6767	6776	6785	6794	6803	1	2	3	4	5	5	6	7	8
48	6812	6821	6830	6839	6848	6857	6866	6875	6884	6893	1	2	3	4	4	5	6	7	8
49	6902	6911	6920	6928	6937	6946	6955	6964	6972	6981	1	2	3	4	4	5	6	7	8
50	6990	6998	7007	7016	7024	7033	7042	7050	7059	7067	1	2	3	3	4	5	6	7	8

LOGARITHMS (Contd.)

	0	1	2	3	4	5	6	7	8	9	Mean Differences								
											1	2	3	4	5	6	7	8	9
51	7076	7084	7093	7101	7110	7118	7126	7135	7143	7152	1	2	3	3	4	5	6	7	8
52	7160	7168	7177	7185	7193	7202	7210	7218	7226	7235	1	2	2	3	4	5	6	7	7
53	7243	7251	7259	7267	7275	7284	7292	7300	7308	7316	1	2	2	3	4	5	6	6	7
54	7324	7332	7340	7348	7356	7364	7372	7380	7388	7396	1	2	2	3	4	5	6	6	7
55	7404	7412	7419	7427	7435	7443	7451	7459	7466	7474	1	2	2	3	4	5	5	6	7
56	7482	7490	7497	7505	7513	7520	7528	7536	7543	7551	1	2	2	3	4	5	5	6	7
57	7559	7566	7574	7582	7589	7597	7604	7612	7619	7627	1	2	2	3	4	5	5	6	7
58	7634	7642	7649	7657	7664	7672	7679	7686	7694	7701	1	1	2	3	4	4	5	6	7
59	7709	7716	7723	7731	7738	7745	7752	7760	7767	7774	1	1	2	3	4	4	5	6	7
60	7782	7789	7796	7803	7810	7818	7825	7832	7839	7846	1	1	2	3	4	4	5	6	6
61	7853	7860	7868	7875	7882	7889	7896	7903	7910	7917	1	1	2	3	4	4	5	6	6
62	7924	7931	7938	7945	7952	7959	7966	7973	7980	7987	1	1	2	3	3	4	5	6	6
63	7993	8000	8007	8014	8021	8028	8035	8041	8048	8055	1	1	2	3	3	4	5	5	6
64	8062	8069	8075	8082	8089	8096	8102	8109	8116	8122	1	1	2	3	3	4	5	5	6
65	8129	8136	8142	8149	8156	8162	8169	8176	8182	8189	1	1	2	3	3	4	5	5	6
66	8195	8202	8209	8215	8222	8228	8235	8241	8248	8254	1	1	2	3	3	4	5	5	6
67	8261	8267	8274	8280	8287	8293	8299	8306	8312	8319	1	1	2	3	3	4	5	5	6
68	8325	8331	8338	8344	8351	8357	8363	8370	8376	8382	1	1	2	3	3	4	4	5	6
69	8388	8395	8401	8407	8414	8420	8426	8432	8439	8445	1	1	2	2	3	4	4	5	6
70	8451	8457	8463	8470	8476	8482	8488	8494	8500	8506	1	1	2	2	3	4	4	5	6
71	8513	8519	8525	8531	8537	8543	8549	8555	8561	8567	1	1	2	2	3	4	4	5	5
72	8573	8579	8585	8591	8597	8603	8609	8615	8621	8627	1	1	2	2	3	4	4	5	5
73	8633	8639	8645	8651	8657	8663	8669	8675	8681	8686	1	1	2	2	3	4	4	5	5
74	8692	8698	8704	8710	8716	8722	8727	8733	8739	8745	1	1	2	2	3	4	4	5	5
75	8751	8756	8762	8768	8774	8779	8785	8791	8797	8802	1	1	2	2	3	3	4	5	5
76	8808	8814	8820	8825	8831	8837	8842	8848	8854	8859	1	1	2	2	3	3	4	5	5
77	8865	8871	8876	8882	8887	8893	8899	8904	8910	8915	1	1	2	2	3	3	4	4	5
78	8921	8927	8932	8938	8943	8949	8954	8960	8965	8971	1	1	2	2	3	3	4	4	5
79	8976	8982	8987	8993	8998	9004	9009	9015	9020	9025	1	1	2	2	3	3	4	4	5
80	9031	9036	9042	9047	9053	9058	9063	9069	9074	9079	1	1	2	2	3	3	4	4	5
81	9085	9090	9096	9101	9106	9112	9117	9122	9128	9133	1	1	2	2	3	3	4	4	5
82	9138	9143	9149	9154	9159	9165	9170	9175	9180	9186	1	1	2	2	3	3	4	4	5
83	9191	9196	9201	9206	9217	9217	9222	9227	9232	9238	1	1	2	2	3	3	4	4	5
84	9243	9248	9253	9258	9263	9269	9274	9279	9284	9289	1	1	2	2	3	3	4	4	5
85	9294	9299	9304	9360	9315	9320	9325	9330	9335	9340	1	1	2	2	3	3	4	4	5
86	9345	9350	9355	9360	9365	9370	9375	9380	9385	9390	1	1	2	2	3	3	4	4	5
87	9395	9400	9405	9410	9415	9420	9425	9430	9435	9440	0	1	1	2	2	3	3	4	4
88	9445	9450	9455	9460	9465	9469	9474	9479	9484	9489	0	1	1	2	2	3	3	4	4
89	9494	9499	9504	9509	9513	9518	9523	9528	9533	9538	0	1	1	2	2	3	3	4	4
90	9542	9547	9552	9557	9562	9566	9571	9576	9581	9586	0	1	1	2	2	3	3	4	4
91	9590	9595	9600	9605	9609	9614	9619	9624	9628	9633	0	1	1	2	2	3	3	4	4
92	9638	9643	9647	9652	9657	9661	9666	9671	9675	9680	0	1	1	2	2	3	3	4	4
93	9685	9689	9694	9699	9703	9708	9713	9717	9722	9727	0	1	1	2	2	3	3	4	4
94	9731	9736	9741	9745	9750	9754	9759	9763	9768	9773	0	1	1	2	2	3	3	4	4
95	9777	9782	9786	9791	9795	9800	9805	9809	9814	9818	0	1	1	2	2	3	3	4	4
96	9823	9827	9832	9836	9841	9845	9850	9854	9859	9863	0	1	1	2	2	3	3	4	4
97	9868	9872	9877	9881	9886	9890	9894	9899	9903	9908	0	1	1	2	2	3	3	4	4
98	9912	9916	9921	9926	9930	9934	9939	9943	9948	9952	0	1	1	2	2	3	3	4	4
99	9956	9961	9965	9969	9974	9978	9983	9987	9991	9996	0	1	1	2	2	3	3	3	4

ANTILOGARITHMS

	0	1	2	3	4	5	6	7	8	9	Mean Differences								
											1	2	3	4	5	6	7	8	9
.00	1000	1002	1005	1007	1009	1012	1014	1016	1019	1021	0	0	1	1	1	1	2	2	2
.01	1023	1026	1028	1030	1033	1035	1038	1040	1042	1045	0	0	1	1	1	1	2	2	2
.02	1047	1050	1052	1054	1057	1059	1062	1064	1067	1069	0	0	1	1	1	1	2	2	2
.03	1072	1074	1076	1079	1081	1084	1086	1089	1091	1094	0	0	1	1	1	1	2	2	2
.04	1096	1099	1102	1104	1107	1109	1112	1114	1117	1119	0	1	1	1	1	2	2	2	2
.05	1122	1125	1127	1130	1132	1135	1138	1140	1143	1146	0	1	1	1	1	2	2	2	2
.06	1148	1151	1153	1156	1159	1161	1164	1167	1169	1172	0	1	1	1	1	2	2	2	2
.07	1175	1178	1180	1183	1186	1189	1191	1194	1197	1199	0	1	1	1	1	2	2	2	2
.08	1202	1205	1208	1211	1213	1216	1219	1222	1225	1227	0	1	1	1	1	2	2	2	3
.09	1230	1233	1236	1239	1242	1245	1247	1250	1253	1256	0	1	1	1	1	2	2	2	3
.10	1259	1262	1265	1268	1271	1274	1276	1279	1282	1285	0	1	1	1	1	2	2	2	3
.11	1288	1291	1294	1297	1300	1303	1306	1309	1312	1315	0	1	1	1	2	2	2	2	3
.12	1318	1321	1324	1327	1330	1334	1337	1340	1343	1346	0	1	1	1	2	2	2	2	3
.13	1349	1352	1355	1358	1361	1365	1368	1371	1374	1377	0	1	1	1	2	2	2	3	3
.14	1380	1384	1387	1390	1393	1396	1400	1403	1406	1409	0	1	1	1	2	2	2	3	3
.15	1413	1416	1419	1422	1426	1429	1432	1435	1439	1442	0	1	1	1	2	2	2	3	3
.16	1445	1449	1452	1455	1459	1462	1466	1469	1472	1476	0	1	1	1	2	2	2	3	3
.17	1479	1483	1486	1489	1493	1496	1500	1503	1507	1510	0	1	1	1	2	2	2	3	3
.18	1514	1517	1521	1524	1528	1531	1535	1538	1542	1545	0	1	1	1	2	2	2	3	3
.19	1549	1552	1556	1560	1563	1567	1570	1574	1578	1581	0	1	1	1	2	2	3	3	3
.20	1585	1589	1592	1596	1600	1603	1607	1611	1614	1618	0	1	1	1	2	2	3	3	3
.21	1622	1626	1629	1633	1637	1641	1644	1648	1652	1656	0	1	1	2	2	2	3	3	3
.22	1660	1663	1667	1671	1675	1679	1683	1687	1690	1694	0	1	1	2	2	2	3	3	3
.23	1698	1702	1706	1710	1714	1718	1722	1726	1730	1734	0	1	1	2	2	2	3	3	4
.24	1738	1742	1746	1750	1754	1758	1762	1766	1770	1774	0	1	1	2	2	2	3	3	4
.25	1778	1782	1786	1791	1795	1798	1803	1807	1811	1816	0	1	1	2	2	2	3	3	4
.26	1820	1824	1828	1832	1837	1841	1845	1849	1854	1858	0	1	1	2	2	3	3	3	4
.27	1862	1866	1871	1875	1879	1884	1888	1892	1897	1901	0	1	1	2	2	3	3	3	4
.28	1905	1910	1914	1919	1923	1928	1932	1936	1941	1945	0	1	1	2	2	3	3	4	4
.29	1950	1954	1959	1963	1968	1972	1977	1982	1986	1991	0	1	1	2	2	3	3	4	4
.30	1995	2000	2004	2009	2014	2018	2023	2028	2032	2037	0	1	1	2	2	3	3	4	4
.31	2042	2046	2051	2056	2061	2065	2070	2075	2080	2084	0	1	1	2	2	3	3	4	4
.32	2089	2094	2099	2104	2109	2113	2118	2123	2128	2133	0	1	1	2	2	3	3	4	4
.33	2138	2143	2148	2153	2158	2163	2168	2173	2178	2183	0	1	1	2	2	3	3	4	4
.34	2188	2193	2198	2203	2208	2213	2218	2223	2228	2234	1	1	2	2	3	3	4	4	5
.35	2239	2244	2249	2254	2259	2265	2270	2275	2280	2286	1	1	2	2	3	3	4	4	5
.36	2291	2296	2301	2307	2312	2317	2323	2328	2333	2339	1	1	2	2	3	3	4	4	5
.37	2344	2350	2355	2360	2366	2371	2377	2382	2388	2393	1	1	2	2	3	3	4	4	5
.38	2399	2404	2410	2415	2421	2427	2432	2438	2443	2449	1	1	2	2	3	3	4	4	5
.39	2455	2460	2466	2472	2477	2483	2489	2495	2500	2506	1	1	2	2	3	3	4	5	5
.40	2512	2518	2523	2529	2535	2541	2547	2553	2559	2564	1	1	2	2	3	4	4	5	5
.41	2570	2576	2582	2588	2594	2600	2606	2612	2618	2624	1	1	2	2	3	4	4	5	5
.42	2630	2636	2642	2649	2655	2661	2667	2673	2679	2685	1	1	2	2	3	4	4	5	6
.43	2692	2698	2704	2710	2716	2723	2729	2735	2742	2748	1	1	2	3	3	4	4	5	6
.44	2754	2761	2767	2773	2780	2786	2793	2799	2805	2812	1	1	2	3	3	4	4	5	6
.45	2818	2825	2831	2838	2844	2851	2858	2864	2871	2877	1	1	2	3	3	4	5	5	6
.46	2884	2891	2897	2904	2911	2917	2924	2931	2938	2944	1	1	2	3	3	4	5	5	6
.47	2951	2958	2965	2972	2979	2985	2992	2999	3006	3013	1	1	2	3	3	4	5	5	6
.48	3020	3027	3034	3041	3048	3055	3062	3069	3076	3083	1	1	2	3	4	4	5	6	6
.49	3090	3097	3105	3112	3119	3126	3133	3141	3148	3155	1	1	2	3	4	4	5	6	6
.50	3162	3170	3177	3184	3192	3199	3206	3214	3221	3228	1	1	2	3	4	4	5	6	7

ANTILOGARITHMS (Contd.)

	0	1	2	3	4	5	6	7	8	9	Mean Differences								
											1	2	3	4	5	6	7	8	9
.51	3236	3243	3251	3258	3266	3273	3281	3289	3296	3304	1	2	2	3	4	5	5	6	7
.52	3311	3319	3327	3334	3342	3350	3357	3365	3373	3381	1	2	2	3	4	5	5	6	7
.53	3388	3396	3404	3412	3420	3428	3436	3443	3451	3459	1	2	2	3	4	5	6	6	7
.54	3467	3475	3483	3491	3499	3508	3516	3524	3532	3540	1	2	2	3	4	5	6	6	7
.55	3548	3556	3565	3573	3581	3589	3597	3606	3614	3622	1	2	2	3	4	5	6	7	7
.56	3631	3639	3648	3656	3664	3673	3681	3690	3698	3707	1	2	3	3	4	5	6	7	8
.57	3715	3724	3733	3741	3750	3758	3767	3776	3784	3793	1	2	3	3	4	5	6	7	8
.58	3802	3811	3819	3828	3837	3846	3855	3864	3873	3882	1	2	3	4	4	5	6	7	8
.59	3890	3899	3908	3917	3926	3936	3945	3954	3963	3972	1	2	3	4	5	5	6	7	8
.60	3981	3990	3999	4009	4018	4027	4036	4046	4055	4064	1	2	3	4	5	6	6	7	8
.61	4074	4083	4093	4102	4111	4121	4130	4140	4150	4159	1	2	3	4	5	6	7	8	9
.62	4169	4178	4188	4198	4207	4217	4227	4236	4246	4256	1	2	3	4	5	6	7	8	9
.63	4266	4276	4285	4295	4305	4315	4325	4335	4345	4355	1	2	3	4	5	6	7	8	9
.64	4365	4375	4385	4395	4406	4416	4426	4436	4446	4457	1	2	3	4	5	6	7	8	9
.65	4467	4477	4487	4498	4508	4519	4529	4539	4550	4560	1	2	3	4	5	6	7	8	9
.66	4571	4581	4592	4603	4613	4624	4634	4645	4656	4667	1	2	3	4	5	6	7	9	10
.67	4677	4688	4699	4710	4721	4732	4742	4753	4764	4775	1	2	3	4	5	7	8	9	10
.68	4786	4797	4808	4819	4831	4842	4853	4864	4875	4887	1	2	3	4	6	7	8	9	10
.69	4898	4909	4920	4932	4943	4955	4966	4977	4989	5000	1	2	3	5	6	7	8	9	10
.70	5012	5023	5035	5047	5058	5070	5082	5093	5105	5117	1	2	4	5	6	7	8	9	11
.71	5129	5140	5152	5164	5176	5188	5200	5212	5224	5236	1	2	4	5	6	7	8	10	11
.72	5248	5260	5272	5284	5297	5309	5321	5333	5346	5358	1	2	4	5	6	7	9	10	11
.73	5370	5383	5395	5408	5420	5433	5445	5458	5470	5483	1	3	4	5	6	8	9	10	11
.74	5495	5508	5521	5534	5546	5559	5572	5585	5598	5610	1	3	4	5	6	8	9	10	12
.75	5623	5636	5649	5662	5675	5689	5702	5715	5728	5741	1	3	4	5	7	8	9	10	12
.76	5754	5768	5781	5794	5808	5821	5834	5848	5861	5875	1	3	4	5	7	8	9	11	12
.77	5888	5902	5916	5929	5943	5957	5970	5984	5998	6012	1	3	4	5	7	8	10	11	12
.78	6026	6039	6053	6067	6081	6095	6109	6124	6138	6152	1	3	4	6	7	8	10	11	13
.79	6166	6180	6194	6209	6223	6237	6252	6266	6281	6295	1	3	4	6	7	9	10	11	13
.80	6310	6324	6339	6353	6368	6383	6397	6412	6427	6442	1	3	4	6	7	9	10	12	13
.81	6457	6471	6486	6501	6516	6531	6546	6561	6577	6592	2	3	5	6	8	9	11	12	14
.82	6607	6622	6637	6653	6668	6683	6699	6714	6730	6745	2	3	5	6	8	9	11	12	14
.83	6761	6776	6792	6808	6823	6839	6855	6871	6887	6902	2	3	5	6	8	9	11	13	14
.84	6918	6934	6950	6966	6982	6998	7015	7031	7047	7063	2	3	5	6	8	10	11	13	15
.85	7079	7096	7112	7129	7145	7161	7178	7194	7211	7228	2	3	5	7	8	10	12	13	15
.86	7244	7261	7278	7295	7311	7328	7345	7362	7379	7396	2	3	5	7	8	10	12	13	15
.87	7413	7430	7447	7464	7482	7499	7516	7534	7551	7568	2	3	5	7	9	10	12	14	16
.88	7586	7603	7621	7638	7656	7674	7691	7709	7727	7745	2	4	5	7	9	11	12	14	16
.89	7762	7780	7798	7816	7834	7852	7870	7889	7907	7925	2	4	5	7	9	11	13	14	16
.90	7943	7962	7980	7998	8017	8035	8054	8072	8091	8110	2	4	6	7	9	11	13	15	17
.91	8128	8147	8166	8185	8204	8222	8241	8260	8279	8299	2	4	6	8	9	11	13	15	17
.92	8318	8337	8356	8375	8395	8414	8433	8453	8472	8492	2	4	6	8	10	12	14	15	17
.93	8511	8531	8551	8570	8590	8616	8630	8650	8670	8690	2	4	6	8	10	12	14	16	18
.94	8710	8730	8750	8770	8790	8810	8831	8851	8872	8892	2	4	6	8	10	12	14	16	18
.95	8913	8933	8954	8974	8995	9016	9036	9057	9078	9099	2	4	6	8	10	12	15	17	19
.96	9120	9141	9162	9183	9204	9226	9247	9268	9290	9311	2	4	6	8	11	13	15	17	19
.97	9333	9354	9376	9397	9419	9441	9462	9484	9506	9528	2	4	7	9	11	13	15	17	20
.98	9550	9572	9594	9616	9638	9661	9683	9705	9727	9750	2	4	7	9	11	13	16	18	20
.99	9772	9793	9817	9840	9863	9886	9908	9931	9954	9977	2	5	7	9	11	14	16	18	20